Analysis and Synthesis of Time Delay Systems

Analysis and Synthesis of Time Delay Systems

H. Górecki, S. Fuksa, P. Grabowski and A. Korytowski

*Institute of Automatic Control,
Technical University of Mining and Metallurgy
Kraków, Poland*

A Wiley-Interscience Publication

John Wiley & Sons

Chichester · New York · Brisbane · Toronto · Singapore

PWN — Polish Scientific Publishers

Warszawa

Graphic design by *Zygmunt Ziemka*

Copyright © 1989 by PWN—Polish Scientific Publishers—Warszawa

All rights reserved.

No part of this book may be reproduced by any means, or transmitted, or translated into a machine language without the written permission of the publisher.

British Library Cataloguing in Publication Data:

Analysis and synthesis of time delay systems.
1. Control theory   2. Functional differential equations—Delay equations
I. Górecki, Henryk
629.8′312   QA402.3

ISBN 0-471-27622-7

Printed in Poland by D.R.P.

Preface

The purpose of this book is to give a comprehensive review of modern theory of control systems described by functional-differential equations. Emphasis is put on the methods and results which are applicable to analysis and synthesis of industrial control systems with time delays. The book is intended for a reader with a knowledge of mathematical methods of control theory though some parts (e.g., Chapters 1, 2, 7) may be easily read by a control engineer without special mathematical background. The first two introductory chapters contain examples of systems with delays and a short presentation of typical control problems. Chapters 3 and 4 give basic results of the mathematical theory of functional-differential and discrete equations. Identification of time-delay systems is treated in Chapter 5. An extensive presentation of stability theory is given in Chapter 6. Chapter 7 will be probably most interesting for a control engineer. It deals with PID and Smith's controllers, determination of stability regions, controller settings, etc. A review of controllability and observability results is presented in Chapter 8. Chapter 9 gives general results referring to optimal control whereas the linear-quadratic problem of optimal control is discussed in detail in Chapter 10. The contributions of the four authors are distributed as follows: H. Górecki—Chapters 1 and 7; S. Fuksa—Chapters 1, 5, 8 and 9; P. Grabowski—Chapter 6; A. Korytowski—Chapters 1, 2, 3, 4, 7 and 10. A large part of the application-oriented material is based on the book by Górecki (1971).

The authors would like to express their gratitude to A. Olbrot for a revision of the manuscript and many valuable suggestions.

Contents

Preface V

Part I Introduction 1

1 Examples of Systems with Time Delays 1
   1.1 Systems with Lumped Delays 2
      1.1.1 Transport processes. Control of coal flow rate on a belt
           conveyor 2
      1.1.2 Metal rolling system 4
      1.1.3 Population models 5
      1.1.4 Economic systems 6
      1.1.5 Control in computerized systems with state reconstruction 7
      1.1.6 Remote control 8
      1.1.7 Urban traffic 10
   1.2 Systems with Distributed Delays 11
      1.2.1 Electric transmission line 12
      1.2.2 Heat exchangers 18

2 Formulation of Control Problems 24
   2.1 Model Construction 24
   2.2 Stability and Level Stabilization 25
   2.3 Terminal Control 26
   2.4 Tracking Problems 26
   2.5 Optimal Control 27
   2.6 The Existence of Solutions to Control Problems 29
   2.7 Feedback Regulator and Connected Problems 29
   2.8 Analysis and Synthesis of Control Systems 31

Part II Models of time-delay systems ... 32

3 Models of Continuous Time-delay Systems ... 32
3.1 Functional-differential Equations ... 32
3.1.1 Classification of functional differential equations ... 32
3.1.2 Existence, continuous dependence and uniqueness of solutions of FDE ... 34
3.1.3 Step method ... 37
3.2 Linear Continuous-time FDE ... 41
3.2.1 Linear models and variation-of-constants formulas ... 41
3.2.2 Step method ... 49
3.2.3 Characteristic equation and exponential solutions ... 53
3.2.4 Laplace transform and series expansions of solutions ... 57
3.3 Input, Output, State and Transfer Functions ... 63
3.3.1 Basic definitions ... 63
3.3.2 State equation for linear systems ... 67
3.3.3 Transfer functions ... 70

4 Models of Discrete Time-delay Systems ... 72
4.1 Discrete-time Equations ... 72
4.2 Linear Discrete Models ... 74
4.3 State and Output Equations; Transfer Functions ... 78
4.4 Approximation of Continuous-time Systems by Discrete-time Systems ... 80
4.4.1 Linear systems ... 80
4.4.2 Nonlinear systems ... 85

5 Identification of Linear Systems ... 87
5.1 Problem Statement ... 87
5.2 Mathematical Methods in Identification Problems ... 89
5.3 Some Particular Identification Problems and Methods ... 95

Part III Control problems ... 100

6 Stability ... 100
6.1 Lyapunov's Second Method for FDE in the Space C ... 100
6.2 Stability Theory of Abstract Linear Systems and its Applications to Systems Involving Delay ... 123

6.2.1 Some basic results in abstract linear systems theory . . 123
 6.2.2 Stability of abstract linear systems 125
 6.2.3 Application of the abstract theory to systems with delay 130
6.3 Stability of Linear Systems with Nonlinear Perturbations
 (Lyapunov's First Method) 154
 6.3.1 Variation-of-constants formula for neutral linear functional-differential equations 155
 6.3.2 Lyapunov's first method for hereditary systems 165
 6.3.3 Stability of a class of abstract linear systems with a nonlinear perturbation 168
 6.3.4 Examples 174
6.4 Criteria for the Location of Characteristic Function Zeros in Left Complex Half-plane 177
 6.4.1 The principle of argument 178
 6.4.2 Application of the principle of argument to the location of characteristic function zeros 179
 6.4.3 Pontryagin's theory of quasipolynomial zero location . . 185

7 Conventional Regulation Problems 200
7.1 Stability Region of a Continuous-time Conventional Regulation System 200
7.2 Parametric Synthesis of Continuous Regulators 215
 7.2.1 Determination of the regulator type 215
 7.2.2 Criteria for regulator setting selection based on characteristic roots 218
 7.2.3 Criteria based on frequency characteristics 221
 7.2.4 Integral criteria based on step response 224
7.3 Smith-type Regulation Systems 232
 7.3.1 The Smith principle 232
 7.3.2 Mismatch, stability and sensitivity 238

8 Controllability, Observability and Related Problems for Linear Systems 245
8.1 Definitions of Basic Notions 245
8.2 R^p-controllability of Linear FDE Systems 251
8.3 The Null-state Controllability of FDE Systems 261
8.4 Controllability, Observability and Pole Assignment 268

9 Optimal Control for Continuous-time Systems 285
9.1 Formulation of Optimal Control Problem 285

- 9.2 Necessary Conditions of Optimality 286
  - 9.2.1 Necessary conditions. General approach 286
  - 9.2.2 Fundamental properties of FDE systems 291
  - 9.2.3 Necessary optimality conditions for FDE system ... 294
- 9.3 Some Particular Optimization Problems 297

10 Linear-quadratic Problem of Optimal Control 303
- 10.1 Problem Formulation 303
- 10.2 Equivalent Linear-quadratic Problems 306
- 10.3 Step Method 309
  - 10.3.1 Problem formulation 309
  - 10.3.2 Variation of the performance functional and the adjoint equation 313
  - 10.3.3 Optimal control and canonical system of equations . 317
  - 10.3.4 Basic algebraic equation and solution of optimal control problem 319
  - 10.3.5 Step method solution in case of a delay-free performance functional 321
  - 10.3.6 Open-loop and closed-loop optimal control 323
  - 10.3.7 Other variants of the step method 327
  - 10.3.8 Linear-quadratic problem with infinite control time . 328
  - 10.3.9 Examples 331
- 10.4 Finite-time Problem with Arbitrary Delays 333
  - 10.4.1 Problem formulation 333
  - 10.4.2 Optimal control and canonical set of equations ... 334
  - 10.4.3 Optimal requlator 336
  - 10.4.4 General characterization of operator $K(s)$ 338
  - 10.4.5 Characterization of components of $K(s)$ by means of differential equations 342
  - 10.4.6 Characterization of the free term of the optimal regulator 345
  - 10.4.7 Generalizations 345
- 10.5 Infinite-time Problem with Arbitrary Delays 347
  - 10.5.1 Problem formulation 347
  - 10.5.2 Optimal regulator 348
  - 10.5.3 Operator Riccati equation and stability of the closed-loop system 350
  - 10.5.4 Characterization of operator K 352
- 10.6 Discrete Approximation of the Linear-quadratic Problem with Finite Control Time 354

10.6.1 Discrete approximation of the optimal control problem 354
10.6.2 Discrete approximation of the optimal regulator . . . 359

References 360

Index 367

PART I
INTRODUCTION

1

Examples of Systems with Time Delays

The rapid development of manufacturing process automatization has been largely due to increasing demands with respect to the rate and quality of production, during and since the Second World War. The high production rates resulted in high speed variation of control variables. The period of time between the measurement of the controlled variable, the control decision and the control signal action therefore grew in importance. The result of a feedback action delay is critical in the case of aeroplanes, rockets, or large remotely controlled plants. The delayed response of the control system to disturbances appearing in the process results in the generation of oscillations in the closed-loop system, and also in the nonstability of this system. When a mathematical model incorporates time delays, equations of a more general form than that of differential equations have to be used. These are difference-differential equations, which are a class of more general functional-differential equations.

Difference-differential equations were initially introduced in the eighteenth century. However, the rapid development of the theory of these equations was prompted by practical needs, and did not come until after the Second World War. The basic theory concerning stability of systems described by equations of this type was developed by Pontryagin (1942). Important works have been written by Bellman and Cooke (1963), Myshkis (1972), Pinney (1958), Halanay (1966), El'sgol'c and Norkin (1971), Yanushevski (1978), Marshall (1979) and Hale (1977). The theory of optimal control of systems with time delays was developed by Krasovski (1962), Oğuztöreli (1966), Salukvadze (1962), Gabasov and Kirillova (1971), Kharatishvili (1966), Chyung and Lee (1966, 1970), Banks and others (1969, 1972, 1973, 1975), Delfour and others (1972, 1974, 1975, 1977a, b, 1980a, b).

A mathematical model in a difference-differential equation form includes as special cases systems described by differential equations, i.e. certain continuous-time control systems, and systems described by difference equations, i.e. certain impulse control systems. Apart from lumped delays, which lead to difference-differential equations, control systems may incorporate so-called distributed delays. These delays occur in distributed parameter systems represented by partial differential equations. In control theory the notion of equivalent delay is also used. This is useful when replacing higher-order differential equations with lower-order difference-differential equations or nonlinear differential equations with linear difference-differential equations.

If we consider the relationship between theory and practice, the question arises as to what sort of phenomena can be represented by a mathematical model which has the form of a difference-differential equation, i.e. where the time delay between the action and the reaction is to be introduced. It appears that this kind of phenomenon occurs not only in technology, but also in biology, economics, etc. The most typical processes of this kind are transport, mixing, burning, evolution, economic fluctuations and bureaucracy. Distributed delays almost always occur in long lines, e.g. electric, heating, hydraulic or pneumatic lines. Long line models, however, can often be well-approximated by means of difference-differential equations.

1.1 Systems with Lumped Delays

1.1.1 Transport processes. Control of coal flow rate on a belt conveyor

The control system (Figure 1.1.1) ensures a constant coal feed rate to a mill. The coal is fed from the container A by the belt conveyor B to the scales C, and from there to the mill D. The required amount of coal w can be fixed by hanging a weight on the left arm of the scales. The difference between the set value w and the actual weight of coal y is fed as an error signal ε to the controller R. The controller forms the control signal u and feeds it to the motor M which drives the slide damper Y.

The change in the amount of coal being transferred from the container A onto the conveyor B at time t is not measured at once, but after a certain period of time h during which the coal is conveyed onto the scales C. The length of this period of time depends on the conveying velocity v and on

1.1 Systems with Lumped Delays

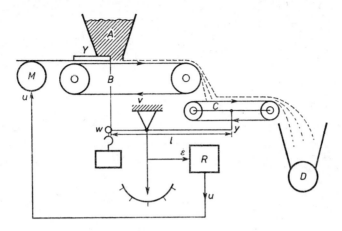

Figure 1.1.1 Control of coal flow rate on a belt conveyor

the distance l between the measurement point and the control point. The time h for the constant velocity v is

$$h = \frac{l}{v}. \quad (1.1.1)$$

The control signal u is transferred through the process without distortions but with a delay h. Assuming a step variation of the signal u, the reaction of the process is illustrated in Figure 1.1.2.

The process is described by the relation between the small changes Δy and Δu around the equilibrium point determined by the set value,

$$\Delta y(t) = K\Delta u(t-h). \quad (1.1.2)$$

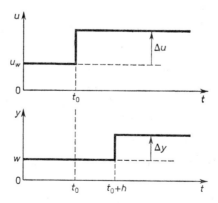

Figure 1.1.2 Step response of a process with delay

The operator transmittance of the control system is

$$G(s) = \frac{Y(s)}{U(s)} = Ke^{-sh}. \tag{1.1.3}$$

Assuming $s = i\omega$, the frequency transmittance is

$$G(i\omega) = Ke^{-i\omega h} = K(\cos\omega h - i\sin\omega h). \tag{1.1.4}$$

A diagram of this is shown in Figure 1.1.3. The relationship (1.1.4) yields the fact that the real and the imaginary frequency characteristics are of a cosine and sine form. The pass band is not limited in this case since the ordinates of

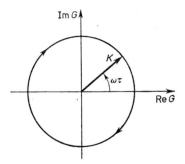

Figure 1.1.3 Frequency characteristic of a process with delay

these characteristics do not approach zero when the frequency ω increases. Of course, the unlimited pass band of the harmonics results from the fact that the delay unit only shifts the input signal and does not introduce any distortions.

1.1.2 Metal rolling system

Metallic tape is put into a pair of reducing rollers (Figure 1.1.4). After rolling, the tape is of thickness x, width w and has a velocity v. Thickness of the reduced tape is measured at a distance d behind the rollers. Measurement of thickness is used to control the distance u between the rollers. Under a simplifying assumption that the width of the material does not change during rolling, we have the following mass conservation law:

$$x(t)v(t) = \text{constant}.$$

Since $v(t)$ is not constant, the delay $\tau(t)$ between the rollers and the thickness sensor is described by the relation

$$d = \int_{t-\tau(t)}^{t} v(s)\,ds.$$

1.1 Systems with Lumped Delays

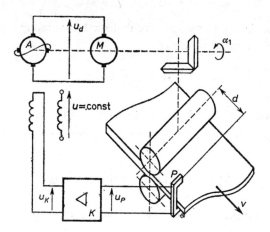

Figure 1.1.4 Metal rolling system

From the two above relations, after differentiating both sides of the latter we obtain

$$\frac{1}{x(t)} - \frac{1-\dot{\tau}(t)}{x(t-\tau(t))} = 0.$$

Assuming that the feedback regulator is realized in the form of an integral controller described by the equation

$$\dot{x}(t) = K[u(t) - x(t-\tau(t))]$$

we obtain the following system of equations:

$$\dot{x}_1(t) = -Kx_1(t-x_2(t)) + Ku(t)$$

$$\dot{x}_2(t) = 1 - \frac{x_1(t-x_2(t))}{x_1(t)}$$

where

$$x_1(t) = x(t)$$
$$x_2(t) = \tau(t).$$

1.1.3 Population models

Let $x(t)$ denote the number of individuals in a population at time t. Let us suppose that every member of the population has a life time T. Assuming that the number of births per unit time is a function of $x(t)$ only, namely $g(x(t))$, we have the simplest model of population growth

$$\dot{x}(t) = g(x(t)) - g(x(t-T)).$$

We can make the above model more realistic by assuming that the number of deaths per unit time at time t is

$$-\int_0^T g(x(t-s))\dot{b}(s)\,ds$$

where $b(s)$ represents the probability of survival to age s.

As in the previous model we can state the following, more complicated relation:

$$\dot{x}(t) = g(x(t)) + \int_0^T g(x(t-s))\dot{b}(s)\,ds$$

which gives a more precise model of population growth.

1.1.4 Economic systems

Delays in economic systems appear in a natural way because decisions and effects are separated by some interval of time.

We shall briefly describe the Kalecki model.

The model assumes that income $Y(t)$ may be split into consumption $C(t)$, investment $I(t)$ and autonomous expenditure A,

$$Y(t) = C(t) + I(t) + A$$

where

$$C(t) = cY(t).$$

Income is determined by investment

$$Y(t) = \frac{I(t)+A}{1-c}. \tag{1.1.5}$$

It is assumed that following the decision to invest $B(t)$ (i.e. to order capital equipment), deliveries are made after a fixed interval of time h, and disbursements $I(t)$ are spread over a period of length h.

If $K(t)$ denotes the stock of capital assets at time t, then $\dot{K}(t)$ is the rate of delivery of the new equipment. The assumption on lag gives

$$\frac{d}{dt}K(t) = B(t-h) \tag{1.1.6}$$

$$I(t) = \frac{1}{h}\int_{t-h}^t B(s)\,ds. \tag{1.1.7}$$

$B(t)$ is determined by the rate of saving $S(t) = (1-c)Y(t)$ and by the stock $K(t)$,

$$B(t) = a(1-c)Y(t) - kK(t) + \varepsilon \tag{1.1.8}$$

1.1 Systems with Lumped Delays

where a, k are positive constants and ε is a trend coefficient.
From (1.1.6) and (1.1.7) we have

$$I(t) = \frac{1}{h}[K(t+h) - K(t)]. \tag{1.1.9}$$

From (1.1.5) and (1.1.9)

$$Y(t) = \frac{1}{h(1-c)}[K(t+h) - K(t)] + \frac{A}{1-c}. \tag{1.1.10}$$

Substituting (1.1.10) and (1.1.7) into (1.1.8) we obtain

$$\frac{d}{dt}K(t+h) = \frac{a}{h}K(t+h) - \left(k + \frac{a}{h}\right)K(t) + (aA + \varepsilon) \tag{1.1.11}$$

which gives the Kalecki model in the form of an equation with lumped delay.

1.1.5 Control in computerized systems with state reconstruction

Suppose we are given a linear system described by the following relations:

$$\dot{x}(t) = Ax(t) + Bu(t)$$
$$y(t) = Cx(t)$$
$$x(t) \in R^n \quad u(t) \in R^m \quad y(t) \in R^k$$
$$A, B, C - \text{matrices}.$$

Let us assume that this system is large and the law of control is complex. Computer control is usually in operation in such cases.

Let us denote the feedback control law by

$$R: R^n \ni x(t) \mapsto u(t) \in R^m.$$

We can distinguish two specific intervals of time:

The first interval of length T_0 is devoted to the reconstruction of the state of the system based on observations of $y(t)$ and $u(t)$ in this interval; reconstructing is not trivial because of the form of matrix C.

The second interval of length T_1 is the delay caused by producing values of control and feedback transmission of information between the computer and the system.

Let us take a linear observer of the system in the form

$$x(t) = \int_0^{T_0} G_1(s) y(t-s) \, ds + \int_0^{T_0} G_2(s) u(t-s) \, ds$$

where G_1, G_2 are $n \times k$ and $n \times m$ matrices, respectively. It is easy to show that G_1, G_2 must fulfil the following equations:

$$I = \int_0^{T_0} G_1(s) C e^{-As} ds$$

$$\int_0^\tau G_1(s) C e^{A(\tau-s)} B ds + G_2(\tau) = e^{A\tau} B \qquad \tau \in [0, T_0].$$

Observers of this integral type are relatively less sensitive with respect to random noise which may appear in the measurement of y and u.

Taking into account the above relations, we construct equations describing the behaviour of the feedback-controlled system:

$$\dot{x}(t) = Ax(t) + Bu(t)$$
$$y(t) = Cx(t)$$
$$\hat{x}(t) = \int_0^{T_0} G_1(s) y(t-s) ds + \int_0^{T_0} G_2(s) u(t-s) ds$$
$$= \int_0^{T_0} G_1(s) Cx(t-s) ds + \int_0^{T_0} G_2(s) u(t-s) ds$$
$$u(t) = R(\hat{x}(t-T_1))$$

where $\hat{x}(t)$ denotes an estimated value of the state of the system at time t on the basis of observations of y and u in the interval $[t-T_0, t]$.

Reducing this system to one equation, we obtain the following functional-differential relation for the closed-loop control system:

$$\dot{x}(t) = Ax(t) + BR\left(\int_0^{T_0} G_1(s) Cx(t-T_1-s) ds + \int_0^{T_0} G_2(s) R(x(t-T_1-s)) ds\right).$$

1.1.6 Remote control

Let us consider the remote control of the motion of an artificial satellite. The satellite can change position by means of jet-propelled motors which permit motion along three independent axes. These motors can be telemetrically controlled by an Earth control centre. This control centre receives telemetric data from the satellite about its current position and velocity. Taking into account the boundedness of the velocity of light and the resulting retardation in feedback transmission of information between the Earth control centre and the satellite, we can write the following equations describing the satellite's motion:

1.1 Systems with Lumped Delays

$$\ddot{x}_1(t) = -\frac{k \cdot x_1(t)}{r^3(t)} + F_1(t) \cdot \frac{1}{m_s}$$

$$\ddot{x}_2(t) = -\frac{k \cdot x_2(t)}{r^3(t)} + F_2(t) \cdot \frac{1}{m_s}$$

$$\ddot{x}_3(t) = -\frac{k \cdot x_3(t)}{r^3(t)} + F_3(t) \cdot \frac{1}{m_s}$$

$$F(t) = G(t-\tau)$$

$$G(t) = R\big(x(t-\tau), \dot{x}(t-\tau)\big)$$

where

$$x(t) = \begin{bmatrix} x_1(t) \\ x_2(t) \\ x_3(t) \end{bmatrix}$$

denotes the current position of the satellite with respect to the centre of the Earth,

$r(t)$—radius of orbit, $r(t) = \|x(t)\|$,

$$F(t) = \begin{bmatrix} F_1(t) \\ F_2(t) \\ F_3(t) \end{bmatrix}$$ —forces produced by jet-propelled motors,

$G(t)$—values of forces sent telemetrically from the Earth control centre to be produced by motors,

$R(x, \dot{x})$—function defined on telemetric data $(x, \dot{x})$ received from the satellite, which are calculated by the Earth control centre; this function might, for example, be the optimal law of control with respect to some goal,

k—gravity constant multiplied by the mass of Earth,

m_s—mass of satellite,

τ—retardation in communication between satellite and Earth centre,

$$\tau = \frac{1}{c}\|x(t) - z\|,$$

c—velocity of light,

z—position of the Earth centre.

Assuming that changes of position are slow as compared with the frequency of feedback communication, we can fix retardation at an appropriate interval of time and obtain the following constant-delay equations describing the feedback control process:

$$\ddot{x}_1(t) = \frac{-kx_1(t)}{r^3(t)} + R_1\big(x(t-2\tau), \dot{x}(t-2\tau)\big) \cdot \frac{1}{m_s}$$

$$\ddot{x}_2(t) = \frac{-kx_2(t)}{r^3(t)} + R_2\big(x(t-2\tau),\ \dot{x}(t-2\tau)\big) \cdot \frac{1}{m_s}$$

$$\ddot{x}_3(t) = \frac{-kx_3(t)}{r^3(t)} + R_3\big(x(t-2\tau),\ \dot{x}(t-2\tau)\big) \cdot \frac{1}{m_s}.$$

1.1.7 Urban traffic

Time delays are a common feature of urban traffic models used for the determination of optimal signal settings at intersections of street networks. The intersections are represented by nodes of the network; they are connected by links with one-way traffic only. Let us denote by $x_i(t)$ the length of the queue of vehicles waiting at the downstream node of the i-th link. The rate of change is given by

$$\dot{x}_i(t) = q_i(t) - r_i(t)$$

where $q_i(t)$ is the rate of arrivals and $r_i(t)$ is the rate of departures. The delays in the vehicle transport process are basically the average times h_i needed to drive through the i-th link from its upstream to downstream node. If we denote the flow that originates in the i-th link (vehicles from parking lots, etc.), we obtain the following formula which expresses the rate of arrivals by the rates of departures from the upstream node:

$$q_i(t) = \sum_{l \in L_i} r_l(t-h_l) a_{li} + z_i(t)$$

where summation is over all links which end at the upstream node of the i-th link, and a_{li} are division factors.

Flow through the i-th link cannot exceed its saturation value s_i. Outflows r_j are controlled by light signals at the intersections: if $u_{ij}(t) =$ green, right of way is given to those vehicles which move from the i-th to the j-th link. If $u_{ij}(t) =$ red, they have no right of way. We neglect here the amber and red-amber signals because for control purposes the normal procedure is to define two light signals only: effective green for the period when the vehicles actually move, and effective red for the rest of the cycle. Let M_i be the set of all links which have their beginnings at the downstream node of the i-th links. Thus,

$$r_i(t) = 0 \quad \text{if} \quad u_{ij}(t) = \text{red for all } j \in M_i.$$

Denote by $G_j(t)$ the set of all links from L_j for which $u_{lj}(t) =$ green, and define

$$R_j(t) = \sum_{l \in G_j(t)} r_l(t) a_{lj}.$$

Then
$$R_j(t) = s_j$$
if for at least one link from $G_j(t)$ we have $x_l(t) > 0$. Otherwise
$$R_j(t) = \min\left\{s_j, \sum_{l \in G_j(t)} q_l(t) a_{lj}\right\}.$$
Introducing constant division factors we obtain
$$r_i(t) = b_{ij} R_j(t).$$

Of course this model of urban traffic is valid only provided $x_i(t)$ are much less than the lengths of links. A more realistic model is obtained if we introduce distributed delays to the formula for the rate of arrivals,
$$q_i(t) = \sum_{l \in L_i} \int_{t-d}^{t+d} r_l(s - h_i) g(s) \, ds \, a_{li} + z_i(t).$$
The parameter d and weighting function g characterize the dispersion of vehicle groups.

The controls in the system are the light signals at intersections, i.e. the functions u_{ij}. In a real control problem many constraints are imposed on these functions. All of them must be periodic with a common period called the cycle, the green light must not normally be displayed more than once per cycle for each approach and the duration times for signal lights are limited. Security demands give rise to important interdependence between light signals.

This model can serve as a basis for the optimization of traffic lights. It allows us to estimate D, the global time lost by drivers while waiting in queues at intersections. A simple calculation yields
$$D = \sum_i \int_0^T q_i(t) t_1 \, dt$$
where T is the period under consideration and t_1 is determined from the equality
$$\int_t^{t_1} r_i(s) \, ds = x_i(t).$$

1.2 Systems with Distributed Delays

The processes described above include lumped delays, and so they can be represented by difference-differential equations. However the majority of thermal processes, together with processes in which the signal is transmitted by long electric, hydraulic or other lines, show a delay distributed along the entire length of the spatial coordinate. This time delay is usually accompanied by disturbances introduced to the transmitted signal. Processes of this type are often represented by partial differential equations.

1.2.1 Electric transmission line

The electric transmission line is a good illustration of distributed parameter systems in which a distributed delay occurs. The signals transmitted by an object of this type are delayed and distorted.

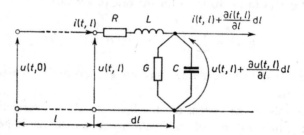

Figure 1.2.1 An infinitesimal part of a long electric line

A homogeneous electric line will be considered, i.e. one where the parameters per unit length (resistance R, inductance L, leakance G, and capacity C) are constant and independent of coordinate l. An infinitesimal part of the long line is shown in Figure 1.2.1. Using loop analysis, Kirchhoff's voltage law gives

$$u(t,l) - \left[u(t,l) + \frac{\partial u(t,l)}{\partial l} dl\right] = R dl\, i(t,l) + L dl \frac{\partial i(t,l)}{\partial t}. \quad (1.2.1)$$

Using node analysis, Kirchhoff's current law gives

$$i(t,l) - \left[i(t,l) + \frac{\partial i(t,l)}{\partial l} dl\right] = G dl\, u(t,l) + C dl \frac{\partial u(t,l)}{\partial t}. \quad (1.2.2)$$

Ordering the above expressions, we obtain the following two basic equations describing the long electric line:

$$-\frac{\partial u(t,l)}{\partial l} = R i(t,l) + L \frac{\partial i(t,l)}{\partial t} \quad (1.2.3)$$

$$\cdot \frac{\partial i(t,l)}{\partial l} = G u(t,l) + C \frac{\partial u(t,l)}{\partial t}. \quad (1.2.4)$$

1.2 Systems with Distributed Delays

The system of partial differential equations (1.2.3), (1.2.4) will be solved using the double Laplace transform. After transformation with respect to variable t we obtain

$$-\frac{dU(s, l)}{dl} = I(s, l)(R+sL) - Li(0, l) \qquad (1.2.5)$$

$$-\frac{dI(s, l)}{dl} = U(s, l)(G+sC) - Cu(0, l). \qquad (1.2.6)$$

Transforming the equations again with respect to variable l and then ordering them, we obtain

$$qU(s, q) + (R+sL)I(s, q) = U(s, 0) + LI(0, q) \qquad (1.2.7)$$
$$(G+sC)U(s, q) + qI(s, q) = CU(0, q) + I(s, 0). \qquad (1.2.8)$$

Hence

$$U(s, q) = \frac{q[U(s, 0) + LI(0, q)] - (R+sL)[CU(0, q) + I(s, 0)]}{q^2 - (G+sC)(R+sL)} \qquad (1.2.9)$$

$$I(s, q) = \frac{q[CU(0, q) + I(s, 0)] - (G+sC)[U(s, 0) + LI(0, q)]}{q^2 - (G+sC)(R+sL)}. \qquad (1.2.10)$$

The characteristic equation is of the form

$$q^2 - (G+sC)(R+sL) = 0. \qquad (1.2.11)$$

Its two roots are

$$q_{1,2} = \pm \gamma \qquad (1.2.12)$$

where

$$\gamma(s) = \sqrt{(G+sC)(R+sL)} \qquad (1.2.13)$$

is termed a wave propagation constant corresponding to a given frequency s.

It is assumed that at the instant $t = 0$ there is no energy stored in the line along its entire length. Thus

$$U(0, q) = 0 \quad \text{and} \quad I(0, q) = 0, \qquad (1.2.14)$$

and equations (1.2.9) and (1.2.10) together with (1.2.13) yield

$$U(s, q) = \frac{q}{q^2 - \gamma^2} U(s, 0) - \frac{R+sL}{q^2 - \gamma^2} I(s, 0) \qquad (1.2.15)$$

$$I(s, q) = \frac{q}{q^2 - \gamma^2} I(s, 0) - \frac{G+sC}{q^2 - \gamma^2} U(s, 0). \qquad (1.2.16)$$

The inverse transform with respect to variable q yields

$$U(s, l) = U(s, 0)\cosh \gamma l - I(s, 0) Z \sinh \gamma l \qquad (1.2.17)$$

$$I(s, l) = I(s, 0)\cosh \gamma l - \frac{U(s, 0)}{Z} \sinh \gamma l \qquad (1.2.18)$$

where

$$Z(s) = \sqrt{\frac{R+sL}{G+sC}} \qquad (1.2.19)$$

is the line wave impedance. From equations (1.2.17) and (1.2.18) the transmittance (or transfer) matrix between the points $l = 0$ and $l = l_k$ of the unloaded long electric line can be determined. Denoting

$$V(s) = \begin{bmatrix} U(s,0) \\ I(s,0) \end{bmatrix} \qquad (1.2.20)$$

$$X(s) = \begin{bmatrix} U(s,l_k) \\ I(s,l_k) \end{bmatrix} \qquad (1.2.21)$$

we have

$$G(s)V(s) = X(s) \qquad (1.2.22)$$

where

$$G(s) = \begin{bmatrix} \cosh \gamma l_k & -Z \sinh \gamma l_k \\ -\dfrac{1}{Z} \sinh \gamma l_k & \cosh \gamma l_k \end{bmatrix}. \qquad (1.2.23)$$

is the transmittance matrix. Thus the line is a double input and double output element. Its block diagram is shown in Figure 1.2.2.

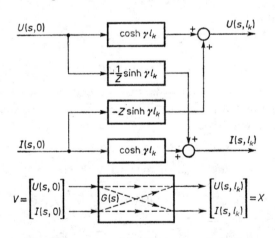

Figure 1.2.2 Block diagram of long electric line

If instead of the conditions $U(s, 0)$ and $I(s, 0)$ at the beginning of the line, the conditions $U(s, l_k)$ and $I(s, l_k)$ are given at the end of the line, equations (1.2.17) and (1.2.18) are transformed by substituting l by $l-l_k$ and changing the signs of the hyperbolic sine terms. Then

1.2 Systems with Distributed Delays

$$U(s, l) = U(s, l_k)\cosh\gamma(l_k-l) + I(s, l_k)Z\sinh\gamma(l_k-l) \quad (1.2.24)$$

$$I(s, l) = I(s, l_k)\cosh\gamma(l_k-l) + \frac{U(s, l_k)}{Z}\sinh\gamma(l_k-l). \quad (1.2.25)$$

We can now consider a case frequently met in practice when there are lumped parameter elements at both ends of the line (Figure 1.2.3). The system is described by the following operator equations:

$$E_0(s) - I(s, 0)Z_1 = U(s, 0) \quad \text{at the beginning of the line} \quad (1.2.26)$$

$$U(s, l_k) = I(s, l_k)Z_z \quad \text{at the end of the line.} \quad (1.2.27)$$

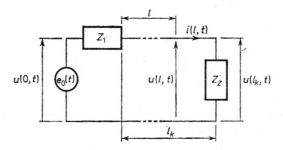

Figure 1.2.3 Long line with lumped parameter elements on input and output

From these relations and from (1.2.17) and (1.2.18) we obtain

$$U(s, l) = \frac{E_0}{Z_z} \frac{Z_z\cosh\gamma(l_k-l) + Z\sinh\gamma(l_k-l)}{\left(1 + \frac{Z_1}{Z_z}\right)\cosh\gamma l_k + \left(\frac{Z}{Z_z} + \frac{Z_1}{Z}\right)\sinh\gamma l_k} \quad (1.2.28)$$

and

$$I(s, l) = \frac{E_0}{Z_z} \frac{\cosh\gamma(l_k-l) + \frac{Z_z}{Z}\sinh\gamma(l_k-l)}{\left(1 + \frac{Z_1}{Z_z}\right)\cosh\gamma l_k + \left(\frac{Z}{Z_z} + \frac{Z_1}{Z}\right)\sinh\gamma l_k}. \quad (1.2.29)$$

If we assume that the load impedance Z_z is equal to the wave impedance Z or that the length of the line is infinite, $l_k = \infty$, expressions (1.2.28) and (1.2.29) become considerably simpler:

$$U(s, l) = \frac{Z}{Z+Z_1}E_0 e^{-\gamma l} \quad (1.2.30)$$

$$I(s, l) = \frac{1}{Z+Z_1}E_0 e^{-\gamma l} \quad (1.2.31)$$

since in this case there is no wave reflection from the end of the line.

Further simplification in formulas (1.2.30) and (1.2.31) can be obtained by assuming the impedance of the supply voltage source equal to zero, $Z_1 = 0$. Then

$$U(s, l) = E_0 e^{-\gamma l} \tag{1.2.32}$$

$$I(s, l) = \frac{E_0}{Z} e^{-\gamma l}. \tag{1.2.33}$$

Now we shall consider a particularly interesting case of the long electric line without distortions. If

$$\frac{R}{L} = \frac{G}{C} = \alpha, \tag{1.2.34}$$

then the line wave impedance does not depend on s,

$$Z = \sqrt{\frac{L}{C}}. \tag{1.2.35}$$

The wave propagation constant is

$$\gamma = \sqrt{LC}(s+\alpha) = \frac{s+\alpha}{v}. \tag{1.2.36}$$

From formulas (1.2.32) and (1.2.33)

$$U(s, l) = E_0(s) e^{-\frac{l}{v}\alpha} e^{-\frac{l}{v}s} \tag{1.2.37}$$

$$I(s, l) = \sqrt{\frac{C}{L}} E_0(s) e^{-\frac{l}{v}\alpha} e^{-\frac{l}{v}s}. \tag{1.2.38}$$

Then

$$\left.\begin{aligned} u(t, l) &= e_0\left(t - \frac{l}{v}\right) \exp\left(-\frac{l}{v}\alpha\right) \\ i(t, l) &= \sqrt{\frac{C}{L}} e_0\left(t - \frac{l}{v}\right) \exp\left(-\frac{l}{v}\alpha\right) \end{aligned}\right\} \text{ for } t \geq \frac{l}{v} \tag{1.2.39}$$

$$u(t, l) = 0 \quad i(t, l) = 0 \quad \text{for } t < \frac{l}{v}. \tag{1.2.40}$$

These formulas describe wave propagation along the long line with constant speed v and with no distortions. However, the wave dies out exponentially with an increase in distance l. The distance decay factor is

$$T_l = \frac{v}{\alpha}. \tag{1.2.41}$$

1.2 Systems with Distributed Delays

Formulas (1.2.37) and (1.2.38) directly yield the line transmittances (with respect to time)

$$G_U(s, l) = \frac{U(s, l)}{E_0(s)} = e^{-\frac{l}{v}\alpha} e^{-\frac{l}{v}s} \qquad (1.2.42)$$

$$G_I(s, l) = \frac{I(s, l)}{E_0(s)} = \sqrt{\frac{C}{L}} e^{-\frac{l}{v}\alpha} e^{-\frac{l}{v}s}. \qquad (1.2.43)$$

The step response of the line (to a constant input voltage E) is shown in Figure 1.2.4a. The appropriate cross-sections are shown in Figure 1.2.4b and c. It can be easily seen that for a given distance l_1 the line behaves as a delay element with a lumped delay $\tau = l_1/v$. The gain (or amplification factor) of this

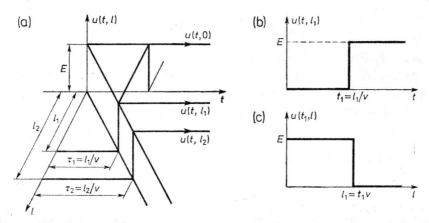

Figure 1.2.4 Step response of the long line with $G = R = 0$ (a); cross-section of the voltage wave at a fixed distance l_1 (b); cross-section of the voltage wave at a fixed moment of time τ_1 (c)

element depends on l_1 and is equal to $\exp(-\alpha l_1/v)$. Only in the case of negligible resistance and leakance, $R = 0$ and $G = 0$, there are no losses in the line and the wave does not decay during its movement along the line.

From formula (1.2.42) the voltage amplitude characteristic can be easily obtained as

$$|G_U(i\omega, l_1)| = e^{-\frac{l_1}{v}\alpha}, \qquad (1.2.44)$$

as can its phase characteristic

$$\varphi(\omega, l_1) = -\omega \frac{l_1}{v}. \qquad (1.2.45)$$

1.2.2 Heat exchangers

Heat exchangers are typical objects with distributed parameters. They will be dealt with on the basis of the works of Nöldus (1967) and Devyatov (1950). The transient state of the heat exchange process in a heat exchanger of the pipe-in-pipe type can be approximately represented by the following system of equations:

$$\frac{\partial T_1}{\partial t} + h_1 \frac{\partial T_1}{\partial l} = k_{s1}(T_s - T_1) \tag{1.2.46}$$

$$\frac{\partial T_2}{\partial t} + h_2 \frac{\partial T_2}{\partial l} = k_{s2}(T_s - T_2) \tag{1.2.47}$$

$$\frac{\partial T_s}{\partial t} = k_{1s}(T_1 - T_s) + k_{2s}(T_2 - T_s) \tag{1.2.48}$$

where T_1 is the temperature of the first medium, T_2 is the second medium temperature, T_s is the partition wall temperature, the k are the medium-to-wall heat exchange coefficients and the h are the coefficients characterizing the direct flow and counterflow cases. In the case of direct flow

$$h_1 = v_1 \quad h_2 = v_2 \tag{1.2.49}$$

and in the case of counterflow

$$h_1 = v_1 \quad h_2 = -v_2 \tag{1.2.50}$$

where v are the velocities of the medium flows.

Transforming equations (1.2.46), (1.2.47) and (1.2.48) we obtain (for simplicity the same symbols are used for Laplace transforms and their originals)

$$(s + k_{s1}) T_1(s, l) + h_1 \frac{\mathrm{d}T_1(s, l)}{\mathrm{d}l} - k_{s1} T_s(s, l) = T_1(0, l) \tag{1.2.51}$$

$$(s + k_{s2}) T_2(s, l) + h_2 \frac{\mathrm{d}T_2(s, l)}{\mathrm{d}l} - k_{s2} T_s(s, l) = T_2(0, l) \tag{1.2.52}$$

$$(s + k_{1s} + k_{2s}) T_s(s, l) - k_{1s} T_1(s, l) - k_{2s} T_2(s, l) = T_s(0, l). \tag{1.2.53}$$

It is assumed that the initial values $T_1(0, l)$, $T_2(0, l)$ and $T_s(0, l)$ satisfy the stationary heat exchange equations which can be derived from (1.2.46), (1.2.47) and (1.2.48) by equating all time derivatives to zero. The equations for the temperature deviations from the steady-state values are therefore of the form

1.2 Systems with Distributed Delays

$$(s+k_{s1})\Delta T_1(s, l)+h_1\frac{\mathrm{d}\Delta T_1(s, l)}{\mathrm{d}l}-k_{s1}\Delta T_s(s, l) = 0 \quad (1.2.54)$$

$$(s+k_{s2})\Delta T_2(s, l)+h_2\frac{\mathrm{d}\Delta T_2(s, l)}{\mathrm{d}l}-k_{s2}\Delta T_s(s, l) = 0 \quad (1.2.55)$$

$$(s+k_{1s}+k_{2s})\Delta T_s(s, l)-k_{1s}\Delta T_1(s, l)-k_{2s}\Delta T_2(s, l) = 0 \quad (1.2.56)$$

where

$$\Delta T_1(s, l) = T_1(s, l)-\frac{T_1(0, l)}{s} \quad (1.2.57)$$

$$\Delta T_2(s, l) = T_2(s, l)-\frac{T_2(0, l)}{s} \quad (1.2.58)$$

$$\Delta T_s(s, l) = T_s(s, l)-\frac{T_s(0, l)}{s}. \quad (1.2.59)$$

Solving the equation system (1.2.54), (1.2.55) and (1.2.56) with respect to the temperature deviations ΔT_1 and ΔT_2 we obtain

$$\Delta T_1(s, l_k) = G_{11}(s)\Delta T_1(s, 0)+G_{21}(s)\Delta T_2(s, 0) \quad (1.2.60)$$

$$\Delta T_2(s, l_k) = G_{12}(s)\Delta T_1(s, 0)+G_{22}(s)\Delta T_2(s, 0) \quad (1.2.61)$$

where l_k is the exchanger length.

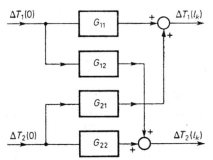

Figure 1.2.5 Block diagram of a heat exchanger

The block diagram of the heat exchanger (Figure 1.2.5) is therefore in full accordance with the diagram obtained for the electric line (Figure 1.2.2). The particular transmittances, however, are of a complicated structure. In the case of direct flow they are

$$G_{11}(s) = e^{A(s)}\cosh B(s)-\frac{C(s)}{B(s)}\sinh B(s) \quad (1.2.62)$$

$$G_{21}(s) = \frac{D_1(s)}{1+\dfrac{s}{k_{2s}}} e^{A(s)}\frac{\sinh B(s)}{B(s)} \quad (1.2.63)$$

and in the case of counterflow

$$G_{11}(s) = \frac{e^{A(s)}}{\cosh B(s) + \dfrac{C(s)}{B(s)} \sinh B(s)} \qquad (1.2.64)$$

$$G_{21}(s) = \frac{D_1(s)}{1 + \dfrac{s}{k_{2s}}} \frac{\sinh B(s)}{B(s)} \frac{1}{\cosh B(s) + \dfrac{C(s)}{B(s)} \sinh B(s)} \qquad (1.2.65)$$

where

$$A(s) = -\frac{s(\tau_1 + \tau_2) + D_1(s) + D_2(s)}{2} \qquad (1.2.66)$$

$$B(s) = \sqrt{C(s)^2 + \frac{D_1(s) D_2(s)}{\left(1 + \dfrac{s}{k_{2s}}\right)\left(1 + \dfrac{s}{k_{1s}}\right)}} \qquad (1.2.67)$$

$$C(s) = \frac{s(\tau_1 - \tau_2) + D_1(s) - D_2(s)}{2} \qquad (1.2.68)$$

$$D_1(s) = \frac{1}{h_1} \frac{k_{2s} k_{s1}}{k_{1s} + k_{2s}} \frac{1 + \dfrac{s}{k_{2s}}}{1 + \dfrac{s}{k_{1s} + k_{2s}}} \qquad (1.2.69)$$

$$D_2(s) = \frac{1}{h_2} \frac{k_{1s} k_{s2}}{k_{1s} + k_{2s}} \frac{1 + \dfrac{s}{k_{1s}}}{1 + \dfrac{s}{k_{1s} + k_{2s}}} \qquad (1.2.70)$$

$$\tau_1 = \frac{l_k}{h_1} \qquad \tau_2 = \frac{l_k}{h_2}. \qquad (1.2.71)$$

The transmittances $G_{12}(s)$ and $G_{22}(s)$ are obtained by exchanging indexes 1 and 2 in the above expressions, since equations (1.2.54), (1.2.55) and (1.2.56) are symmetric with respect to these indexes. Assuming that k_{1s} and k_{2s} approach infinity, i.e. that the influence of the partition wall is negligible (thin-wall exchanger) one obtains a remarkable simplification (Nöldus, 1967). The examples of logarithmic amplitude and phase characteristics shown in Figures 1.2.6 and 1.2.7 are taken from Nöldus (1967). The oscillations seen in these characteristics are typical for heat exchangers. An approximation of the heat exchanger model by a transmittance containing only a rational function of the operator s and a term with a lumped delay may lead to considerable errors, since the time constant of the partition wall is disregarded.

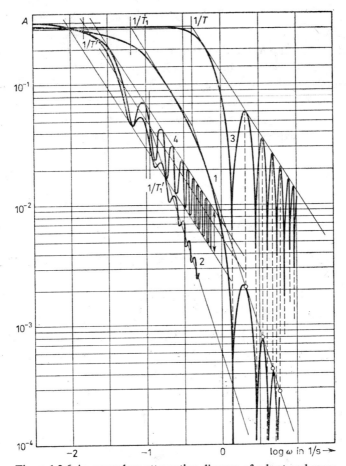

Figure 1.2.6 An exemplary attenuation diagram of a heat exchanger

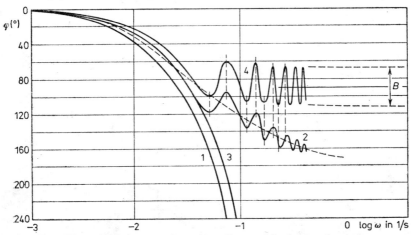

Figure 1.2.7 An exemplary phase diagram of a heat exchanger

It is particularly important that this time constant be taken into account in the case of direct-flow heat exchangers.

In practice, however, such approximations using rational transmittances and lumped elements are often the only modelling possibility. As has been pointed out by Devyatov, in the case of counterflow exchangers $G_{11}(s)$ for small s can be expressed approximately in the form

$$G_{11}(s) = \frac{G_{11}(0)}{1+as} e^{-s\tau_1} \qquad (1.2.72)$$

where

$$a = -\tau_1 - \frac{d}{ds} \ln G_{11}(s) \bigg|_{s=0}. \qquad (1.2.73)$$

The approximation of a heat exchanger within the entire frequency range can be realized by the relationship

$$G_{11}(s) = G_{11}(0) \frac{(1+a_1 s)(1+a_2 s)}{(1+a_3 s)(1+a_4 s)} e^{-s\tau_1}. \qquad (1.2.74)$$

a_i, $i = 1, \ldots, 4$, are calculated on the basis of the first two terms of the expansion of $G_{11}(s)$ for low frequencies (small s), and on the basis of the first three terms of the expansion for high frequencies.

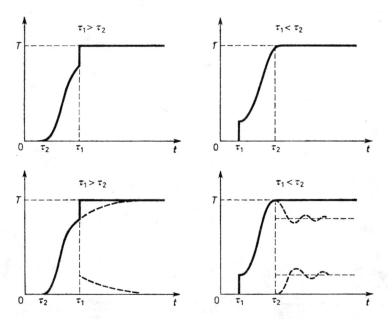

Figure 1.2.8 Step responses of heat exchanger

1.2 Systems with Distributed Delays

The remaining transmittances G_{12}, G_{21} and G_{22} are approximated in a similar way.

To determine the approximate transmittances of direct-flow heat exchangers, one should proceed as in the case of counterflow exchangers. However, certain singularities of the step response shape should be taken into account. The step responses for different τ_1 and τ_2 are shown in Figure 1.2.8.

If $\tau_1 > \tau_2$, the approximate transmittance is

$$G_{11}(s) = c_1 \frac{a_2 s^2 + a_1 s + 1}{b^2 s^2 + 2\xi bs + 1} e^{-s\tau_1} + c_2 \frac{e^{-s\tau_2}}{b^2 s^2 + 2\xi bs + 1} \qquad (1.2.75)$$

and if $\tau_1 < \tau_2$,

$$G_{11}(s) = \left[a_0 + c_1 \frac{a_3 s + 1}{b^2 s^2 + 2\xi bs + 1} \right] e^{-s\tau_1} + c_2 \frac{e^{-s\tau_2}}{b^2 s^2 + 2\xi bs + 1}. \qquad (1.2.76)$$

The number of parameters is greater than in the case of counterflow. These can be determined on the basis of the step responses, taking into account the following:
(a) the magnitude a_0 of the initial jump at instant $t = \tau_1$,
(b) the first derivative at $t = \tau_1 + 0$,
(c) the second derivative at $t = \tau_1 + 0$,
(d) the magnitude of the step response in the steady state,
(e) the magnitude of the step response at $t = \tau_2$.

2

Formulation of Control Problems

2.1 Model Construction

In a model, which is a formal description of a system, we wish to incorporate information necessary for the control of this system. Essentially control a purposeful action, and we must have a sufficiently accurate means to forecast its effect. This is what the model is for.

The construction of a model consists of two stages:
(a) choice of a class of models, and
(b) selection of a particular model from this class.

At stage (a) two different, frequently intermingling and complementary approaches can be distinguished. The choice of class of models may be based on a profound and detailed knowledge of the physical, chemical, economic or other phenomena which occur in the controlled process. The disadvantage of this approach is that it often leads to unnecessarily complicated models. The other approach, which is closer to phenomenological or behavioural methods of analysis, merely takes the effectiveness of the model into account, and not its justification. The simplest possible class of models (from the mathematical point of view) is chosen, provided they describe the system or process with sufficient accuracy. Short-sightedness is a real danger here. Models constructed in this way are only to be used strictly within the range of conditions for which they have been created.

Stage (b), called identification, is based on measurements on the real system (process). In order to choose a particular model from a given class, a number of numerical or functional parameters must be determined so as to achieve consistency between model forecasts and measurements.

Identification is a basic element of several other notions of control theory which are distinguished for traditional or formal reasons, or because of the specificity of technical realization. Adaptation consists in periodically updating model parameters according to changes in environment. The determination of the state of the process from measurements is traditionally called observation or state reconstruction. In the framework of probabilistic and stochastic models the word 'estimation' is used for identification which is based on measurements with stochastic errors. In the theory of stochastic processes, filtration is used for the estimation of the current value of a parameter or state based on output measurements performed up to the moment of estimation. The similar estimation of a future value is called prediction and that of a past value, interpolation.

2.2 Stability and Level Stabilization

It is possible to control a process on the basis of its model only if it allows the effects of control to be predicted with sufficient accuracy. To that end it is necessary that small errors in the determination of initial conditions, model parameters and the circumstances under which the process proceeds, do not cause large errors in the future results of control action. In other words, the system or process should not be too sensitive to these factors. One of the basic properties of a system which ensures this is stability. Of course an object or process which we intend to control need not be stable beforehand. As a rule, one of the fundamental tasks of an automatic control system is to generate a control signal (on the basis of measurement of suitably chosen output quantities) such that the whole control system, consisting of the controlled process and the controller (the control generating device), is stable.

A system is said to be at its equilibrium point if, provided there are no external influences, the quantities characterizing the behaviour of the system do not change in time. Usually this means that the state of the system is constant. Postponing precise mathematical formulation we can say that an equilibrium point is (asymptotically) stable if any small deviation caused by a perturbation of short duration vanishes automatically with time. Therefore an asymptotically stable control system automatically maintains, or stabilizes, its equilibrium state. This manifests itself in practice by the fact that the observed and measured output quantities which characterize the system's behaviour oscillate irregularly with small amplitude around constant values.

One of the most common tasks of automatic control systems is level stabilization which consists in maintaining given constant values of some output quantities, no matter what the external influences, model inaccuracy or stochastic disturbances are. In practice this problem is solved by the construction of a controller that creates an asymptotically stable equilibrium point of the control system where the respective output quantities assume the desired values.

2.3 Terminal Control

Let us imagine a traveller who has to get to a place A by a particular time despite problems of transport. To this end he undertakes a number of actions, preceded by decisions, which we may call control. From the traveller's point of view reaching A in time is crucial for quality of this control. Such factors as comfort, route or to some extent also cost, are clearly of minor importance. Therefore it is the final state of the process, its termination which is most significant, and not the course. Another typical example can be found in traffic control. Dispersing a traffic jam during rush hours is a serious control problem. Here again the process of dispersing itself is not of prime importance. The control purpose is to restore normal traffic conditions as soon as possible. In a planned economy, enterprises are evaluated at the end of each planning period. Therefore the control task here is to achieve a predetermined terminal state in the system.

In general, control problems where the aim is to be achieved at the end of the process are called terminal control problems.

2.4 Tracking Problems

A wide class of control problems can be formulated as tracking problems. It is postulated that the output trajectory of the controlled process (i.e. the trajectory of the measurable quantities characterizing the process) follow a given function of time called the reference trajectory. For example, control of a space vehicle consists in correcting its velocity by activating appropriate rocket motors so that the vehicle trajectory be close to the planned one. Let us note that if the reference trajectory is constant the problem of level stabilization arises as a special case.

Also for the control a reference function is often introduced. For example, it might be a control which generates the reference output trajectory (provided there are no disturbances), or a control function for which energy consumption is minimal. Frequently the model used to control a process is a result of the linearization of a more complicated nonlinear model. In this case the reference control and reference trajectory are determined by the point in the function space at which the model was linearized.

2.5 Optimal Control

It is worth noting that in many cases it is convenient to treat terminal control problems as tracking problems. In order to do this the final fragment of output trajectory, or even just its final point, must be assigned much more weight than the rest of it—in the sense that a high level of accuracy of the overlapping of the real and the reference trajectory is required only at the end.

2.5 Optimal Control

In theoretical physics there is a tendency to represent processes as solutions to variational problems of the minimization of appropriately constructed functionals. This approach makes it possible for laws of Nature to be formulated concisely. Also in technology and economy optimization is increasingly used as a universal method of analysis and design. In static problems methods based on optimization are already well established. In dynamic problems, among which control problems are usually reckoned to belong, this is less evident because of the theoretical, computational and implementational difficulties which are met in the case of more complicated control systems.

A typical formulation of an optimal control (or dynamic optimization) problem contains as its basic elements a model of the controlled dynamic process and a performance functional. Within the model a space of controls $\mathcal{U}$ and a set of admissible controls contained in $\mathcal{U}$ must be determined. A space of parameters $\mathcal{P}$ is also defined whose elements in the optimization problem can be scalar decision variables such as time of termination of the process, its initial state or gain coefficients. To each control of $\mathcal{U}$ and each parameter of $\mathcal{P}$ the model assigns an output trajectory which belongs to a given space $\mathcal{Y}$. Thus a function $g: \mathcal{P} \times \mathcal{U} \to \mathcal{Y}$ is defined. In the widely used state space approach a state space $\mathcal{X}$ is introduced, $\mathcal{X} \subset \mathcal{P}$, and a state equation which determines the function $g_1: \mathcal{P} \times \mathcal{U} \to \mathcal{X}$. The relation between the current state and control values on the one hand and the current value of output (that is, the current values of measured or observed quantities) on the other, $g_2: \mathcal{X} \times U \to Y$, is given by an output equation, where U is a space of momentary control values, and Y a space of momentary output values. Thus for each moment t we have $y(t) = g_2(x(t), u(t))$, provided the system is stationary (constant in time).

The constraints which determine the sets of admissible controls and admissible parameters are an important part of the model. They may be given directly or in the form of constraints imposed on the states or outputs obtained from the system equations.

Within the constraints imposed on control there are usually many possible algorithmic and constructional solutions, yielding different admissible controls. In most cases a control designer is able to build up a criterion that allows him to order the admissible controls according to their quality, that is to assign numerical estimates to them. A performance functional defined in this way may be related to some easily measurable physical or economic quantities such as energy consumption, cost, income, time taken to reach the goal or tracking accuracy. We should stress that in an optimization problem the purpose of the control, and the demands and conditions imposed on the control and the parameters, can be expressed both in the performance functional and in the constraints, and these are to some extent equivalent. It is convenient to express those requirements that must be strictly satisfied in the form of constraints, whereas those which are intuitively related to loss or profit and with respect to which there is more tolerance may be expressed in a performance functional.

From the formal point of view an optimal control problem consists in finding an admissible control u_0 and possibly an admissible parameter p_0 which minimize the performance functional $J: \mathscr{P} \times \mathscr{U} \to R$. The elements u_0 and p_0 are called optimal. If the state space approach is used, the performance functional is defined directly on the pair (control, output trajectory), nevertheless with the aid of the state equation and the output equation the functional J can be easily determined. Of course in many cases J depends on controls only, and then we have $J: \mathscr{U} \to R$.

Practical implementations of optimal control theory—in the form of optimally controlled systems—are still rare. At present the role of the theory is rather to produce patterns which should be aimed for in real systems. If a performance functional value obtained in a real system is close to the optimal one obtained from a model, the existing controller may be considered satisfactory. If not, the theoretically optimal controls and trajectories give the designer some clue how to improve the control system, and thus they help to construct suboptimal control systems.

The rapid development of computer technology, especially the common use of microcomputers in process control, leads us to expect that the range of practical applications of optimal control will also expand rapidly. To this end a further development of the theory is necessary so that it may be widely used in analysis of realistic, complex problems.

The present theory of optimal control is an extension of the classical calculus of variations which has proved ineffective in many of the problems of modern technology. The landmark work of Pontryagin and his group, and also of Bellman in the 1950's gave birth to two new approaches to variational problems, one based on the maximum principle and the other on the optimality principle. Today theoretical work concentrates on the extension of the class of models (discrete-continuous models), multicriterial optimization where control quality is estimated by means of many independent performance functionals, and on optimal control in large-scale systems, in multilevel, multilayer and dispersed structures. The sensitivity of optimal control systems with respect to model inaccuracy, disturbances, etc. becomes very important. The problem of optimization of the measurement or observation process together with control, receives much attention. Considerable development has been seen in the theory of adaptive systems. Qualitatively new methods of analysis and design of optimal systems, based on man–computer dialogue, are constantly being developed.

2.6 The Existence of Solutions to Control Problems

A fundamental question in every control problem is one of goal attainability. In other words, we ask if there exists an admissible control such that the control system will accomplish its task. This question is dealt with theoretically in the problems of controllability and stabilizability, and from another point of view it is involved in the problem of optimal control existence. Generally speaking, the controllability of a system refers to the question as to what output or state trajectories are available. In most cases the question is whether the system can be steered to a certain state or set in the state space, or whether an output variable can reach a given value at some moment (or a value from a given set). Stabilizability is a more specific feature of a system. In practice it means that certain predetermined quantities characterizing the system's behaviour can be maintained at a given level with good accuracy. As the term itself implies, stabilizability is necessary for a solution of the stabilization problem to exist.

Frequently controllability or stabilizability are necessary conditions of existence of a solution of an optimal control problem. However, in many cases these conditions are not sufficient. A more thorough investigation often requires the solution of difficult mathematical problems concerning the existence of a functional minimum in a functional space.

Along with existence, the uniqueness of optimal control is established. This is important since if in a real problem many optimal solutions exist, there is a chance to impose additional conditions or to make demands which would make the solution unique.

2.7 Feedback Regulator and Connected Problems

Feedback occurs in all control systems where control is determined or where corrections of control are made on the basis of output measurements.

Thus a closed feedback loop is formed which works in the following way. The output is measured, which permits estimating whether the control aim is being realized with sufficient accuracy. If it is not, a new corrected control is determined using a model and this in turn generates new output values. Measurement of these values closes the loop.

If the measurements are used to modify or correct the model, we call the control system adaptive. We call control in a closed loop (automatic) regulation if the loop works automatically, permanently (in the sense that it is always active or ready to act), and if it maintains a one-directional flow of interactions from the output towards the input of the system.

Formally, a regulator determined on state is defined as a function that at every moment assigns a control value to the state of the system. This function may depend on time. An analogous function which assigns control values to output values is called a regulator determined on output.

In optimal control systems the regulator determined on output is constructed as a series connection of two devices—an observer which estimates the current state on the basis of the output trajectory and a regulator determined on state.

Where we have at our disposal a perfectly accurate model of the control system, closed-loop control is exactly equivalent to open-loop control, where control is determined once and for all at the beginning of control process and is not subject to modifications resulting from output measurements. However, because of its lesser sensitivity with respect to measurement and modelling errors and to stochastic disturbances, feedback control shows many advantages in real systems due to which it is in many cases the only way of accomplishing control tasks. As a rule stabilization and tracking systems contain feedback loops. Generally speaking, feedback is used in all cases where the influence of stochastic factors such as disturbances, measurement errors, model inaccuracy, small and irregular external interactions must be reduced.

As has already been said, regulation, and especially optimal regulation, demands current state reconstruction. This creates another existence problem in the investigation of optimal feedback control systems. Besides the questions posed in Section 2.6, we must find out whether the choice of measured output quantities is sufficient in order to obtain the information concerning system behaviour needed to determine optimal control. This is the general problem of observability; usually it refers to the reconstruction of the current state of the system on the basis of knowledge of the output trajectory and control up to the present moment.

2.8 Analysis and Synthesis of Control Systems

To round off this introduction we shall briefly describe how the problems formulated above are interconnected in the general methodology of analysis and synthesis of control systems.

In analysis we deal with a given control system. Depending on our objective we can either search for a model (or improve an existing one), or try to answer questions concerning selected qualitative features of the system, e.g. whether it is optimal according to some criterion, what its sensitivity is, whether it is controllable, observable, etc.

Synthesis, or the theoretical design of control systems, is a more complicated task. The following elements can be distinguished here:

— Determination of the aim of control.

— Construction of a model of the controlled process with selection of measurable output quantities on the basis of which it is possible to check or estimate the realization of the control aim and the state of the process, as well as input quantities which are treated as causes of output changes.

— Division of the input quantities into manipulable (those which we can steer or change purposefully) and non-manipulable. The latter, if their influence is harmful, are called disturbances.

— Characterization of disturbances (by building a model).

— Selection of control quantities and possibly other decision variables. These must be relatively easily manipulable and output sensitivity to them should be appropriately great.

— Precise characterization of the set of admissible controls (and possibly the set of admissible parameters).

— Choice of control system structure, with special attention to the question as to which control actions may be realized in feedback form, and which in open loops.

— Assumptions as to the type and structure of the regulator and observer (measurement system)—for closed-loop controls.

— Synthesis criterion, for example a performance functional in the case of optimal control systems. Often classical criteria such as overshoot magnitude, properties of a frequency characteristic, phase and amplitude margins, etc. are also used.

These elements can be found in practically all control synthesis problems, but not always in the same order. As a rule, synthesis is an iterative process in which the design cycle is passed through many times. The reason for this is that analysis of further steps increases the amount of information available about the system, and this makes it possible for earlier steps of the procedure to be reconsidered with a greater degree of precision.

PART II
MODELS OF TIME-DELAY SYSTEMS

3

Models of Continuous Time-delay Systems

3.1 Functional-differential Equations

3.1.1 Classification of functional differential equations

Let $x(t)$ be an n-dimensional variable describing the behaviour of a process under control u in time interval $t_0 \leqslant t \leqslant t_1$. All the following remarks and theorems will also be valid, after obvious modifications, in the case of an infinite time interval $[t_0, \infty)$. We say that x satisfies a functional-differential equation (FDE) if for $t \in [t_0, t_1]$ the derivative $\dot{x}(t)$ depends on values of the variable x, its derivative $\dot{x}$ or control u at times not necessarily equal to t. In order to solve an FDE, the functions x and u must often be determined outside the interval $[t_0, t_1]$ in the form of boundary conditions.

Most generally an FDE is formulated as follows. Let $T_1(t)$, $T_2(t)$ and $T_3(t)$ be time-dependent sets of real numbers, defined for all $t \in [t_0, t_1]$. Let us assume that x is an absolutely continuous function in $[t_0, t_1]$, and that outside this interval it is a predetermined function. The control u is here treated as given in the whole necessary time interval. We shall use the convention that $\dot{x}(t)$ for $t \in [t_0, t_1)$ denotes the right-hand derivative of x and for other t it is a predetermined function which may be in no relation to the derivative of x. For each $t \in [t_0, t_1]$ we define x_t, $\dot{x}_t$ and u_t as shifted restrictions of x, $\dot{x}$ and u to $T_1(t)$, $T_2(t)$ and $T_3(t)$ respectively: $x_t(s) = x(t+s)$, $s \in T_1(t)$, $\dot{x}_t(s) = \dot{x}(t+s)$, $s \in T_2(t)$, $u_t(s) = u(t+s)$, $s \in T_3(t)$. Let f be a function $(x_t, \dot{x}_t, u_t, t) \mapsto R^n$.

3.1 Functional-differential Equations

The function x satisfies an FDE in $[t_0, t_1]$ if for almost every $t \in [t_0, t_1]$ the following equality holds:
$$\dot{x}(t) = f(x_t, \dot{x}_t, u_t, t), \qquad (3.1.1)$$
or in an integral form
$$x(t) = x(t_0) + \int_{t_0}^{t} f(x_s, \dot{x}_s, u_s, s)\,ds. \qquad (3.1.2)$$

The dependence of $\dot{x}(t)$ on control values at times different from t is less essential in the analysis of control systems as it does not change the type of the equation from a mathematical point of view. However, such dependence may cause serious difficulties in control synthesis and in order to overcome this special methods are used similar to those applied in the case of system equations with a deviation of argument in the function x. That is why in control theory, systems with deviations of argument in control function are treated as functional-differential systems.

Since models of dynamic systems are always constructed with the causality principle more or less explicitly taken into account, very few models are encountered in which the rate of change of the variable describing the process at time t depends on future values of variables or of control. For this reason, as well as the ease of formulation of Cauchy problems, the prevailing majority of theoretical results obtained hitherto concern two types of FDE: retarded FDE and neutral FDE.

An FDE is retarded if $T_2(t) = \emptyset$, $\forall t \in [t_0, t_1]$, therefore the right-hand side of (3.1.1) does not depend on the derivative $\dot{x}$, and
$$\forall t \in [t_0, t_1] \qquad T_1(t) \cup T_3(t) \subset (-\infty, 0]. \qquad (3.1.3)$$
An FDE is neutral if inclusion (3.1.3) holds and
$$\forall t \in [t_0, t_1] \qquad T_2(t) \subset (-\infty, 0). \qquad (3.1.4)$$

This distinction has far-reaching consequences for the properties of solutions. In applications, sets $T_1(t)$, $T_2(t)$, $T_3(t)$ are usually uniformly bounded, i.e. there is a constant $h \geq 0$ such that
$$\forall t \in [t_0, t_1] \qquad T_1(t) \cup T_2(t) \cup T_3(t) \subset [-h, 0].$$
If sets $T_1(t)$, $T_2(t)$, $T_3(t)$ are finite, a retarded FDE is called an FDE with lumped delays. If these sets are a continuum we say that an FDE is an equation with distributed delays. An example of a system with a lumped delay is
$$\dot{x}(t) = f\bigl(x(t), x(t-h(t)), u(t)\bigr) \qquad h(t) > 0 \qquad (3.1.5)$$
and of a system with distributed delay
$$\dot{x}(t) = \int_{t-h}^{t} g\bigl(x(s), u(s), t, s\bigr)\,ds.$$

Sometimes a lumped delay depends on variable $x(t)$ or control value $u(t)$, e.g.

$$\dot{x}(t) = f\Big(x(t), x\big(t-h(x(t), u(t), t)\big), u(t)\Big). \tag{3.1.6}$$

Other criteria of classification are similar to those used for ordinary differential equations. If function f in (3.1.1) is affine with respect to the three first arguments (or the two first arguments in the case of a retarded equation, when $\dot{x}_t$ does not appear in the right-hand side), the equation is called linear. If function f does not depend explicitly on t, equation (3.1.1) is autonomous or stationary.

3.1.2 Existence, continuous dependence and uniqueness of solutions of FDE

First we shall formulate, after Hale (1977), some basic theorems on the existence and properties of solutions for retarded FDE. Since in the considerations which follow the control u is treated as a given, fixed function of time, it will not appear in the right-hand sides of the equations.

Let h be a nonnegative real number. By R^n we denote the n-dimensional linear space over reals with the euclidean norm $|\cdot|$. $C([a, b], R^n)$ is a space of continuous functions $[a, b] \to R^n$ supplied with the norm $|\varphi| = \sup_{a \leqslant t \leqslant b} |\varphi(t)|$. For any $x \in C([t_0-h, t_1], R^n)$, $t_1 > t_0$, we introduce the notation $x_t \in C([-h, 0], R^n)$ where $x_t(s) = x(t+s)$, $s \in [-h, 0]$. Let D be an open subset of $C([-h, 0], R^n)$ and a function $f: D \times [t_0, t_1] \to R^n$ be given. A retarded FDE on $D \times [t_0, t_1]$ is the relation

$$\dot{x}(t) = f(x_t, t) \quad t \in [t_0, t_1] \quad x_t \in D. \tag{3.1.7}$$

A function $x: [t_0-h, t_1] \to R^n$ is a solution of equation (3.1.7) if $x \in C([t_0-h, t_1], R^n)$, $x_t \in D$, and equality (3.1.7) holds for almost all $t \in [t_0, t_1]$. If we additionally assume that $x_{t_0} = \varphi$, $\varphi \in C([-h, 0], R^n)$, then x is a solution of equation (3.1.7) with an initial condition φ. In order to express the dependence on the initial condition we shall also use the notation $x(\cdot, \varphi)$ and $x(\cdot, t_0, \varphi)$. After replacement of t_1 by ∞, a solution of equation (3.1.7) on an infinite interval $[t_0-h, \infty)$ can be defined in a similar way.

To formulate the theorems on existence, continuous dependence and uniqueness, the so-called Carathéodory condition will be put on function f. Let Ω be an open subset of $C([-h, 0], R^n) \times R$. A function $f: \Omega \to R^n$ satisfies the Carathéodory condition on Ω if function $f(\varphi, \cdot)$ is measurable for every fixed φ, function $f(\cdot, t)$ is continuous for every fixed t and for every point $(\varphi, t) \in \Omega$ there is a neighbourhood $V(\varphi, t)$ and a Lebesgue integrable function m such that

$$|f(\psi, s)| \leqslant m(s) \quad (\psi, s) \in V(\varphi, t). \tag{3.1.8}$$

3.1 Functional-differential Equations

The reader who is not familiar with measurable functions or Lebesgue integrals, is referred to Rudin (1964) and to Dunford and Schwartz (1967). For practical purposes one can replace measurability of function $f(\varphi, \cdot)$ by pointwise continuity, and Lebesgue integrability by Riemann integrability.

Theorem 3.1.1 (Existence) *Assume that Ω is an open subset of $C([-h, 0], R^n) \times R$. For every point $(\varphi, t_0) \in \Omega$ there exists a solution of equation (3.1.7) passing through (φ, t_0), i.e. $x_{t_0} = \varphi$.*

Theorem 3.1.2 (Continuous dependence) *Let Ω be an open subset of $C([-h, 0], R^n) \times R$, $(\varphi^0, t_0^0) \in \Omega$, $f^0 \in C(\Omega, R^n)$, and let x^0 be a solution of equation (3.1.7) with the right-hand side equal to f^0, passing through (φ^0, t_0^0). We assume that x^0 exists and is unique in an interval $[t_0^0 - h, b]$, $b > t_0^0$. Let*

$$W^0 = \{(x_t^0, t) : t \in [t_0^0, b]\}$$

and let V^0 be a neighbourhood of W^0, in which function f^0 is bounded. Assume also that (φ^k, f^k, t_0^k), $k = 1, 2, \ldots$, is an arbitrary sequence such that $\varphi^k \to \varphi^0$, $|f^k - f^0|_{V^0} \to 0$, and $t_0^k \to t_0^0$ when $k \to \infty$.

There then exists k^0, such that for every $k \geq k^0$ there is a solution x^k of (3.1.7) with the right-hand side f^k, passing through (φ^k, t_0^k) and determined in $[t_0^k - h, b]$. Moreover $x^k \to x^0$ uniformly in $[t_0^0 - h, b]$. Since not all solutions x^k must be determined on $[t_0^0 - h, b]$, the uniform convergence is understood in the following way. For every $\varepsilon > 0$ there is $k_1(\varepsilon)$ such that $x^k(t)$, $k \geq k_1(\varepsilon)$, is determined on $[t_0^0 - h + \varepsilon, b]$ and $x^k \to x^0$ uniformly on $[t_0^0 - h + \varepsilon, b]$.

Continuous dependence of the solution on initial conditions (that is the initial moment of time and the initial function) and the right-hand side of the equation is a very important property of a differential equation which is a model of a real process. Only due to such continuity may we expect that increasing model accuracy, e.g. by more accurate measurements of initial conditions and other parameters, will result in more accurate model trajectories, converging to the actual trajectories of the real system.

Theorem 3.1.3 (Uniqueness) *Suppose that Ω is an open subset of $C([-h, 0], R^n) \times R$, and $f(\cdot, t)$ satisfies the Lipschitz condition in every bounded subset of Ω, that is, for every bounded $V \subset \Omega$ there is a real $k > 0$ such that*

$$|f(x_t, t) - f(y_t, t)| \leq k|x_t - y_t| \quad (x_t, t) \in V \quad (y_t, t) \in V. \quad (3.1.9)$$

Then a unique solution of equation (3.1.7) passes through every point of Ω.

The theory of neutral FDE is much less developed than the theory of retarded FDE. We shall confine ourselves to a special type of neutral equations, namely

$$\dot{x}(t) = \sum_{i=1}^{N} A_i(t)\dot{x}(t-h_i(t)) + f(x_t, t) \quad t \in [t_0, t_1] \quad (3.1.10)$$

$$x(t) \in R^n \quad \min_i \inf_t h_i(t) > 0 \quad \max_i \sup_t h_i(t) = h < \infty.$$

A_i are essentially bounded measurable matrix functions, the delays h_i are continuous. Moreover we assume that function f satisfies the Carathéodory condition. Symbol $\dot{x}(t)$ denotes the right-hand derivative for $t \geq t_0$ and an arbitrary given function $\mu(t)$ for $t < t_0$. The initial conditions for equation (3.1.10) are given in the form

$$\begin{aligned} x_{t_0} &= \varphi \in C([-h, 0], R^n) \\ \dot{x}_{t_0} &= \mu \in L^\infty([-h, 0], R^n) \end{aligned} \quad (3.1.11)$$

where L^∞ denotes the space of all measurable essentially bounded functions. Let D be an open subset of $C([-h, 0], R^n)$ such that function f is determined on $D \times [t_0, t_1]$. A function $x: [t_0-h, t_1] \to R^n$ is a solution of equation (3.1.10) if $x \in C([t_0-h, t_1], R^n)$, $x_t \in D$, and equality (3.1.10) holds for almost all $t \in [t_0, t_1]$. If (3.1.11) is also satisfied, then x is a solution of equation (3.1.10) with initial condition (3.1.11). In the case where function φ is absolutely continuous, i.e. it has a derivative almost everywhere, and $\mu = \dot{\varphi}$, we say that x is a solution of equation (3.1.10) with initial condition φ. We shall use also the notation $x(\cdot, \varphi, \mu)$, $x(\cdot, t_0, \varphi, \mu)$, and in the last case $x(\cdot, \varphi)$, $x(\cdot, t_0, \varphi)$. These definitions can be easily extended to the case $t_1 = \infty$.

Under the above assumptions for the neutral equation (3.1.10), theorems on existence, continuous dependence and uniqueness of solutions strictly analogous to Theorems 3.1.1, 3.1.2 and 3.1.3 are valid.

Of course neutral equations include retarded equations as a special case. To prove this it is sufficient to substitute $A_i = 0$, $i = 1, \ldots, N$, into (3.1.10). Let us note that equation (3.1.10) may be regarded as a generalization of a certain type of difference equation. Let us consider a difference equation

$$x(t) = \sum_{i=1}^{N} B_i(t) x(t-h_i) + g(t) \quad (3.1.12)$$

with differentiable coefficient matrices and a differentiable function g. After differentiation of both sides and an appropiate change in notation we obtain a special form of equation (3.1.10).

The dependence of the right-hand side of equations (3.1.7) or (3.1.10) on control is expressed by replacing function $f(x_t, t)$ by $f(x_t, u_t, t)$, where u_t is a shifted restriction of function u to interval $[-h, 0]$. The control function u is an element of the space of admissible controls $\mathcal{U}$. As a rule we shall

3.1 Functional-differential Equations

assume that $\mathscr{U}$ is a set of measurable essentially bounded functions determined in $[t_0-h, t_1]$ and taking their values from a given set $U \subset R^m$, or that $\mathscr{U}$ is a set of square integrable functions $[t_0-h, t_1] \to U \subset R^m$. It will everywhere be assumed that for every $u \in \mathscr{U}$, the function $(\xi, t) \mapsto f(\xi, u_t, t)$ satisfies the Carathéodory condition. Then Theorems 3.1.1, 3.1.2 and 3.1.3 remain valid for FDE with control.

3.1.3 Step method

The step method, called also the method of sequential integration, is a basic method for solving FDE with lumped delays. It can be applied to those equations which can be treated as ordinary equations without deviations of argument in a certain interval $[t, t+d]$ provided their solution is known up to the moment t. A solution is found on successive intervals, one after another, by solving an ordinary equation without delays in each interval. We shall present this method on a simple equation with a constant lumped delay,

$$\begin{aligned} \dot{x}(t) &= x(t-1) \quad t \geq 0 \\ x(t) &= \varphi(t) \quad t \in [-1, 0] \end{aligned} \qquad (3.1.13)$$

where φ is a given integrable function. For $0 \leq t \leq 1$ we have

$$\dot{x}(t) = \varphi(t-1)$$

then

$$x(t) = \varphi(0) + \int_0^t \varphi(s-1)\,ds \quad t \in [0, 1].$$

Having solved the equation in interval $[0, 1]$ we can extend the solution into interval $[1, 2]$

$$x(t) = x(1) + \int_1^t x(s-1)\,ds \quad t \in [1, 2].$$

By this procedure we are able to construct the solution in any finite interval.

We shall in turn apply the step method to an equation with time-varying delay $h(t) = \tfrac{1}{2}t+1$,

$$\begin{aligned} \dot{x}(t) &= -x(t) + x(\tfrac{1}{2}t-1) + u(t) \quad t \geq 0 \\ x(t) &= \varphi(t) \quad t \in [-1, 0]. \end{aligned} \qquad (3.1.14)$$

We assume that the control is constant, $u(t) = -1$, $t \geq 0$, as is the initial function $\varphi(t) = 1$, $t \in [-1, 0]$. In interval $[0, 2]$ equation (3.1.14) can be written as

$$\dot{x}(t) = -x(t),$$

hence
$$x(t) = e^{-t} \quad t \in [0, 2].$$
Next, in interval [2, 6] we have
$$\dot{x}(t) = -x(t) + \exp(\tfrac{1}{2}t - 1) - 1,$$
and so
$$x(t) = e^{-t+2}x(2) + \int_2^t e^{-t+s}[x(\tfrac{1}{2}s - 1) + u(s)]ds$$
$$= e^{-t}(1 - e^2) + 2\exp(-\tfrac{1}{2}t + 1) - 1.$$

Here we stop, but let us note that due to the fact that the delay varies with time, the lengths of the intervals in which we integrate ordinary equations also vary with time. These intervals are [0, 2], [2, 6], [6, 14], etc.

Let us consider in more detail the sequential integration of the equation
$$\dot{x}(t) = f\big(x(t), x(t - h(t)), t\big) \quad t \geq t_0, \qquad (3.1.15)$$
where h is a continuous function with nonnegative values. The initial function φ must be determined at time t_0 and at all moments $s = t - h(t) \leq t_0$ for $t \geq t_0$. The set of all such s we call the initial set. If $t \mapsto t - h(t)$ is a strictly increasing function and $\inf h(t) = \delta > 0$, then equation $t - h(t) = t_0$ has a unique solution with respect to t which we denote by $g(t_0)$. Obviously $g(t_0) > t_0$. Succeeding intervals in which equation (3.1.15) may be considered as an equation without delay, are $[t_0, g(t_0)]$, $[g(t_0), g^2(t_0)]$, $[g^2(t_0), g^3(t_0)]$, etc., where $g^m(t_0) = g(g^{m-1}(t_0))$, $g^1(t_0) = g(t_0)$.

For applicability of the step method it is not necessary that $t \mapsto t - h(t)$ is strictly increasing. This only serves to make it easier to determine the lengths of succeeding 'steps' of integration. The condition $\inf_t h(t) = \delta > 0$ is more essential. The step method can be applied to equations which contain delays depending on $x(t)$ and $u(t)$. Consider an FDE with a lumped delay
$$\dot{x}(t) = f\big(x(t), x(t - h(x(t), u(t), t)), u(t), t\big) \quad t \geq t_0. \qquad (3.1.16)$$
A sufficient condition under which this equation may be solved with the use of the step method is
$$\inf_{\substack{t \geq t_0 \\ \xi \in D}} h(\xi, u(t), t) = \delta > 0$$
where D is an open subset of R^n containing the solution. Equation (3.1.16) can be solved step by step on succeeding intervals of length not less than δ.

In the case $\inf h = 0$ the step method may fail as can be seen from the example

3.1 Functional-differential Equations

$$\dot{x}(t) = x\left(\frac{t}{2}\right) \quad x(0) = 1 \quad t \geq 0.$$

To solve this equation by the step method we have at least to know its solution on an arbitrarily small interval $[0, a)$, $a > 0$.

For retarded FDE with lumped delays to which the step method is applicable, theorems on the existence, continuous dependence and uniquenes of solutions may be formulated under much weaker assumptions than those in Theorems 3.1.1, 3.1.2 and 3.1.3. Since in each step the FDE is transformed into an equation without delays, we can use the well-known theorem on existence, continuous dependence and uniqueness for ordinary differential equations. For the equation

$$x(t) = f\big(x(t), x(t-h(x(t), t)), t\big) \quad t \in [t_0, t_1] \quad (3.1.17)$$

$$x(t) = \varphi(t) \text{ on the initial set}$$

with $\inf_{\xi, t} h(\xi, t) > 0$ we obtain in the first step

$$\dot{x}(t) = f\big(x(t), \varphi(t-h(x(t), t)), t\big) \quad (3.1.18)$$
$$x(t_0) = \varphi(t_0).$$

Let us assume that the function

$$F(y, t) = f\big(y, \varphi(t-h(y, t)), t\big) \quad (3.1.19)$$

satisfies the Carathéodory condition in a nonempty set $D \times [t_0, t_0+a)$ ($a > 0$ and D an open subset of R^n containing $\varphi(t_0)$). This means that $F(y, \cdot)$ is measurable for every fixed y, $F(\cdot, t)$ is continuous for every fixed t and there is a Lebesgue integrable function m such that

$$|F(z, s)| \leq m(s) \quad (z, s) \in D \times [t, t+a). \quad (3.1.20)$$

A solution of equation (3.1.17) then passes through the initial point (φ, t_0). Moreover, if we assume additionally that function F satisfies the Lipschitz condition with respect to its first argument uniformly in the second argument $t \in [t_0, t_0+a)$, the solution is unique and depends continuously on initial condition $\varphi(t_0)$ and function F, in a similar sense as in Theorem 3.1.2.

The requirement that initial function φ be continuous is often too strong for applications. Therefore formulations of the Cauchy problem with a square integrable initial function may be encountered, and in many cases it is assumed that the initial function φ is continuous for $t < t_0$ with possible discontinuity at t_0, i.e. $\lim_{t \to t_0^-} \varphi(t) \neq \varphi(t_0)$. It is easy to see that in the case of equations with lumped delays to which the step method can be applied, continuity of function φ at t_0 is irrelevant to the existence and uniqueness of solutions.

Much the same the step method can be applied to FDE with many lumped delays, e.g.

$$\dot{x}(t) = f\big(x(t), x(t-h_1(t)), \ldots, x(t-h_k(t)), t\big) \qquad t \in [t_0, t_1] \qquad (3.1.21)$$

where $\min_i \inf_t h_i(t) = \delta > 0$. The initial set on which initial function φ must be determined is in this case defined as

$$\{t_0\} \cup \{s \leqslant t_0 \colon \exists t \in [t_0, t_1], \exists i, \ 1 \leqslant i \leqslant k, s = t - h_i(t)\}.$$

Now let us consider a neutral FDE with a lumped delay

$$\dot{x}(t) = f\big(x(t), x(t-h(t)), \dot{x}(t-h(t)), t\big) \qquad t \in [t_0, t_1]. \qquad (3.1.22)$$

Initial conditions are determined on an initial set, defined similarly to the case of retarded FDE

$$x(t) = \varphi(t) \qquad t \leqslant t_0$$
$$\dot{x}(t) = \mu(t) \qquad t < t_0$$

(in the case of $\dot{x}$ we exclude time t_0 from the initial set). Function μ may be different from the derivative of φ; moreover, function φ need not be differentiable (similarly to the case of retarded FDE with lumped delays). Equation (3.1.22) may be solved by the step method provided $\inf_t h(t) = \delta > 0$.

Under this assumption the conditions of the existence, continuous dependence and uniqueness of solution can be obtained from the well-known theorems concerning ordinary differential equations applied to the function

$$F(y, t) = f\big(y, \varphi(t-h(t)), \mu(t-h(t)), t\big) \qquad t \in [t_0, t_0+\delta]. \qquad (3.1.23)$$

A characteristic feature that distinguishes retarded FDE from neutral ones is the smoothness of solutions. If the right-hand side of a retarded FDE with lumped delays is regular enough, its solution becomes more and more smooth with time. Solutions of neutral equations do not have this property. We shall show this on examples. First, let us consider the retarded FDE

$$\dot{x}(t) = x(t-1) \qquad t \geqslant 0$$
$$x(t) = 0 \qquad t < 0$$
$$x(0) = 1.$$

Its solution is

$$x(t) = \sum_{i=0}^{[t]} \frac{(t-i)^i}{i!} \qquad t \geqslant 0$$

where $[t]$ is the greatest integer less than or equal to t. This solution is analytic in every open interval $(i, i+1)$, $i = 0, 1, \ldots$ At $t = 0$ it is discontinuous, at $t = 1$ its first derivative is discontinuous; generally, at $t = i$ its i-th derivative has a discontinuity. Therefore, as time goes by the solution becomes smoother and smoother.

Now consider the neutral equation
$$\dot{x}(t) = -\dot{x}(t-1) \quad t \geq 0$$
$$\dot{x}(t) = 1 \quad t < 0$$
$$x(0) = 1.$$

Its solution is
$$x(t) = \begin{cases} [t]+1-t & t \in [2i, 2i+1] \\ t-[t] & t \in [2i-1, 2i) \end{cases} \quad (i = 0, 1, \ldots).$$

The first derivative of this function is discontinuous at each point $t = i$, $i = 0, 1, \ldots$ Therefore in spite of the regularity of the right-hand side, the solution is not smoother than C^0 on any half-line $[a, \infty)$, $a \geq 0$.

3.2 Linear Continuous-time FDE

3.2.1 Linear models and variation-of-constants formulas

The functional-differential equation (3.1.1) is linear if its right-hand side is affine with respect to the state, its derivative (in the case of neutral systems) and to the control. A quite general form of the neutral equation known to be useful in the applications is

$$\sum_{i=0}^{N} [A_i(t)\dot{x}(t-h_i) + B_i(t)x(t-h_i) + C_i(t)u(t-h_i)] +$$
$$+ \int_{-h}^{0} B(t, \tau)x(t+\tau)d\tau + \int_{-h}^{0} C(t, \tau)u(t+\tau)d\tau = f(t) \quad t \geq t_0. \quad (3.2.1)$$

Here $x(t) \in R^n$, $u(t) \in R^m$, $f(t) \in R^n$, A_i, B_i, B, C_i, C are matrix-valued functions of compatible dimensions, $A_0(t)$ has an inverse $A_0(t)^{-1}$ bounded for $t \geq t_0$, $0 = h_0 < h_1 < \ldots < h_N = h$. In problems of control synthesis we usually treat the restriction $u|_{[t_0-h, t_0)}$ as a given initial condition. It is convenient (and more general) to consider $\dot{x}(t)$ for $t < t_0$ as a given function, not necessarily equal to the derivative of x. Therefore the initial conditions for the system equation are
$$x(t) = \varphi(t) \quad t \in [t_0-h, t_0]$$
$$\dot{x}(t) = \mu(t) \quad t \in (t_0-h, t_0) \quad (3.2.2)$$
$$u(t) = \nu(t) \quad t \in [t_0-h, t_0)$$
where φ, μ and ν are predetermined functions.

The theorems on the existence, uniqueness and continuous dependence of the solution of Section 3.1 can obviously be applied to system (3.2.1), (3.2.2) under appropriate assumptions. Now we shall give a slightly different formulation which will be useful for the linear-quadratic problem of optimal control. We assume that φ, μ and ν are square integrable on (t_0-h, t_0), function f and control u are square integrable on every interval $[t_0, t]$, $t > t_0$, A_i, B_i, C_i, B, C and $t \mapsto A_0(t)^{-1}$ are bounded and integrable on every finite subset of their domain. An important feature of this formulation is that φ need not be continuous and as a result, the solution x is required to be continuous only for $t > t_0$ with right-hand continuity at t_0. Discontinuous initial functions are often encountered and in particular the initial instantaneous state $x(t_0)$ may be in no relation to $x(t)$, $t < t_0$. This, together with the fact that $x(t_0)$ and $x(t)$, $t < t_0$ affect the solution in different ways—$x(t_0)$ by a matrix operator whereas $x(t)$, $t < t_0$ by an integral operator—make it natural that the Cartesian product

$$H = R^n \times L^2(-h, 0; R^n) \qquad (3.2.3)$$

be used as the space of initial data. This space applies to retarded systems with no deviations of argument in the control. If we want to include the initial conditions for $\dot{x}$ and u in the initial data we must extend the Cartesian product; then

$$H = R^n \times L^2(-h, 0; R^n) \times L^2(-h, 0; R^n) \times L^2(-h, 0; R^m). \qquad (3.2.4)$$

The neutral system (3.2.1) becomes a retarded one if $A_i = 0$, $i = 1, \ldots, N$, and $A_0(t) = I$:

$$\dot{x}(t) + \sum_{i=0}^{N} [B_i(t)x(t-h_i) + C_i(t)u(t-h_i)] +$$

$$+ \int_{-h}^{0} B(t, \tau)x(t+\tau)d\tau + \int_{-h}^{0} C(t, \tau)u(t+\tau)d\tau = f(t) \qquad t \geq t_0 \qquad (3.2.5)$$

We shall also consider systems (3.2.1) and (3.2.5) with the starting moment shifted to a certain $s \geq t_0$. Of course the shift will then refer to the initial condition too.

Theorem 3.2.1 For every t_1, s, $t_1 \geq s \geq t_0$, every initial condition $\xi \in H$, every control $u \in L^2(s, t_1; R^m)$ and every $f \in L^2(s, t_1; R^n)$, equations (3.2.1) and (3.2.5) have unique solutions $x(\cdot, \xi, u, f)$ in $[s-h, t_1]$. Moreover, the function $(\xi, u, f) \mapsto x(\cdot, \xi, u, f)|_{[s, t_1]} \in C(s, t_1)$ is linear and continuous.

3.2 Linear Continuous-time FDE

In the retarded case (3.2.5) the solution has the form:

$$x(t, \xi, u, f) = \Phi(t, s)x(s) - \int_{-h}^{0} \Phi^1(t, s, \tau)x(t+\tau)d\tau -$$

$$- \int_{-h}^{0} \Phi^2(t, s, \tau)u(t+\tau)d\tau - \int_{s}^{t} \Phi(t, \tau)[v(\tau) - f(\tau)]d\tau \quad t \geq s \quad (3.2.6)$$

where

$$v(t) = \sum_{i=0}^{N} C_i(t) \begin{cases} u(t-h_i) & t-h_i \geq s \\ 0 & \text{otherwise} \end{cases} +$$

$$+ \int_{-h}^{0} C(t, \tau) \begin{cases} u(t+\tau) & t+\tau \geq s \\ 0 & \text{otherwise} \end{cases} d\tau. \quad (3.2.7)$$

Φ is a matrix-valued function called the fundamental solution of equation (3.2.5). It is defined by

$$\frac{\partial}{\partial t} \Phi(t, s) + \sum_{i=0}^{N} B_i(t)\Phi(t-h_i, s) + \int_{-h}^{0} B(t, \tau)\Phi(t+\tau, s)d\tau = 0 \quad t > s$$

$$\Phi(s, s) = I \quad \Phi(t, s) = 0 \quad \text{for} \quad t < s. \quad (3.2.8)$$

Furthermore

$$\Phi^1(t, s, \tau) = \sum_{i=1}^{N} \begin{cases} \Phi(t, s+\tau+h_i)B_i(s+\tau+h_i) & \tau+s-t < -h_i \leq \tau \\ 0 & \text{otherwise} \end{cases} +$$

$$+ \int_{-h}^{\tau} \Phi(t, s+\tau-\vartheta)B(s+\tau-\vartheta, \vartheta)d\vartheta. \quad (3.2.9)$$

Φ^2 is obtained from this formula after replacing B_i and B by C_i and C respectively.

This theorem can be proved by substitution of (3.2.6) into equation (3.2.5). Equality (3.2.6), called the variation-of-constants formula, plays an important role in the theory of linear systems with delays. It explicitly shows the linear and continuous dependence of the trajectory $x(\cdot, \xi, u, f)$ on the initial conditions, control and function f, which sometimes represents disturbances. An analogous variation-of-constants formula can be derived for neutral linear FDE (Bellman and Cooke, 1963; Hale, 1977). In this case, however, we shall confine ourselves to stationary systems, namely

$$\sum_{i=0}^{N} [A_i \dot{x}(t-h_i) + B_i x(t-h_i) + C_i u(t-h_i)] + \int_{-h}^{0} C(\tau) u(t+\tau) d\tau = f(t) \quad t \geq 0.$$

(3.2.10)

Assumptions on the coefficients, initial conditions, etc. are as in (3.2.1). The variation-of-constants formula is

$$x(t, \xi, u, f) = \sum_{i=0}^{N} \Phi(t-h_i) A_i x(0) + \int_0^t \Phi(t-\tau) v(\tau) d\tau -$$

$$- \sum_{i=1}^{N} \int_{-h_i}^{0} \Phi(t-h_i-\tau)[A_i \mu(\tau) + B_i \varphi(\tau)] d\tau \quad (3.2.11)$$

where

$$v(t) = f(t) - \sum_{i=0}^{N} C_i u(t-h_i) - \int_{-h}^{0} C(\tau) u(t+\tau) d\tau$$

and the fundamental solution Φ is a matrix-valued piecewise continuous function, uniquely characterized by the conditions:

(1) function $t \mapsto \sum_{i=0}^{N} A_i \Phi(t-h_i)$ is continuous for $t \geq 0+$
(2) $\Phi(0) = A_0^{-1}$ and $\Phi(t) = 0$ for $t < 0$
(3) for almost every $t > 0$, Φ is differentiable and

$$\sum_{i=0}^{N} [A_i \dot{\Phi}(t-h_i) + B_i \Phi(t-h_i)] = 0. \quad (3.2.12)$$

The reader familiar with the control theory of finite-dimensional systems described by ordinary differential equations, will realize how helpful the variation-of-constants formula is in solving different problems of synthesis and analysis. In time-delay systems the situation is more difficult because even in the stationary case the fundamental solution is not analytic and can be expressed by explicit formulas only for very simple equations.

Let us consider two examples. In the first one the system is described by the equation

$$\dot{x}(t) = x(t-1) + u(t) \quad t \geq 0.$$
$$x(t) = \varphi(t) \quad t \leq 0.$$

Its fundamental solution satisfies

$$\dot{\Phi}(t) = \Phi(t-1) \quad t \geq 0$$
$$\Phi(0) = 1 \quad \Phi(t) = 0 \quad t < 0.$$

3.2 Linear Continuous-time FDE

Hence (see Section 3.1.3)

$$\Phi(t) = \sum_{i=0}^{[t]} \frac{(t-i)^i}{i!} \qquad t \geq 0$$

and the solution of the system equation has the form

$$x(t) = \Phi(t)\varphi(0) + \int_{-1}^{0} \Phi^1(t,\tau)\varphi(\tau)d\tau + \int_{0}^{t} \Phi(t-\tau)u(\tau)d\tau$$

where

$$\Phi^1(t,\tau) = \begin{cases} \Phi(t-\tau-1) & \tau-t < -1 < \tau \\ 0 & \text{otherwise.} \end{cases}$$

Now, let the system equation be

$$\dot{x}(t) + \dot{x}(t-1) = f(t)$$
$$x(0) = \varphi(0) \quad \dot{x}(t) = \mu(t) \quad t < 0.$$

The fundamental solution almost everywhere satisfies the equation

$$\dot{\Phi}(t) + \dot{\Phi}(t-1) = 0 \quad \Phi(0) = 1 \quad \text{and} \quad \Phi(t) = 0 \quad \text{for} \quad t < 0.$$

Hence, taking into account that $t \mapsto \Phi(t) + \Phi(t-1)$ must be continuous for $t \geq 0+$, we obtain

$$\Phi(t) = \begin{cases} 0 & t \in [2k-1, 2k) \\ 1 & t \in [2k, 2k+1) \end{cases} \quad (k = 0, 1, 2, \ldots).$$

Then

$$x(t) = [\Phi(t) + \Phi(t-1)]\varphi(0) + \int_{0}^{t} \Phi(t-\tau)f(\tau)d\tau - \int_{-1}^{0} \Phi(t-1-\tau)\mu(\tau)d\tau$$

$$= \varphi(0) + \begin{cases} \sum_{i=0}^{k-1} \int_{t-2i-1}^{t-2i} f(\tau)d\tau + \int_{0}^{t-2k} f(\tau)d\tau - \int_{-1}^{t-2k-1} \mu(\tau)d\tau & t \in [2k, 2k+1) \\ \sum_{i=0}^{k} \int_{t-2i-1}^{t-2i} f(\tau)d\tau - \int_{t-2k-2}^{0} \mu(\tau)d\tau & t \in [2k+1, 2k+2) \end{cases}$$

$$(k = 0, 1, 2, \ldots).$$

The formulation of the Cauchy problem for neutral systems given by (3.2.1) and (3.2.2) has one serious shortcoming, no matter how natural it looks. Suppose the physical properties of the modelled system imply that for every differentiable initial function φ the initial condition for $\dot{x}$ is compatible, i.e. μ is the derivative of φ. Both for theoretical and practical reasons it is essential to know the reaction of such a system to a discontinuous initial function φ. Of special importance are discontinuities of the type $\lim_{t \to t_0-} \varphi(t) \ne \varphi(t_0)$. Of course description (3.2.1), (3.2.2) is not suitable for dealing with such cases unless we decide to introduce distributions or to study the solution by means of limit procedures. We shall show one way of overcoming this difficulty (Hale, 1977). First, let us consider the example

$$\dot{x}(t) + \dot{x}(t-1) = 0 \quad t \geq 0$$
$$x(t) = \varphi(t) \quad t \leq 0. \quad (3.2.13)$$

For sufficiently smooth initial functions this equation is equivalent to

$$\frac{d}{dt}[x(t) + x(t-1)] = 0 \quad t \geq 0 \quad x(t) = \varphi(t) \quad t \leq 0. \quad (3.2.14)$$

However, the latter equation also makes sense for discontinuous initial conditions. To avoid distributional differentiation one must only require that $t \mapsto x(t) + x(t-1)$ is absolutely continuous for $t \geq 0$. The idea of the approach is to replace description (3.2.13) by (3.2.14). A solution of initial-value problem (3.2.14) is defined similarly as that of (3.2.13) with one major difference: continuity of $x|_{[t_0, \infty)}$ is replaced by continuity of $t \mapsto x(t) + x(t-1)$, $t \geq t_0$.

Now we shall apply the idea to more general neutral systems. Let us assume that matrix-valued functions $A_i: [t_0, \infty) \to L(R^n)$, $i = 0, \ldots, N$, are absolutely continuous and the inverse $A_0(t)^{-1}$ exists everywhere. A linear neutral equation has the form

$$\frac{d}{dt}\left[\sum_{i=0}^{N} A_i(t) x(t-h_i)\right] + \sum_{i=0}^{N} [B_i(t) x(t-h_i) + C_i(t) u(t-h_i)] +$$
$$+ \int_{-h}^{0} B(t, \tau) x(t+\tau) d\tau + \int_{-h}^{0} C(t, \tau) u(t+\tau) d\tau = f(t) \quad t \geq t_0. \quad (3.2.15)$$

The initial conditions are

$$x(t) = \varphi(t) \quad t \in [t_0-h, t_0] \quad u(t) = v(t) \quad t \in [t_0-h, t_0). \quad (3.2.16)$$

With other assumptions as in (3.2.1) one can easily prove that equation (3.2.15) has a unique solution for a wide class of discontinuous initial functions (e.g. piecewise continuous with possible discontinuity at t_0). Of course the solution x is no longer required to be continuous. This requirement is replaced by continuity of $t \mapsto \sum_{i=0}^{N} A_i(t) x(t-h_i)$ for $t > t_0$ and right-hand continuity at t_0.

The description of control systems by sets of differential equations of first order is popular, but it is not the only one used in control theory.

3.2 Linear Continuous-time FDE

It is sometimes convenient to describe a system with one output y by one scalar equation of higher order, e.g.

$$\sum_{i=0}^{N}\left[\sum_{j=0}^{n}a_{ij}(t)y^{(j)}(t-h_i)+b_i(t)u(t-h_i)\right]+\sum_{j=0}^{n-1}\int_{-h}^{0}a_j(t,\tau)y^{(j)}(t+\tau)d\tau+$$

$$+\int_{-h}^{0}b(t,\tau)u(t+\tau)d\tau = f(t), \qquad (3.2.17)$$

where $a_{0n}(t)$ has a bounded inverse for every t, $a_{ij}(t)$ and $a_j(t, \tau)$ are scalars $y^{(j)}$ denotes the j-th derivative of the dependent scalar variable y, $b_i(t)$ and $b(t, \tau)$ are m-dimensional row vectors and the control $u(t)$ is an m-dimensional column vector. The form of description (3.2.17) is less general than the vector first-order one (3.2.1), because it is always possible to construct an equation of the form (3.2.1) which is equivalent to (3.2.17), while the converse is obviously untrue. In a standard technique for transforming equation (3.2.17) into (3.2.1) new variables are introduced: $x_1 = y$, $\dot{x}_i(t) = x_{i+1}(t)$, $i = 1, \ldots, n-1$. We then have from (3.2.17)

$$\sum_{i=0}^{N}\left[a_{in}(t)\dot{x}_n(t-h_i)+\sum_{j=1}^{n}a_{i,j-1}(t)x_j(t-h_i)+b_i(t)u(t-h_i)\right]+$$

$$+\sum_{j=1}^{n}\int_{-h}^{0}a_{j-1}(t,\tau)x_j(t+\tau)d\tau+\int_{-h}^{0}b(t,\tau)u(t+\tau)d\tau = f(t). \qquad (3.2.18)$$

This, together with the definitions of the new variables, yields a set of n first-order equations equivalent to (3.2.17). The task of rewriting this in vector-matrix notation is left to the reader. Notice that the same technique may be applied to more general nonlinear FDE of higher order with time-dependent and distributed delays.

For many linear systems model (3.2.17) is not adequate unless it contains some terms with derivatives of control. Stationary control systems are often described by equations of the form

$$\sum_{i=0}^{N}\left[\sum_{j=0}^{n}a_{ij}y^{(j)}(t-h_i)+\sum_{j=0}^{n-1}b_{ij}u^{(j)}(t-h_i)\right] = f(t). \qquad (3.2.19)$$

a_{ij} are constant reals, $a_{0n} \neq 0$, b_{ij} are constant rows of dimension m, $u^{(j)}(t)$ denotes the j-th derivative of the m-dimensional control $u(t)$ (column vector). Equation (3.2.19) can be transformed in many ways to the vector form (3.2.1). We shall show two such transformations. In the first one we assume $f = 0$ and introduce new variables $x_1, \ldots, x_n$ satisfying $\dot{x}_i(t) = x_{i+1}(t)$, $i = 1, \ldots, n-1$, and

$$\sum_{i=0}^{N}\left[a_{in}\dot{x}_n(t-h_i)+\sum_{j=1}^{n}a_{i,j-1}x_j(t-h_i)\right]+u(t) = 0. \qquad (3.2.20)$$

Then

$$y(t) = \sum_{i=0}^{N}\sum_{j=1}^{n}b_{i,j-1}x_j(t-h_i). \qquad (3.2.21)$$

The reader should rewrite this system of equations in vector-matrix notation. Now let $x_1 = y$,

$$\sum_{i=0}^{N}[a_{in}\dot{x}_1(t-h_i)+a_{i,n-1}x_1(t-h_i)+b_{i,n-1}u(t-h_i)] = x_2(t)$$

$$\dot{x}_2(t)+\sum_{i=0}^{N}[a_{i,n-2}x_1(t-h_i)+b_{i,n-2}u(t-h_i)] = x_3(t)$$

$$\cdots\cdots\cdots\cdots\cdots\cdots\cdots\cdots\cdots\cdots\cdots\cdots\cdots\cdots$$

$$\dot{x}_{n-1}(t)+\sum_{i=0}^{N}[a_{i1}x_1(t-h_i)+b_{i1}u(t-h_i)] = x_n(t). \qquad (3.2.22)$$

From (3.2.19) we get

$$\dot{x}_n(t)+\sum_{i=0}^{N}[a_{i0}x_1(t-h_i)+b_{i0}u(t-h_i)] = f(t). \qquad (3.2.23)$$

The first-order systems thus obtained are equivalent to (3.2.19), provided the control function is sufficiently smooth. However, often equation (3.2.19) is used as a model of a system controlled by a piecewise continuous function u. Then it makes sense only if understood in a distributional sense, whereas the first-order vector equation may be understood as a functional identity, i.e. in the 'normal' sense. Again we come to the conclusion that equation (3.2.1) is more general.

In some models the highest order of derivative of u is n. Let us consider a system with scalar control

$$\sum_{i=0}^{N}\sum_{j=0}^{n}[a_{ij}y^{(j)}(t-h_i)+b_{ij}u^{(j)}(t-h_i)] = 0. \qquad (3.2.24)$$

The above transformations may sometimes also be used here, e.g. if $a_{in} = b_{in}$, $i = 0,\ldots,N$, we put $\dot{x}_i(t) = x_{i+1}(t), i = 1,\ldots,n-1$,

$$\sum_{i=0}^{N}\left[a_{in}\dot{x}_n(t-h_i)+\sum_{j=1}^{n}a_{i,j-1}x_j(t-h_i)\right]+u(t) = 0, \qquad (3.2.25)$$

3.2 Linear Continuous-time FDE

whence

$$\sum_{i=0}^{N}\sum_{j=1}^{n}(b_{i,j-1}-a_{i,j-1})x_j(t-h_i)-u(t) = y(t). \qquad (3.2.26)$$

3.2.2 Step method

We shall consider retarded and neutral systems with lumped, constant and commensurate delays, in the most general case described by the equation

$$\sum_{i=0}^{N}[A_i(t)\dot{x}(t-i\varrho)+B_i(t)x(t-i\varrho)+C_i(t)u(t-i\varrho)] = f(t) \quad t \in [0, T] \qquad (3.2.27)$$

with initial conditions

$$\begin{aligned} x(t) &= \varphi(t) & t \in [-N\varrho, 0] \\ \dot{x}(t) &= \mu(t) & t \in (-N\varrho, 0) \\ u(t) &= \nu(t) & t \in [-N\varrho, 0). \end{aligned}$$

We assume that ϱ is a positive constant, the function $t \mapsto A_0(t)^{-1}$ is measurable and bounded, A_i, $i = 0, \ldots, N$, are bounded and integrable and μ is square integrable. Other assumptions are as in (3.2.1). For such systems the step method yields analytic formulas for solutions. The vector-matrix approach, first studied by Olbrot (1972, 1973a) consists in the construction of a set of equations equivalent to (3.2.27), of greater dimension and with split boundary conditions, but without any deviations of argument and considered only on interval $[0, \varrho]$.

We shall begin with a simple example of a stationary system with one delay (Olbrot, 1973a):

$$\dot{x}(t) = Ax(t)+Bx(t-1)+Cu(t) \quad t \geq 0 \qquad (3.2.28)$$
$$x(t) = \varphi(t) \quad t \in [-1, 0] \quad x(t) \in R^n \quad u(t) \in R^m.$$

Let

$$x^i(t) = x((i-1)\varrho+t) \quad u^i(t) = u((i-1)\varrho+t)$$
$$\tilde{x}^j(t) = \text{col}(x^1(t), \ldots, x^j(t)) \quad \tilde{u}^j(t) = \text{col}(u^1(t), \ldots, u^j(t))$$

$$\tilde{A}_j = \begin{bmatrix} A & 0 & \cdots & 0 \\ B & A & \cdots & 0 \\ & \ddots & \ddots & \\ 0 & \cdots & B & A \end{bmatrix} \quad \tilde{C}_j = \text{diag}(C, \ldots, C)$$

$$\tilde{K}_j = \begin{bmatrix} 0 & \cdots & & 0 \\ I & \ddots & & \\ & \ddots & \ddots & \\ 0 & \cdots & I & 0 \end{bmatrix}$$

$$y_j(t) = B\text{col}(\varphi(t-1), 0, \ldots, 0) \quad y^0 = \text{col}(\varphi(0), 0, \ldots, 0). \qquad (3.2.29)$$

Then for any positive integer j the following system is equivalent to equation (3.2.28) on interval $[0,j]$:

$$\dot{\tilde{x}}^j(t) = \tilde{A}_j\tilde{x}^j(t)+\tilde{C}_j\tilde{u}^j(t)+y_j(t) \quad t \in [0, 1] \tag{3.2.30}$$

$$\tilde{x}^j(0) = y^0+\tilde{K}_j\tilde{x}^j(1). \tag{3.2.31}$$

Every solution of (3.2.28), corresponding to any initial condition φ, satisfies the system (3.2.30), (3.2.31) with regard to the defining equalities (3.2.29). Conversely, every solution of the latter system satisfies, in virtue of definition (3.2.29) applied 'in the other direction', equation (3.2.28).

Denoting

$$\Pi_j(t) = \exp(\tilde{A}_j t), \tag{3.2.32}$$

we have the variation-of-constants formula

$$\tilde{x}^j(t) = \Pi_j(t)\tilde{x}^j(0)+\int_0^t \Pi_j(t-s)[\tilde{C}_j\tilde{u}^j(s)+y_j(s)]\mathrm{d}s. \tag{3.2.33}$$

From this and (3.2.31)

$$[I-\tilde{K}_j\Pi_j(1)]\tilde{x}^j(0) = y^0+\int_0^1 \tilde{K}_j\Pi_j(1-s)[\tilde{C}_j\tilde{u}^j(s)+y_j(s)]\mathrm{d}s.$$

Since $\tilde{K}_j\Pi_j(1) = \Pi_j(1)\tilde{K}_j$ and $(\tilde{K}_j)^j = 0$, it is easy to check that matrix $I-\tilde{K}_j\Pi_j(1)$ is nonsingular and

$$(I-Y)^{-1} = I+Y+ \ldots +Y^{j-1} \tag{3.2.34}$$

where $Y = \tilde{K}_j\Pi_j(1)$. Then

$$\tilde{x}^j(0) = (I-Y)^{-1}\left\{y^0+\int_0^1 \tilde{K}_j\Pi_j(1-s)[\tilde{C}_j\tilde{u}^j(s)+y_j(s)]\mathrm{d}s\right\}. \tag{3.2.35}$$

Substituting this expression into (3.2.33) we obtain a formula for the solution of system (3.2.30) with boundary conditions (3.2.31). Equalities (3.2.29) give the corresponding solution of (3.2.28) for any $t \in [0,j]$.

Now we shall solve the general system (3.2.27) by means of a more complicated variant of the step method (Korytowski, 1976). This approach will prove useful when we consider the linear-quadratic problem of optimal control. Let σ be a positive integer such that $T+\varrho > \sigma\varrho \geqslant T$, and $\Theta = \sigma\varrho-T$. For any function of time F the following notation will be used:

$$F^i(t) = F(T-i\varrho+t)$$
$$\tilde{F}^j(t) = \mathrm{col}(F^1(t), \ldots, F^j(t)). \tag{3.2.36}$$

3.2 Linear Continuous-time FDE

For simplicity we assume $A_j, B_j, C_j = 0$ for $j > N$. Let also $\eta = \mathrm{col}(\dot{x}^{\sigma+1},\ldots$
$\ldots, \dot{x}^{\sigma+N}, x^{\sigma+1}, \ldots, x^{\sigma+N}, u^{\sigma+1}, \ldots, u^{\sigma+N})$,

$$\tilde{A}_i = \begin{bmatrix} A_0^1 & \cdots & A_{i-1}^1 \\ & \ddots & \vdots \\ 0 & & A_0^i \end{bmatrix}. \tag{3.2.37}$$

$\tilde{B}_i$ and $\tilde{C}_i$ are defined in a similar way, letter A being replaced by B and C respectively. It follows from the nonsingularity of $A_0(t)$ that $\tilde{A}_i(s)$ is nonsingular everywhere. Then we may define

$$\hat{B}_i(t) = -\tilde{A}_i(t)^{-1}\tilde{B}_i(t) \quad \hat{C}_i(t) = -\tilde{A}_i(t)^{-1}\tilde{C}_i(t),$$

$$a_i^j(t) = -\tilde{A}_i(t)^{-1}\begin{bmatrix} A_{j+i}^1(t) \\ \vdots \\ A_{j+1}^i(t) \end{bmatrix} \quad b_i^j(t) = -\tilde{A}_i(t)^{-1}\begin{bmatrix} B_{j+i}^1(t) \\ \vdots \\ B_{j+1}^i(t) \end{bmatrix}$$

$$c_i^j(t) = -\tilde{A}_i(t)^{-1}\begin{bmatrix} C_{j+i}^1(t) \\ \vdots \\ C_{j+1}^i(t) \end{bmatrix} \quad \hat{f}^j(t) = \tilde{A}_j(t)^{-1}\tilde{f}^j(t)$$

$$\delta_i = (a_i^0 \ldots a_i^{N-1} \; b_i^0 \ldots b_i^{N-1} \; c_i^0 \ldots c_i^{N-1})$$

$$e_\sigma = (I\,0) \quad d_\sigma = (0\,I) \quad ((\sigma-1)n \times \sigma n)\text{-matrices.} \tag{3.2.38}$$

The following set of equations is equivalent to (3.2.27):

$$\dot{\tilde{x}}^{\sigma-1}(t) = \hat{B}_{\sigma-1}(t)\tilde{x}^{\sigma-1}(t) + \hat{C}_{\sigma-1}(t)\tilde{u}^{\sigma-1}(t) + \delta_{\sigma-1}(t)\eta(t+\varrho) + \hat{f}^{\sigma-1}(t) \quad t \in [0, \Theta]$$

$$\dot{\tilde{x}}^{\sigma}(t) = \hat{B}_\sigma(t)\tilde{x}^\sigma(t) + \hat{C}_\sigma(t)\tilde{u}^\sigma(t) + \delta_\sigma(t)\eta(t) + \hat{f}^\sigma(t) \quad t \in [\Theta, \varrho] \tag{3.2.39}$$

$$\tilde{x}^{\sigma-1}(\Theta-) = e_\sigma \tilde{x}^\sigma(\Theta+) \quad \tilde{x}^{\sigma-1}(0+) = d_\sigma \tilde{x}^\sigma(\varrho-)$$

$$x^\sigma(\Theta) = x(0).$$

If $\sigma = 1$, only the second differential equation of system (3.2.39) remains valid. The proof is straightforward and we leave it to the reader. Since the case $\sigma = 1$ is trivial (no delays), from now on we shall assume $\sigma > 1$.

Let $\Pi_i(t,s), i = \sigma-1, \sigma$, be the fundamental solutions of equations (3.2.39), that is

$$\frac{\partial}{\partial t}\Pi_i(t,s) = \hat{B}_i(t)\Pi_i(t,s) \quad \Pi_i(s,s) = I. \tag{3.2.40}$$

The variation-of-constants formula applied to (3.2.39) eventually gives

$$\tilde{x}^{\sigma-1}(t) = \Pi_{\sigma-1}(t,\Theta)\xi + \int_\Theta^t \Pi_{\sigma-1}(t,\tau)\gamma_{\sigma-1}(\tau)\,d\tau \quad t \in [0,\Theta]$$

$$\tilde{x}^\sigma(t) = \Pi_\sigma(t,\Theta)e_\sigma^T\xi + Y_\sigma(t)x(0) + \int_\Theta^t \Pi_\sigma(t,\tau)\gamma_\sigma(\tau)\,d\tau \quad t \in [\Theta,\varrho]$$

$$\tag{3.2.41}$$

where $Y_\sigma(t)$ is a submatrix of $\Pi_\sigma(t, \theta)$ consisting of its last n columns,

$$\xi = [\Pi_{\sigma-1}(0, \Theta) - d_\sigma \Pi_\sigma(\varrho, \Theta) e_\sigma^T]^{-1} \cdot \Big[d_\sigma Y_\sigma(\varrho) x(0) +$$
$$+ d_\sigma \int_\Theta^\varrho \Pi_\sigma(\varrho, \Theta) \gamma_\sigma(\tau) d\tau + \int_0^\Theta \Pi_{\sigma-1}(0, \tau) \gamma_{\sigma-1}(\tau) d\tau \Big]$$
$$\gamma_{\sigma-1}(t) = \hat{C}_{\sigma-1}(t) \tilde{u}^{\sigma-1}(t) + \delta_{\sigma-1}(t) \eta(t+\varrho) + \hat{f}^{\sigma-1}(t) \quad (3.2.42)$$
$$\gamma_\sigma(t) = \hat{C}_\sigma(t) \tilde{u}^\sigma(t) + \delta_\sigma(t) \eta(t) + \hat{f}^\sigma(t).$$

We shall analyse an example following the pattern of reasoning used in the above derivations. Let the system be described by a neutral equation

$$\dot{x}(t) + \dot{x}(t-1) + x(t) + x(t-1) = 1 \quad t \in [0, \tfrac{4}{3}]$$

with initial conditions $x(t) = 1$, $t \in [-1, 0]$, and $\dot{x}(t) = 1$, $t < 0$. We recall that $\dot{x}(t)$ for $t < 0$ does not necessarily denote the derivative of x. Of course $T = \tfrac{4}{3}$, $\sigma = 2$ and $\Theta = \tfrac{2}{3}$.

Defining $x^1(t) = x(\tfrac{4}{3} - 1 + t)$, $x^2(t) = x(\tfrac{4}{3} - 2 + t)$ we obtain immediately from the system equation

$$\dot{x}^1(t) + 1 + x^1(t) + 1 = 1 \quad t \in [0, \tfrac{2}{3}]$$
$$\dot{x}^1(t) + \dot{x}^2(t) + x^1(t) + x^2(t) = 1 \quad t \in [\tfrac{2}{3}, 1]$$
$$\dot{x}^2(t) + 1 + x^2(t) + 1 = 1 \quad t \in [\tfrac{2}{3}, 1].$$

The continuity of the solution and the initial conditions imply

$$x^1(\tfrac{2}{3}+) = x^1(\tfrac{2}{3}-), \quad x^2(\tfrac{2}{3}+) = 1, \quad x^1(0+) = x^2(1-).$$

In vector-matrix notation $\dot{x}^1 = -x^1 - 1$, $t \in [0, \tfrac{2}{3}]$,

$$\begin{bmatrix} 1 & 1 \\ 0 & 1 \end{bmatrix} \begin{bmatrix} \dot{x}^1 \\ \dot{x}^2 \end{bmatrix} = \begin{bmatrix} -1 & -1 \\ 0 & -1 \end{bmatrix} \begin{bmatrix} x^1 \\ x^2 \end{bmatrix} + \begin{bmatrix} 1 \\ -1 \end{bmatrix} \quad t \in [\tfrac{2}{3}, 1]$$

$$x^1(\tfrac{2}{3}-) = (1 \ 0) \begin{bmatrix} x^1(\tfrac{2}{3}+) \\ x^2(\tfrac{2}{3}+) \end{bmatrix}$$

$$x^1(0+) = (0 \ 1) \begin{bmatrix} x^1(1-) \\ x^2(1-) \end{bmatrix}$$

$$x^2(\tfrac{2}{3}) = 1.$$

Hence equations (3.2.39) for our particular case take the form

$$\dot{\tilde{x}}^1 = -\tilde{x}^1 - 1 \quad t \in [0, \tfrac{2}{3}]$$
$$\dot{\tilde{x}}^2 = -\tilde{x}^2 + \begin{bmatrix} 2 \\ -1 \end{bmatrix} \quad t \in [\tfrac{2}{3}, 1]$$
$$\tilde{x}^1(\tfrac{2}{3}-) = e_2 \tilde{x}^2(\tfrac{2}{3}+) \quad \tilde{x}^1(0+) = d_2 \tilde{x}^2(1-),$$
$$x^2(\tfrac{2}{3}) = 1.$$

3.2 Linear Continuous-time FDE

The fundamental solutions are $\Pi_1(t, s) = e^{-t+s}$, $\Pi_2(t, s) = e^{-t+s}I$. Then

$$\tilde{x}^1(0) = \exp(\tfrac{2}{3})\tilde{x}^1(\tfrac{2}{3}) - \int_{2/3}^{0} \exp(\tau)d\tau = \exp(\tfrac{2}{3})\tilde{x}^1(\tfrac{2}{3}) - 1 + \exp(\tfrac{2}{3})$$

$$\tilde{x}^2(1) = \exp(-\tfrac{1}{3})\tilde{x}^2(\tfrac{2}{3}) + \int_{2/3}^{1} \exp(\tau - 1)\begin{bmatrix} 2 \\ -1 \end{bmatrix}d\tau$$

$$= \exp(-\tfrac{1}{3})\tilde{x}^2(\tfrac{2}{3}) + (1 - \exp(-\tfrac{1}{3}))\begin{bmatrix} 2 \\ -1 \end{bmatrix}.$$

Multiplying the second equality by (0 1) and comparing it with the first one we obtain from the continuity condition

$$\exp(\tfrac{2}{3})\tilde{x}^1(\tfrac{2}{3}) - 1 + \exp(\tfrac{2}{3}) = \exp(-\tfrac{1}{3})x^2(\tfrac{2}{3}) - 1 + \exp(-\tfrac{1}{3}).$$

Since $x^2(\tfrac{2}{3}) = 1$ we have

$$x^1(\tfrac{2}{3}) = 2\exp(-1) - 1$$

and

$$\tilde{x}^1(t) = \exp(-t + \tfrac{2}{3})\tilde{x}^1(\tfrac{2}{3}) - \int_{2/3}^{t} \exp(-t + \tau)d\tau = 2\exp(-t - \tfrac{1}{3}) - 1$$

$$\tilde{x}^2(t) = \exp(-t + \tfrac{2}{3})\tilde{x}^2(\tfrac{2}{3}) + \int_{2/3}^{t} \exp(-t + \tau)\begin{bmatrix} 2 \\ -1 \end{bmatrix}d\tau$$

$$= \exp(-t + \tfrac{2}{3})\begin{bmatrix} 2\exp(-1) - 3 \\ 2 \end{bmatrix} + \begin{bmatrix} 2 \\ -1 \end{bmatrix}.$$

Finally, we return to the original notation

$$x(t) = \begin{cases} 2e^{-t} - 1 & t \in [0, 1] \\ e^{-t}(2 - 3e) + 2 & t \in [1, \tfrac{4}{3}]. \end{cases}$$

Certainly, in this trivial problem one can obtain the result in a much simpler way.

3.2.3 Characteristic equation and exponential solutions

Let us consider the stationary homogeneous equation

$$\sum_{i=0}^{N} [A_i \dot{x}(t - h_i) + B_i x(t - h_i)] + \int_{-h}^{0} B(\tau) x(t + \tau) d\tau = 0 \quad t \geq 0 \quad (3.2.43)$$

with assumptions as in (3.2.1) (hence $\det A_0 \neq 0$) and with unspecified initial conditions. Let λ be a complex number and let c be a complex n-vector different from zero. Equation (3.2.43) has an exponential solution $e^{\lambda t}c$ in $[-h, \infty)$ if and only if

$$\Delta(\lambda)c = 0 \qquad (3.2.44)$$

where

$$\Delta(\lambda) = \sum_{i=0}^{N}(\lambda A_i + B_i)e^{-\lambda h_i} + \int_{-h}^{0} B(\tau)e^{\lambda \tau}d\tau. \qquad (3.2.45)$$

A nonzero vector c satisfying (3.2.44) exists if and only if matrix $\Delta(\lambda)$ is singular, that is

$$\det \Delta(\lambda) = 0. \qquad (3.2.46)$$

This is the so-called characteristic equation. The determinant $\det \Delta(\cdot)$ is sometimes called the characteristic quasipolynomial. As we shall see further on, the roots of equation (3.2.46) (characteristic roots) contain important information on system (3.2.43). It is evident that explicit formulas for these can only be found in trivial cases.

Let us denote $g(s) = \det \Delta(s)$. All complex zeros of g are pairwise conjugated. A simple evaluation shows that all zeros of g are located in a left half-plane $\text{Re}\, s \leq a$ for some real a. If g is not a polynomial, the set of zeros is infinite and countable. To give a more precise characterization we need additional assumptions.

Theorem 3.2.2 In equation (3.2.43) let $A_i = 0$, $i > 0$ (retarded case), and let g not be a polynomial. Then all zeros of g can be ordered in a sequence $s_1, s_2, \ldots$, such that $|s_k| \to \infty$ when $k \to \infty$. The real parts of the zeros are uniformly bounded above and for any real a, the number of zeros whose real parts are greater than a is at most finite. All zeros have finite multiplicities.

For proof see Hale (1977) and Myshkis (1972). We recall that a zero s of g is said to be of multiplicity p if $g^{(j)}(s) = 0, j = 0, \ldots, p-1$, and $g^{(p)}(s) \neq 0$.

Although probably it is impossible to express the roots of the characteristic equation by elementary operations and functions, asymptotic formulas for roots with large moduli are available. Following El'sgol'c and Norkin (1971) we shall show a method for obtaining them in case $B = 0$ and $N = 1, h = h_1$. We assume that no powers of the term e^{-sh} of order higher than one occur in the determinant (for this, it is sufficient that $\text{rank}\, A_1 + \text{rank}\, B_1 = 1$). Then

$$g(s) = \det(sA_0 + B_0 + sA_1 e^{-sh} + B_1 e^{-sh}) = \sum_{i=0}^{n} a_i s^i + \sum_{i=0}^{p} b_i s^i e^{-sh} \qquad (3.2.47)$$

3.2 Linear Continuous-time FDE

where a_i, b_i are real coefficients, $a_n \neq 0$, $b_p \neq 0$, $n \geq p$. The predominant terms are $a_n s^n$ and $b_p s^p e^{-sh}$. Therefore for large $|s|$ the roots of the characteristic equation may be approximated by the roots of the equation

$$a_n s^n + b_p s^p e^{-sh} = 0. \tag{3.2.48}$$

First, let $n = p$. Then

$$e^{-sh} = -\frac{a_n}{b_n}$$

and the asymptotic formula for the roots is

$$s_k \approx \begin{cases} -\dfrac{1}{h} \ln \left|\dfrac{a_n}{b_n}\right| + \dfrac{2k\pi}{h} i & \dfrac{a_n}{b_n} > 0 \\ -\dfrac{1}{h} \ln \left|\dfrac{a_n}{b_n}\right| + \dfrac{(2k+1)\pi}{h} i & \dfrac{a_n}{b_n} < 0 \end{cases} \quad (k = 0, \pm 1, \pm 2, \ldots). \tag{3.2.49}$$

With increasing $|k|$ these expressions become increasingly accurate, that is the differences between the corresponding roots of (3.2.46) and (3.2.48) tend to zero. The absolute value of the error is bounded above by K/k for some $K > 0$.

Now let $n > p$ (retarded case). Then from (3.2.48)

$$e^{sh} s^r = a \tag{3.2.50}$$

where $r = n - p > 0$ and $a = -\dfrac{b_p}{a_n}$. Putting $s = s_k = x_k + i y_k$ and raising the moduli of both sides to power $2/r$ we obtain

$$\exp\left(\frac{2x_k}{r} h\right) (x_k^2 + y_k^2) = |a|^{2/r}. \tag{3.2.51}$$

Since there is only a finite number of roots in every right half-plane, we have $x_k \to -\infty$ and

$$x_k^2 \exp \frac{2x_k h}{r} \to 0 \quad \text{when } k \to \infty. \tag{3.2.52}$$

Then

$$y_k^2 \exp \frac{2x_k h}{r} \to |a|^{2/r} \quad (k \to \infty). \tag{3.2.53}$$

As $a \neq 0$, this implies

$$\left|\frac{y_k}{x_k}\right| \to \infty \quad (k \to \infty).$$

The complex roots are pairwise conjugate, so we may choose the roots with $y_k > 0$. Then

$$\arg s_k \to \frac{\pi}{2} \quad (k \to \infty).$$

Taking the logarithm of both sides of (3.2.50) and comparing the imaginary parts we obtain

$$y_k h + r \arg s_k = \arg a + 2k\pi,$$

and

$$y_k = \begin{cases} \dfrac{\pi}{h}\left(2k - \dfrac{r}{2}\right) + \varepsilon_1 & a > 0 \\ \dfrac{\pi}{h}\left(2k + 1 - \dfrac{r}{2}\right) + \varepsilon_1 & a < 0 \end{cases} \quad (3.2.54)$$

$\varepsilon_1 \to 0$ when $k \to \infty$. From (3.2.53) we compute

$$x_k = \frac{1}{h}(\ln|a| - r\ln y_k) + \varepsilon_2,$$

$\varepsilon_2 \to 0$ when $k \to \infty$. Hence

$$s_k \approx \begin{cases} \dfrac{1}{h}\left[\ln|a| - r\ln\left(2k - \dfrac{r}{2}\right)\dfrac{\pi}{h}\right] \pm \left(2k - \dfrac{r}{2}\right)\dfrac{\pi i}{h} & a > 0 \\ \dfrac{1}{h}\left[\ln|a| - r\ln\left(2k + 1 - \dfrac{r}{2}\right)\dfrac{\pi}{h}\right] \pm \left(2k + 1 - \dfrac{r}{2}\right)\dfrac{\pi i}{h} & a < 0 \end{cases}$$

$$(k = 0, 1, 2, \ldots). \quad (3.2.55)$$

The absolute value of the approximation error in (3.2.55) is bounded above by $K\dfrac{\ln k}{k}$, for some constant $K > 0$.

The asymptotic formulas (3.2.49) and (3.2.55) give qualitative information on the location of characteristic roots. In simple cases they may be sufficiently accurate even for not very large k; if not, they may serve as the starting point for a numerical procedure. Similar results can be obtained for systems with many delays—for details see Bellman and Cooke (1963).

Now we return to the problem of solutions connected with characteristic roots. We already know that for every such root λ there is a (nonzero) vector c such that $e^{\lambda t}c$ is a solution of equation (3.2.43) with unspecified initial conditions. If λ is a root of multiplicity p, $p > 1$, there can also exist solutions of the form $e^{\lambda t}w(t)$ where w is a polynomial $R \ni t \mapsto w(t) \in C^n$ of degree less than p (C^n is the space of complex n-vectors). The maximal number of linearly independent solutions corresponding to the root λ is equal

3.2 Linear Continuous-time FDE

to its multiplicity. Since the system is linear and we have not fixed the initial conditions, any linear combination of solutions is also a solution. Therefore we can always construct a real solution by pairing functions of the form $e^{\lambda t}w(t)$. Adding to each its conjugate and dividing by two we obtain $\mathrm{Re}\,[e^{\lambda t}w(t)]$. By subtraction we get another real solution, $\mathrm{Im}\,[e^{\lambda t}w(t)]$.

Let us suppose that the characteristic equation has an infinite number of roots and let us arrange these in a sequence λ_i, $i = 1, 2, \ldots$, in such a way that $|\lambda_i| \leq |\lambda_{i+1}|$. Let p_i be the multiplicity of λ_i and $w_{ij}(t)e^{\lambda_i t}$, $j = 1, \ldots, p_i$, be linearly independent solutions of (3.2.43). Then the infinite linear combination

$$x(t) = \sum_{i=1}^{\infty} \sum_{j=1}^{p_i} a_{ij} w_{ij}(t) e^{\lambda_i t} \qquad (3.2.56)$$

is a solution of the system equation (3.2.43) if it is convergent and term-by-term differentiable. It is important for us to construct solutions in this series form because in practice we are interested in solving Cauchy problems rather than in finding any solution with unspecified initial conditions. As we shall see in the next section, by an appropriate choice of coefficients a_{ij} it is possible to fit solutions of the form (3.2.56) to a wide class of predetermined initial conditions. The series (3.2.56) may be also suitable for expressing solutions of nonhomogeneous Cauchy problems (i.e. with nonzero control and function f).

3.2.4 Laplace transform and series expansions of solutions

We consider a stationary system

$$\sum_{i=0}^{N} [A_i \dot{x}(t-h_i) + B_i x(t-h_i) + C_i u(t-h_i)] +$$

$$+ \int_{-h}^{0} B(\tau) x(t+\tau) \, d\tau + \int_{-h}^{0} C(\tau) u(t+\tau) \, d\tau = f(t) \qquad t \geq 0 \qquad (3.2.57)$$

with initial conditions

$$x(t) = \varphi(t) \quad t \in [-h, 0]$$
$$\dot{x}(t) = \mu(t) \quad t \in (-h, 0)$$
$$u(t) = \nu(t) \quad t \in [-h, 0)$$

and assumptions as in (3.2.1). In Section 3.2.1 equalities are given which determine the relation between the input of the system, which may consist of control u, possibly function f and initial conditions, and the trajectory x. The form of the relation is not very simple; it is given by integral operators whose kernels are determined by an FDE. In many applications a control engineer prefers an algebraic, multiplicative form of this relation, though it is available only in the domain of complex variable, between the Laplace transforms of the input and the trajectory. An important source of motivation for the use of Laplace transforms is tradition. The classical analysis and synthesis of control systems is based on frequency techniques and practitioners are accustomed to think in those terms, no matter how limited they are. We should also stress that from the theoretical point of view, the Laplace transform is a convenient tool for investigation of different properties of linear systems.

Let F be a function such that $(0, \infty) \to R^n$. Its Laplace transform $\hat{F}$, denoted also by LF, is a function of a complex variable defined by the formula

$$\hat{F}(s) = \int_0^\infty e^{-st} F(t) \, dt. \tag{3.2.58}$$

Lemma 3.2.1 (i) *If F is measurable and $|F(t)| \leqslant ae^{bt}$, $t \in [0, \infty)$ for some positive constants a and b, then its Laplace transform exists and is an analytic function of s for $\operatorname{Re} s > b$.*

(ii) *Now suppose that $F: [0, \infty) \to R^n$ is a given function of bounded variation on any compact set and $b > 0$ is a real such that function $t \mapsto F(t) e^{-bt}$ is Lebesgue integrable on $[0, \infty)$. Then for any $c > b$,*

$$\frac{1}{2\pi i} \lim_{\omega \to \infty} \int_{c-i\omega}^{c+i\omega} e^{st} \hat{F}(s) \, ds = \begin{cases} \tfrac{1}{2}[F(t+) + F(t-)] & t > 0 \\ \tfrac{1}{2} F(0+) & t = 0. \end{cases} \tag{3.2.59}$$

The transformation of $\hat{F}$ determined by (3.2.59) may be regarded as the inverse of the Laplace transform and is denoted by L^{-1}.

We shall apply the Laplace transform to both sides of equation (3.2.57) assuming that u and f satisfy the conditions of Lemma 3.2.1. We denote by $\hat{x}$, $\hat{u}$, and $\hat{f}$ the Laplace transforms of $x|_{[0, \infty)}$, $u|_{[0, \infty)}$ and f, respectively. Let

$$P(s) = \sum_{i=0}^{N} e^{-sh_i} \left[A_i x(0) - \int_{-h_i}^{0} e^{-st}(A_i \mu(t) + B_i \varphi(t) + C_i \nu(t)) \right] dt -$$

$$- \int_{-h}^{0} \int_{\tau}^{0} e^{s(\tau-t)} [B(\tau) \varphi(t) + C(\tau) \nu(t)] \, dt \, d\tau$$

$$\Delta(s) = \sum_{i=0}^{N} e^{-sh_i}(sA_i + B_i) + \int_{-h}^{0} e^{st} B(t) \, dt \tag{3.2.60}$$

$$H(s) = \sum_{i=0}^{N} e^{-sh_i} C_i + \int_{-h}^{0} e^{st} C(t) \, dt.$$

3.2 Linear Continuous-time FDE

Then
$$\Delta(s)\hat{x}(s) = P(s) - H(s)\hat{u}(s) + \hat{f}(s) \tag{3.2.61}$$
and
$$\hat{x}(s) = \Delta(s)^{-1}[P(s) - H(s)\hat{u}(s) + \hat{f}(s)]. \tag{3.2.62}$$

It is noteworthy that the matrix $\Delta(s)$ is identical with the characteristic matrix (3.2.45). Therefore $\Delta(s)^{-1}$ exists if and only if s is not a characteristic root. It is also evident that all characteristic roots which are not equal to zeros of $P(s) - H(s)\hat{u}(s) + \hat{f}(s)$, are poles of the right-hand side of (3.2.62).

For $t > 0$, x is a continuous function, and so by virtue of (3.2.59)
$$x(t) = \frac{1}{2\pi i} \lim_{\omega \to \infty} \int_{c-i\omega}^{c+i\omega} e^{st} \Delta(s)^{-1}[P(s) - H(s)\hat{u}(s) + \hat{f}(s)] ds \tag{3.2.63}$$

where c is a sufficiently large real.

Now we shall show a method of expanding the solution x of (3.2.57) into a series of the form (3.2.56) on the basis of the Cauchy theorem. Let C_k, $k = 1, 2, \ldots$, be a sequence of closed contours on the complex plane, $C_k = C_k^1 \cup C_k^2$. C_k^1 is a segment of the straight line, $C_k^1 = \{s, \mathrm{Re}\, s = c, \mathrm{Im}\, s \in [-\omega_k, \omega_k]\}$, $\omega_{k+1} > \omega_k$ for every k, and $\omega_k \to \infty$ when $k \to \infty$. C_k^2 lies on the left of C_k^1 and is chosen in such a way that C_{k+1} contains C_k. Each contour C_k contains zero in its interior and does not pass through the poles of $\hat{x}(s)$; the distance between zero and C_k^2 tends to infinity as $k \to \infty$. The details of construction depend on the particular problem involved (see Bellman and Cooke, 1963).

From the Cauchy theorem
$$\int_{C_k} e^{st} \hat{x}(s) ds = 2\pi i \sum_k \tag{3.2.64}$$

where $\sum_k$ stands for the sum of all residues of $e^{st}\hat{x}(s)$ within C_k.

Hence
$$\frac{1}{2\pi i} \int_{C_k^1} e^{st} \hat{x}(s) ds = \sum_k - \frac{1}{2\pi i} \int_{C_k^2} e^{st} \hat{x}(s) ds. \tag{3.2.65}$$

When $k \to \infty$, the integral on the left tends to $x(t)$ for any $t > 0$. If the integral in the right-hand side tends to zero, we obtain a series representation of the solution
$$x(t) = \text{sum of all residues of } e^{st}\hat{x}(s). \tag{3.2.66}$$

More definite results are available for uncontrolled homogeneous systems (Bellman and Cooke, 1963). Then

$$\hat{x}(s) = \Delta(s)^{-1} P(s)$$

$$P(s) = \sum_{i=0}^{N} e^{-sh_i} \left[A_i x(0) - \int_{-h_i}^{0} e^{-st}(A_i \mu(t) + B_i \varphi(t)) \, dt \right] -$$

$$- \int_{-h}^{0} \int_{\tau}^{0} e^{s(\tau-t)} B(\tau) \varphi(t) \, dt \, d\tau.$$

Assume that functions μ and φ are continuous and bounded in $(-h, 0)$. Let $\lambda_1, \lambda_2, \ldots$ be all characteristic roots arranged in order of nondecreasing absolute value, and let

$$\inf_{i,j} |\lambda_i - \lambda_j| > 0 \quad (i \neq j). \tag{3.2.67}$$

(This condition is always satisfied in retarded systems.) Formula (3.2.66) then holds for $t > nh$ and $x(t)$ can be expressed in the form (3.2.56).

For neutral systems formula (3.2.66) is valid for $t > 0$ if $\det A_N \neq 0$.

A group of methods for series expansion of the solution in the time domain are based on the series expansion of the Laplace transform of the solution in the domain of complex variable. To this series the inverse Laplace transform (3.2.59) is applied term by term yielding a function of time. This technique must be used with particular care because convergence of the two series is not sufficient for the result to be correct. A general theorem which is often useful will be given (Korn and Korn, 1961).

Theorem 3.2.3 *Let the function $s \mapsto F(s)$ of a complex variable be determined by a series*

$$F(s) = \sum_{i=1}^{\infty} F_i(s) \tag{3.2.68}$$

convergent in the half-plane $\operatorname{Re} s \geq a$. *Let f_i, $i = 1, 2, \ldots$, be a sequence o functions of time such that for all s, $\operatorname{Re} s \geq a$, we have*

$$F_i(s) = (Lf_i)(s) \quad (i = 1, 2, \ldots).$$

Moreover we assume that all functions $t \mapsto |f_i(t)| e^{-at}$ are integrable on $[0, \infty)$ and the series

$$\sum_{i=1}^{\infty} \int_{0}^{\infty} |f_i(t)| e^{-at} \, dt$$

is convergent. Then the series $\sum_{i=1}^{\infty} f_i(t)$, $t \in [0, \infty)$, is convergent to a function f and

3.2 Linear Continuous-time FDE

$$(Lf)(s) = \left(L\sum_{i=1}^{\infty} f_i\right)(s) = \left(\sum_{i=1}^{\infty} Lf_i\right)(s) = \sum_{i=1}^{\infty} F_i(s) = F(s) \quad \operatorname{Re} s \geq a.$$
(3.2.69)

We shall mention briefly two particular approaches. If the Laplace transform $\hat{f}$ of a real function f is represented by a series of the form

$$\hat{f}(s) = \sum_{i=1}^{\infty} \frac{a_i}{s^i} \tag{3.2.70}$$

convergent for $|s| > r > 0$, then for all $t > 0$

$$f(t) = \sum_{i=0}^{\infty} \frac{a_{i+1}}{i!} t^i. \tag{3.2.71}$$

Now assume that the Laplace transform $\hat{f}$ of f, $\hat{f}(s) = (Lf)(s)$, $\operatorname{Re} s > 0$, is analytic at $s = \infty$. It can then be expressed by a series

$$\hat{f}(s) = (1-z)\sum_{i=0}^{\infty} c_i z^i$$

$$z = \frac{s-0.5}{s+0.5} \quad \operatorname{Re} s > a$$

$$c_i = \sum_{j=0}^{i} C_i^j \frac{1}{j!} \hat{f}^{(j)}\left(\frac{1}{2}\right) \quad (i = 0, 1, \ldots) \tag{3.2.72}$$

convergent for $|z| < 1$. If the assumptions of Theorem 3.2.3 are satisfied then

$$f(t) = e^{-\frac{t}{2}} \sum_{i=0}^{\infty} \frac{c_i L_i(t)}{i!} \tag{3.2.73}$$

for almost all $t \geq 0$, where L_i is the i-th Laguerre polynomial,

$$L_0 = 1, \quad L_{i+1}(t) = (2i+1-t)L_i(t) - i^2 L_{i-1}(t) \quad (i = 1, 2, \ldots). \tag{3.2.74}$$

In further discussion we shall confine ourselves to scalar systems described by equation (3.2.19). Applying the Laplace transform to both sides we obtain

$$\sum_{i=0}^{N} e^{-sh_i} \left\{ \sum_{j=0}^{n} a_{ij} \left[s^j \hat{y}(s) + \sum_{p=1}^{j} s^{j-p} y^{(p-1)}(0+) + \int_{-h_i}^{0} e^{-st} y^{(j)}(t)\,dt \right] + \right.$$
$$\left. + \sum_{j=0}^{n-1} b_{ij} \left[s^j \hat{u}(s) + \sum_{p=1}^{j} s^{j-p} u^{(p-1)}(0+) + \int_{-h_i}^{0} e^{-st} u^{(j)}(t)\,dt \right] \right\} = \hat{f}(s).$$

Hence
$$\hat{y}(s) = \Delta(s)^{-1}[P(s) - H(s)\hat{u}(s) + \hat{f}(s)] \qquad (3.2.75)$$
where
$$\Delta(s) = \sum_{i=0}^{N} \sum_{j=0}^{n} a_{ij} e^{-sh_i} s^j$$

$$H(s) = \sum_{i=0}^{N} \sum_{j=0}^{n-1} b_{ij} e^{-sh_i} s^j \qquad (3.2.76)$$

$$P(s) = \sum_{i=0}^{N} e^{-sh_i} \left\{ \sum_{j=0}^{n} a_{ij} \left[\sum_{p=1}^{j} s^{j-p} y^{(p-1)}(0+) + \int_{-h_i}^{0} e^{-st} y^{(j)}(t) dt \right] + \right.$$
$$\left. + \sum_{j=0}^{n-1} b_{ij} \left[\sum_{p=1}^{j} s^{j-p} u^{(p-1)}(0+) + \int_{-h_i}^{0} e^{-st} u^{(j)}(t) dt \right] \right\}.$$

Let us consider the homogeneous case with zero initial conditions. Then $P = 0$ and
$$\hat{y}(s) = -\Delta(s)^{-1} H(s) \hat{u}(s). \qquad (3.2.77)$$
Denote
$$G(s) = \Delta(s)^{-1} H(s). \qquad (3.2.78)$$
In the applications (feedback control) we encounter an important case when $G(s)$ is expressed in the form
$$G(s) = \frac{G_0(s)}{1 + G_0(s)} \qquad (3.2.79)$$
where
$$G_0(s) = F_0(s) e^{-sh} = \frac{\sum_{j=0}^{n} b_j s^j}{\sum_{j=0}^{n} a_j s^j} e^{-sh} \qquad h > 0 \quad a_n \neq 0. \qquad (3.2.80)$$
Assume that $\hat{u}$ is a rational function
$$\hat{u}(s) = \frac{\sum_{j=0}^{n-1} \beta_j s^j}{\sum_{j=0}^{n} \alpha_j s^j} \qquad \alpha_n \neq 0. \qquad (3.2.81)$$
Formally we may write
$$G(s) = \frac{G_0(s)}{1 + G_0(s)} = \sum_{i=1}^{\infty} (-1)^{i+1} G_0(s)^i \qquad (3.2.82)$$

whence

$$\hat{y}(s) = \sum_{i=1}^{\infty} (-1)^i F_0(s)^i e^{-ish} \hat{u}(s). \qquad (3.2.83)$$

To justify this formula observe that for sufficiently large c

$$\lim_{k \to \infty} \lim_{\omega \to \infty} \int_{c-i\omega}^{c+i\omega} \left[G(s) + \sum_{i=1}^{k} (-1)^i G_0(s)^i \right] \hat{u}(s) e^{st} ds = 0$$

which proves that $L^{-1}\hat{y}$ can be calculated in a term-by-term fashion. As for each i, $F_0(s)^i \hat{u}(s)$ are rational functions, there exist functions of time x_i, $x_i = 1, \ldots$, such that

$$(Lx_i)(s) = (-F_0(s))^i \hat{u}(s).$$

Therefore functions $s \mapsto (-F_0(s))^i \hat{u}(s) e^{-ish}$, $i = 1, 2, \ldots$, are Laplace transforms of functions y_i, $i = 1, \ldots$,

$$y_i(t) = \begin{cases} x_i(t - ih) & t \geq ih \\ 0 & t < ih. \end{cases} \qquad (3.2.84)$$

Finally we obtain

$$y(t) = \sum_{i=1}^{k(t)} y_i(t) \qquad (3.2.85)$$

where $k(t)$ is the greatest integer such that $k(t)h \leq t$.

3.3 Input, Output, State and Transfer Functions

3.3.1 Basic definitions

This section will be devoted to the formalization of several basic notions of control theory such as input, control, state and output. The modern control philosophy, most elegantly expressed in the state-space approach, has its roots in classical physics which in turn is based on the category of causality. Although it can also be applied to nonlinear FD systems, so far the state-space theory proved most effective in the linear case.

In all descriptions of dynamical systems time appears as an independent variable. The set of all moments of time T may be continuous or discrete, finite or not. Let us introduce the spaces of input values W and output values Y. The input is a function of time $w: T \to W$ which represents the cause, something that makes the system react. The measurable or observable reactions of the system, or the effect, are described by the output $y: T \to Y$. The idea is that the output contains all the available information about the system which is necessary to evaluate the system's performance or accomplishment of the control task. The output may also have components which are not causally connected with the input and, on the other hand, it may include some components of the input—if they are measurable and essential for the system's performance.

The causality principle says that for every dynamical system there exists a function f that assigns a unique output y to every input w. Thus there is a description of the dynamical system

$$y(t) = f(w, t) \quad t \in T. \tag{3.3.1}$$

Since an effect cannot precede its cause, function f must be non-anticipative:

$$\forall t \in T \quad (w(s) = w'(s), s \leqslant t) \Rightarrow (f(w, t) = f(w', t)). \tag{3.3.2}$$

In the applications an explicit form of function f occurs very rarely. Usually it is determined by a differential equation.

Now we shall analyse the input in more detail. It is convenient to treat its components which do not change in time separately—these will be called parameters. If we denote by $\mathscr{W}$ the space of all input functions $w: T \to W$, we have $\mathscr{W} = \mathscr{W}_1 \times P$ where $\mathscr{W}_1$ is a space of functions of time and P is a space of (constant) parameters. In control systems a part of input represents the purposeful actions of a human (or automatic) operator. The parameters whose values are chosen by the operator so that the task of the system might be accomplished, are called preset or adjustable parameters. The functions of time $u: T \to U$ which describe the purposeful input actions are called controls. U is the space of control values, and the space of all controls is denoted by $\mathscr{U}$. In real systems controls are subject to different restrictions, many of them due to technological limitations. This fact is taken into account by selecting from $\mathscr{U}$ the subset of admissible controls $\mathscr{U}_{ad}$. Frequently this is done by imposing constraints on control values; an admissible control takes its values only from a given set $U_{ad} \subset U$. According to control systems philosophy the control space $\mathscr{U}$ cannot be quite arbitrary. It is stipulated that a control should represent the will of the operator, his idea of control rather than its technical realization burdened with unavoidable inertia. Therefore at each moment of time the operator should be able to switch from one control to another and also obtain a control. This stipulation is formally expressed by the implication

$$(u, u' \in \mathscr{U}) \Rightarrow (\forall t \in T, u_t \in \mathscr{U}) \tag{3.3.3}$$

where

$$u_t(s) = \begin{cases} u(s) & s < t \\ u'(s) & s \geqslant t. \end{cases}$$

3.3 Input, Output, State and Transfer Functions

We see that some very popular functional spaces, such as C^0 or C^1 are not good control spaces. The spaces of piecewise continuous or square integrable functions satisfy (3.3.3). Since in the applications it is sometimes convenient to deal with continuous or more regular controls, it is advisable in this kind of situation to choose $\mathscr{U}$ so that (3.3.3) holds and the required set of controls is a subset of $\mathscr{U}$. Then we can always determine $\mathscr{U}_{\text{ad}}$ according to our needs.

In real systems there usually are input components harmful or dangerous to the system's performance. This may be due to the fact that it is impossible to fully compensate for this harm by use of control because of the functional character of such components or because they cannot be measured accurately enough. Such components are called disturbances. One of the standard approaches is to treat them within the framework of the theory of stochastic processes. A common practice is to represent as disturbances all factors which cause errors in model performance (i.e. those which make model output differ from the real one), such as errors of modelling, inaccuracy in model parameters, etc.

The state is one of central notions of modern control theory. We introduce a space X—called the state space—and two functions f_1 and f_2. As a rule in dynamical systems the space of parameters P can be decomposed into a Cartesian product of X, T and some space P_1, $P = P_1 \times X \times T$. The X-component of an element of P is called the initial state and the T-component, the initial moment of time. Function f_1 maps $\mathscr{W} \times T$ into X and function f_2 maps $X \times U \times T$ into Y. In some descriptions $f_2: X \times U \times Z \times T \to Y$, where Z is the space of output disturbances. Function f, which describes the dependence between input and output, must satisfy the equality $f = f_2 \circ f_1$. The equation

$$x(t) = f_1(w, t) \quad t \in T \tag{3.3.4}$$

is called the state equation. The output equation is defined as

$$y(t) = f_2\big(x(t), u(t), t\big) \quad t \in T. \tag{3.3.5}$$

Control theory imposes several requirements on function f_1. As the input space can be decomposed into the Cartesian product $\mathscr{W} = \mathscr{W}_1 \times P_1 \times X \times T$ we have $f_1: \mathscr{W}_1 \times P_1 \times X \times T \times T \to X$. The state at moment t represents complete information on the system and f_1 must be non-anticipative, hence f_1 must satisfy the implication

$$(w|_{[t,t')} = w'|_{[t,t')}) \Rightarrow f_1(w, p, x, t, t') = f_1(w', p, x, t, t') \tag{3.3.6}$$

for every $w, w' \in \mathscr{W}_1$, $x \in X$, $t, t' \in T$, $t < t'$. Since the third argument represents the initial state, and the fourth the initial moment of time, we require that for every $w \in \mathscr{W}_1$, $x \in X$, $p \in P_1$, $t \in T$

$$f_1(w, p, x, t, t') = x. \tag{3.3.7}$$

We impose the translation property on the state trajectory,

$$f_1(w, p, x, t, t') = f_1(w, p, f_1(w, p, x, t, t''), t'',)$$

$$\forall w \in \mathscr{W}_1 \quad p \in P_1 \quad x \in X \quad t, t', t'' \in T \quad t \leqslant t'' \leqslant t'. \quad (3.3.8)$$

Other conditions on f_1 require topology in the considered spaces. An important property usually assumed is continuity with respect to the first three arguments. If set T is a continuum it is also assumed that f_1 is continuous with respect to the fourth and fifth arguments. The reader should verify that conditions (3.3.6), (3.3.7), (3.3.8) are independent, that is, none of them is implied by the others.

Lastly, there is still another requirement in state theory which should not, however, be treated too formally or rigorously. In most applications the state of a dynamical system can be interpreted as a certain set (relation, function, vector, sequence, etc.). It is stipulated that the state is a minimal set (in the sense of inclusion) with the above properties, which contains all the information necessary to describe a given process.

It is evident that for every dynamical system with a given input–output relation (3.3.1) one can construct a state space and a state equation in many ways. A trivial and practically useless but formally correct construction is to contain in the state all the information (not necessarily available from the output) on input, that is $\mathscr{W} = X \times T$, $\mathscr{W}_1 = \emptyset$, $P_1 = \emptyset$. Then in the state equation we put $f_1(x, t, t') = x$. One of the sources of motivation for the state space approach is that in most technological processes it is possible to find a simple state space whose elements are much 'smaller' than in the above proposition. Often the state is a finite-dimensional vector; in this book we are interested in a more complicated case where the states contain vector functions. Another source of motivation is that the state space approach gives a formal, theoretical tool for systems description in causality categories, and for using the intuition that there exist no non-local interactions. The latter means that if we can separate a region Ω in the three-dimensional space and a time interval θ such that the whole process we are interested in occurs in $\Omega \times \theta$, its course is uniquely determined by information 'contained' in Ω at the beginning of θ (the initial state) and later interactions through the boundary of Ω.

Now we shall consider the consequences of state properties on the description of system dynamics (3.3.4), (3.3.5). Assume first that the set of time moments is discrete, $T = \{t_i, i = 0, 1, 2, \ldots\}$. By virtue of (3.3.4) and (3.3.8) we get

$$x(t_{i+1}) = f_1(w_1, p, x(t_i), t_i, t_{i+1}) \quad (i = 0, 1, 2, \ldots)$$

3.3 Input, Output, State and Transfer Functions

and because of (3.3.6) we can be sure that there exists a function g such that
$$x(t_{i+1}) = g(w_1(t_i), p, x(t_i), t_i) \quad (i = 0, 1, 2, \ldots). \tag{3.3.9}$$
We have obtained a difference equation which is an equivalent description of the dynamical system. If T is a continuum, an analogous procedure is possible, provided f_1 is regular enough. For a positive Δt we have
$$x(t+\Delta t) = f_1(w_1, p, x(t), t, t+\Delta t).$$
If x is differentiable (i.e. f_1 continuous and almost everywhere differentiable with respect to the last argument) there exists a function g such that
$$\dot{x}(t) = g(w_1(t), p, x(t), t) \quad t \in T. \tag{3.3.10}$$
Unless the state space is finite-dimensional, such a differential equation may be valid only on trajectories starting from a dense subset of initial states (see the next section). Unfortunately, an effective theory of state differential equations for functional-differential systems exists only in the linear case.

In conclusion, we see that the state space approach naturally leads to a local description of system dynamics. This means that the direction and rate of state change are determined only by the state and the input value at the same moment of time. This is a rather attractive feature—such models have been successfully used in physics for a long time.

3.3.2 State equation for linear systems

The state of system (3.2.1) may be determined in many different ways within the framework of the formal definitions. A well established tradition is to take the space of initial conditions H as the state space. In this case the state at moment t consists of four elements
$$\{x(t), x^t, \dot{x}, u^t\}$$
where $x(t) \in R^n$
$$\begin{aligned} x^t &\in L^2(-h, 0; R^n) & x^t(s) &= x(t+s) \\ \dot{x}^t &\in L^2(-h, 0; R^n) & \dot{x}^t(s) &= \dot{x}(t+s) \\ u^t &\in L^2(-h, 0; R^m) & u^t(s) &= u(t+s). \end{aligned} \tag{3.3.11}$$

If there are no delays in the trajectory, its derivative or control, the state space is reduced similarly as the space of initial conditions. For example, in retarded systems the state space H consists of triplets
$$\{x(t), x^t, u^t\}. \tag{3.3.12}$$
If there are no delays in control we have for the state
$$\{x(t), x^t\}. \tag{3.3.13}$$

Although the choice of the state space as H is formally simple and convenient, we must realize that states (3.3.11)–(3.3.13) usually contain much superfluous information. For example, in the system $\dot{x}_1(t) = x_2(t-1)$, $\dot{x}_2(t) = x_1(t)$, the information on $x_1(s)$, $s < t$, is irrelevant after the moment t. Here again we touch on the question of the state being a minimal representation of all the information contained in the system necessary for determining its future trajectory. From the practical point of view it is advisable to store in the state only that information to be brought up to date in order to correct the forecasted output trajectory. The information that does not lose value during the run of the process may, if necessary, be represented by parameters. Thus in most cases the derivative component $\dot{x}^t$ may be removed from the state. For neutral system (3.2.15) the state is defined as

$$\{v(t), x^t, u^t\} \tag{3.3.14}$$

where

$$v(t) = \sum_{i=0}^{N} A_i(t) x(t-h_i).$$

Now we shall construct and discuss a state equation for a retarded system without delays in control,

$$\dot{x}(t) + \sum_{i=0}^{N} B_i(t) x(t-h_i) + \int_{-h}^{0} B(t,\tau) x(t+\tau) \, d\tau + C(t) u(t) = f(t) \quad t \geq 0 \tag{3.3.15}$$

with assumptions as in (3.2.5). The approach will be essentially the same as in Delfour (1977a). The state space is $H = R^n \times L^2(-h, 0; R^n)$, the state of the system at moment t is defined as a pair

$$\tilde{x}(t) = \{\tilde{x}(t)^0, \tilde{x}(t)^1\} \tag{3.3.16}$$

where

$$\tilde{x}(t)^0 = x(t)$$
$$\tilde{x}(t)^1(s) = x(t+s) \quad s \in [-h, 0).$$

H is a Hilbert space with inner product

$$(y|z) = (y^0|z^0)_{R^n} + (y^1|z^1)_{L^2} \quad y, z \in H.$$

The construction of the state differential equation is based on the Hille-Phillips theory (Delfour, 1977a). We introduce a space $W^{1,2}(-h, 0; R^n)$ of all absolutely continuous functions $[-h, 0] \to R^n$ with derivatives in $L^2(-h, 0; R^n)$. Let

$$V = \{(\varphi(0), \varphi): \varphi \in W^{1,2}(-h, 0; R^n)\}. \tag{3.3.17}$$

3.3 Input, Output, State and Transfer Functions

The norm in V is defined by

$$||\varphi||^2 = \int_{-h}^{0} \left(||\varphi(t)||^2 + ||\dot{\varphi}(t)||^2\right) dt. \tag{3.3.18}$$

We introduce mappings $\tilde{B}(t): V \to H$, $\tilde{C}(t): R^m \to H, \tilde{f},$

$$\tilde{B}(t)\varphi = \left(\sum_{i=0}^{N} B_i(t)\varphi(-h_i) + \int_{-h}^{0} B(t,\tau)\varphi(\tau) d\tau, -\dot{\varphi}\right) \tag{3.3.19}$$

$$\tilde{C}(t)w = (C(t)w, 0) \in H \quad \tilde{f}(t) = (f(t), 0) \in H.$$

By Λ we denote a continuous and dense injection $V \to H$. If $v = (\varphi(0), \varphi) \in V$, then $(\Lambda v)^0 = \varphi(0)$ and $(\Lambda v)^1$ is the equivalence class of φ in L^2. The following theorem (Delfour, 1977a) gives the state space description of system (3.3.15).

Theorem 3.3.1 $\forall s, T, 0 \leqslant s \leqslant T, \forall \varphi \in V, \forall u \in L^2(s, T; R^m), \forall f \in L^2(s, T; R^n)$
(i) *The equation*

$$\dot{z}(t) + \tilde{B}(t)z(t) + \tilde{C}(t)u(t) = \tilde{f}(t) \quad t \in [s, T] \tag{3.3.20}$$
$$z(s) = \varphi$$

has a unique solution

$$z \in W(s, T) = \{\zeta \in L^2(s, T; V): \dot{\zeta} \in L^2(s, T; H)\}. \tag{3.3.21}$$

(ii) *The mapping*

$$(\varphi, u, f) \mapsto z(\cdot, \varphi, u, f): V \times L^2(s, T; R^m) \times L^2(s, T; R^n) \to W(s, T) \tag{3.3.22}$$

is linear and continuous.

(iii) *Mapping (3.3.22) can be lifted to a unique linear and continuous function*

$$H \times L^2(s, T; R^m) \times L^2(s, T; R^n) \to C(s, T; H). \tag{3.3.23}$$

There exists a family of operators (called the evolution operators)

$$\tilde{\Phi}: \{(t, s): s \in [0, \infty), t \geqslant s\} \to L(H)$$

such that $\forall \varphi \in H$ *function* $(t, s) \mapsto \tilde{\Phi}(t, s)\varphi$ *is continuous,*

$$\forall \tau, s, t \quad 0 \leqslant \tau \leqslant s \leqslant t < \infty \quad \tilde{\Phi}(t, \tau) = \tilde{\Phi}(t, s)\tilde{\Phi}(s, \tau) \tag{3.3.24}$$

and the solution of (3.3.20) satisfies

$$\Lambda z(t) = \tilde{\Phi}(t, s)\Lambda\varphi - \int_{s}^{t} \tilde{\Phi}(t, \tau)[\tilde{C}(\tau)u(\tau) - \tilde{f}(\tau)] d\tau. \tag{3.3.25}$$

(iv) *Equation* (3.3.20) *is the state equation corresponding to* (3.3.15) *which means that if the initial conditions are the same and relations* (3.3.19) *hold, then state* $\tilde{x}(t)$ *obtained from the trajectory* (3.3.15) *according to rule* (3.3.16) *is identical with* $z(t)$—*the solution of* (3.3.20).

Let us note that in definition (3.3.21) function ζ takes values in V whereas its derivative in H. To clarify this definition, consider an example. Let $x(t) = 0$ for $t \leq 1$ and $x(t) = t-1$ for $t \geq 1$. We construct $x \in L^2(0, 2; V)$ according to (3.3.16), with V given by (3.3.17), $h = 1$, $n = 1$. It is easy to check that derivative $\dot{x}$ does not exist in $L^2(0, 2; V)$, but can be identified with an element of $L^2(0, 2; H)$. We have $\dot{\tilde{x}} = \tilde{\xi}$ where $\tilde{\xi}$ corresponds through (3.3.16) to the function ξ, $\xi(t) = 0$ for $t < 1$ and $\xi(t) = 1$ for $t \geq 1$.

3.3.3 Transfer functions

We consider again the stationary system (3.2.57):

$$\sum_{i=0}^{N} [A_i \dot{x}(t-h_i) + B_i x(t-h_i) + C_i u(t-h_i)] + \int_{-h}^{0} B(\tau) x(t+\tau) d\tau +$$

$$+ \int_{-h}^{0} C(\tau) u(t+\tau) d\tau = f(t) \quad t \geq 0$$

with a linear output equation

$$y(t) = \sum_{i=0}^{N} [D_i x(t-h_i) + E_i u(t-h_i)] + \int_{-h}^{0} D(\tau) x(t+\tau) d\tau + \int_{-h}^{0} E(\tau) u(t+\tau) d\tau,$$

(3.3.26)

where $y(t) \in R^p$, D and E are bounded and measurable. Assuming zero initial conditions we obtain for the Laplace transforms (see (3.2.62))

$$\hat{x}(s) = \Delta(s)^{-1}[-H(s)\hat{u}(s) + \hat{f}(s)] \quad (3.3.27)$$

and

$$\hat{y}(s) = R(s)\hat{x}(s) + S(s)\hat{u}(s)$$

$$R(s) = \sum_{i=0}^{N} e^{-sh_i} D_i + \int_{-h}^{0} e^{st} D(t) dt \quad (3.3.28)$$

$$S(s) = \sum_{i=0}^{N} e^{-sh_i} E_i + \int_{-h}^{0} e^{st} E(t) dt.$$

3.3 Input, Output, State and Transfer Functions

Substituting (3.3.27) into (3.3.28) we obtain

$$\hat{y}(s) = G_u(s)\hat{u}(s) + G_f(s)\hat{f}(s) \tag{3.3.29}$$

$$G_u(s) = -R(s)\Delta(s)^{-1}H(s) + S(s) \tag{3.3.30}$$

$$G_f(s) = R(s)\Delta(s)^{-1}. \tag{3.3.31}$$

The transfer function (or trasmittance) is a basic tool used for the description of the linear stationary system (3.2.57), (3.3.26). It allows us to express in a multiplicative form the dependence of the Laplace transform of the output, $\hat{y}$ on the Laplace transform of some input component $\hat{w}$. More precisely let the initial conditions be zero; then the output depends linearly on the vector $[w_1, \ldots, w_k]$ of all input functions:

$$\hat{y}(s) = \sum_{i=1}^{k} G_i(s)\hat{w}_i(s). \tag{3.3.32}$$

The matrix-valued function of a complex variable G_i is called the transfer function between the i-th input component and the output (or w_i-transfer function). Comparing this with (3.3.29) we see that G_u is the control transfer function and, if f represents disturbances, G_f is the disturbance transfer function.

It is evident from (3.3.30) and (3.3.31) that the only possible poles of G_u and G_f are those of $s \mapsto \Delta(s)^{-1}$, thus both G_u and G_f can have poles only at points which are characteristic roots.

4

Models of Discrete Time-delay Systems

4.1 Discrete-time Equations

In this chapter we consider systems of the form

$$x(t+1) = f(x_t, u_t, t) \qquad t = t_0, t_0+1, t_0+2, \ldots \qquad (4.1.1)$$

where $x(t) \in R^n$ is the dependent variable (instantaneous state), the control u takes values from R^m,

$$\begin{aligned} x_t &= \{x(t-h_i), i = 0, \ldots, N\} \\ u_t &= \{u(t-h_i), i = 0, \ldots, N\} \end{aligned} \qquad (4.1.2)$$

h_i are nonnegative integers,

$$0 = h_0 < h_1 < \ldots < h_N.$$

The initial conditions are

$$\begin{aligned} x(t) &= \varphi(t) \qquad t = t_0 - h_N, t_0 - h_N + 1, \ldots, t_0 \\ u(t) &= v(t) \qquad t = t_0 - h_N, t_0 - h_N + 1, \ldots, t_0 - 1 \end{aligned} \qquad (4.1.3)$$

where φ and v are given functions.

For a given control u, a solution of (4.1.1) with initial conditions (4.1.3) is defined as a sequence $\{x(t), t = t_0 - h_N, t_0 - h_N + 1, \ldots, t_0, t_0 + 1, \ldots\}$ such that $x(t)$ satisfies (4.1.3) for $t \leq t_0$ and (4.1.1) for $t > t_0$. It is evident that we do not need any specific assumptions on f to obtain the existence and uniqueness of the solution. It is also easy to check that the solution depends continuously on initial condition φ if f is continuous in its first argument for any choice of last two arguments. Continuity of the solution with respect to control u requires additionally that f be continuous in its second argument.

4.1 Discrete-time Equations

From the formal standpoint, equation (4.1.1) is a discrete-time difference equation of order h_N+1. By analogy with Section 3.1.1 it may be called a discrete retarded equation. In practice there is no need to introduce discrete equations analogous to neutral ones with continuous time. It is worth noting that the assumption of the set of discrete time moments in the form $\{t_0, t_0+1, \ldots\}$ is not restrictive since for any discrete sequence $\{t_i, i = 0, 1, \ldots\}$ and any function $F\colon \{t_i, i = 0, 1, \ldots\} \to R^k$ we can define $\tilde{F}(t_0+i) = F(t_i)$, $i = 0, 1, \ldots$ An important feature of system (4.1.1) is that by introducing new dependent variables it can be transformed into a first-order difference system or a system without time delays in x. To express this formally, let $\tilde{x}_{k_0}(t) = x(t)$, $k_i = h_N - h_i + 1$, and

$$\begin{aligned}
\tilde{x}_1(t+1) &= \tilde{x}_2(t) \\
\tilde{x}_2(t+1) &= \tilde{x}_3(t) \\
&\cdots\cdots\cdots\cdots \\
\tilde{x}_{k_0}(t+1) &= f\bigl(\tilde{x}_{k_0}(t), \tilde{x}_{k_1}(t), \ldots, \tilde{x}_{k_N}(t), u(t), \ldots, u(t-h_N), t\bigr).
\end{aligned} \qquad (4.1.4)$$

Notice that in continuous-time systems this possibility of removing delays does not exist.

Although, as may be deduced, time delays in discrete-time systems do not create a qualitatively new problem, there are several reasons for our interest in them. Let us recall that when seeking a numerical solution to any control problem involving FDE, sooner or later we must pass to a finite-dimensional model which in most cases is formulated in terms of difference equations. Thus, equation (4.1.1) is regarded as an approximation of a FDE with continuous time. In applications there is even a strong tendency to formulate control problems by means of discrete equations from the very beginning. Apart from the reason we have already mentioned, this is supported by two facts. First, in many processes the values of the process variables are important only at certain discrete moments of time (computers, relay and digital devices, finite automata)—a similar situation arises in continuous-time processes controlled by computers. Second, the discrete-time models appeal to the control engineer because of their simplicity; this makes him hope that mathematical problems will be avoided. This usually proves to be a vain hope since new and complicated questions of approximation must be investigated. However, it cannot be denied that difference schemes used in control theory are hardly specific for particular types of systems described by ordinary, partial or functional differential equations, therefore discrete equations offer a unified approach to most control problems.

4.2 Linear Discrete Models

Linear discrete-time models have the form

$$x(t+1) = \sum_{i=0}^{N} [A_i(t)x(t-h_i) + B_i(t)u(t-h_i)] + f(t)$$

$$t = t_0, t_0+1, t_0+2, \ldots \quad (4.2.1)$$

where $A_i(t)$ and $B_i(t)$ are $(n \times n)$- and $(n \times m)$-matrices, respectively. Initial conditions and other assumptions are as for (4.1.1).

Denote by $x(\cdot, \varphi, \nu, u, f)$ the solution of (4.2.1) corresponding to initial conditions (4.1.3), control $\{u(t), t \geq t_0\}$ and free term f. The following variation-of-constants formula holds for system (4.2.1):

$$x(t_0+j+1, \varphi, \nu, u, f) = \sum_{k=0}^{h_N} \Phi_1(j,k)\varphi(t_0-k) + \sum_{k=1}^{h_N} \Phi_2(j,k)\nu(t_0-k) +$$

$$+ \sum_{k=0}^{j} \Phi_3(j,k)u(t_0+k) + \sum_{k=0}^{j} \Phi(j,k)f(t_0+k) \quad (4.2.2)$$

where Φ is determined by

$$\Phi(j,k) = \sum_{i=0}^{N} A_i(t_0+j)\Phi(j-h_i-1, k) \quad (k \leq j)$$

$$\Phi(j,j) = I \quad \Phi(j,k) = 0 \quad (k > j) \quad (4.2.3)$$

and

$$\Phi_1(j,k) = \sum_{\substack{i \\ k \leq h_i \leq j+k}} \Phi(j, h_i-k)A_i(t_0-k+h_i)$$

$$\Phi_2(j,k) = \sum_{\substack{i \\ k \leq h_i \leq j+k}} \Phi(j, h_i-k)B_i(t_0-k+h_i) \quad (4.2.4)$$

$$\Phi_3(j,k) = \sum_{\substack{i \\ h_i \leq j-k}} \Phi(j, h_i+k)B_i(t_0-k+h_i).$$

Formula (4.2.2) can be verified by direct substitution into the system equation (4.2.1).

The analogy with the continuous-time case which is readily seen in the variation-of-constants formula can be taken much further. The sequel of this section will be devoted to stationary systems

4.2 Linear Discrete Models

$$x(t+1) = \sum_{i=0}^{N} [A_i x(t-h_i) + B_i u(t-h_i)] + f(t) \quad t = t_0, t_0+1, \ldots \quad (4.2.5)$$

A_i and B_i are constant matrices of compatible dimensions. We shall study exponential solutions of the uncontrolled, homogeneous equation

$$x(t+1) = \sum_{i=0}^{N} A_i x(t-h_i) \quad t = t_0, t_0+1, \ldots \quad (4.2.6)$$

with arbitrary initial conditions. Let $\lambda \neq 0$ be a complex number and c a complex nonzero n-vector. Equation (4.2.6) has an exponential solution $\lambda^t c$, $t = t_0 - h_N, t_0 - h_N + 1, \ldots$ if and only if

$$\Delta(\lambda) c = 0 \quad (4.2.7)$$

where

$$\Delta(\lambda) = I \lambda^{h_N+1} - \sum_{i=0}^{N} \lambda^{h_N - h_i} A_i. \quad (4.2.8)$$

A vector $c \neq 0$ satisfying (4.2.7) exists if and only if the characteristic equation holds

$$\det \Delta(\lambda) = 0. \quad (4.2.9)$$

The determinant $\det \Delta(\lambda)$ is a polynomial in λ of degree $n(h_N+1)$; it is called the characteristic polynomial. The roots of (4.2.9) are called the characteristic roots of system (4.2.6) or (4.2.5). Of course all complex roots are pairwise conjugate, since A_i are real. Now we recall a few elementary facts from the theory of linear difference equations. Let λ_i, $i = 1, \ldots, k$, be the sequence of all distinct characteristic roots, each with multiplicity p_i. Then $\sum_{i=1}^{k} p_i = n(h_N+1)$. For each $\lambda_i \neq 0$ there are exactly p_i linearly independent solutions of (4.2.6) of the form $w_{ij}(t) \lambda_i^t$ where w_{ij} is a polynomial of order less than p_i, taking values in the n-dimensional complex space. It is easy to check that if $w_{ij}(t) \lambda_i^t$ is a solution then its conjugate $\bar{w}_{ij}(t) \bar{\lambda}_i^t$ is also a solution, and $\text{Re}[w_{ij}(t) \lambda_i^t]$, $\text{Im}[w_{ij}(t) \lambda_i^t]$ are solutions too. Thus, if all λ_i, $i = 1, \ldots, k$, are different from zero, we can create a set of $n(h_N+1)$ linearly independent real solutions of (4.2.6) with unspecified initial conditions. The solution of (4.2.6) corresponding to any fixed initial function φ can be constructed as a real linear combination of elements of this set. Of course, such a (real) solution can also be obtained as a complex linear combination of the $n(h_N+1)$ linearly independent complex solutions of the form $w_{ij}(t) \lambda_i^t$. Now, if there is a characteristic root equal to zero, we can only say that there is a moment of time T, $T \leq t_0 + n(h_N + 1)$, such that for $t > T$ any solution of (4.2.6) is a linear combination of the solutions $w_{i,j}(t) \lambda_i^t$.

Let us consider a simple example

$$x(t+1) = \begin{bmatrix} 1 & 1 \\ 0 & 0 \end{bmatrix} x(t-1) \quad t = 0, 1, \ldots$$

Here $n = 2$, $h_N = 1$, $A_0 = 0$, $N = 1$,

$$\Delta(\lambda) = \begin{bmatrix} \lambda^2 - 1 & -1 \\ 0 & \lambda^2 \end{bmatrix}$$

The characteristic polynomial $\lambda^2(\lambda^2 - 1)$ has three distinct roots; $\lambda_1 = 0$ with multiplicity 2, $\lambda_2 = 1$ and $\lambda_3 = -1$ both with multiplicity 1. The eigenvectors corresponding to λ_2 and λ_3 are

$$c_2 = \begin{bmatrix} 1 \\ 0 \end{bmatrix} \quad c_3 = \begin{bmatrix} 1 \\ 0 \end{bmatrix}.$$

Therefore for sufficiently large t any solution can be expressed in the form

$$x(t) = \begin{bmatrix} a_1 + a_2(-1)^t \\ 0 \end{bmatrix}$$

where a_1 and a_2 are real constants depending on initial conditions. By simple calculations we check that this formula is valid for all $t > 0$ and

$$a_1 = \tfrac{1}{2}[\varphi_1(0) + \varphi_2(0) + \varphi_1(-1) + \varphi_2(-1)]$$
$$a_2 = \tfrac{1}{2}[\varphi_1(0) + \varphi_2(0) - \varphi_1(-1) - \varphi_2(-1)].$$

An analogy of the Laplace transform in discrete-time systems is the so-called Z-transform. Let F be a function $\{0, 1, 2, \ldots\} \to R^n$. Its Z-transform, denoted by ZF or $\hat{F}$, is a function of complex variable defined by the formula

$$\hat{F}(s) = \sum_{t=0}^{\infty} F(t) s^{-t}. \qquad (4.2.10)$$

We assume that the series in the right-hand side is convergent for $|s| > r$ where r is some positive real. The inverse Z-transform of $\hat{F}$, denoted by $Z^{-1}\hat{F}$, can be found in the following way. Let C be a circle with centre at the origin and radius sufficiently great to contain all poles of $\hat{F}$, and positively oriented. We then have from (4.2.10)

$$\int_C \hat{F}(s) s^{m-1} ds = \sum_{t=0}^{\infty} F(t) \int_C s^{m-t-1} ds$$

4.2 Linear Discrete Models

and

$$(Z^{-1}\hat{F})(m) = F(m) = \frac{1}{2\pi i}\int_C \hat{F}(s)s^{m-1}ds. \qquad (4.2.11)$$

Hence, by virtue of the Cauchy theorem $F(m)$ is equal to the sum of all residues of the function $s \mapsto \hat{F}(s)s^{m-1}$.

Now we shall apply the Z-transform to the stationary system (4.2.5) with $t_0 = 0$ assuming that all the relevant series are convergent in a certain region of complex variables. By $\hat{x}(s)$ we denote the Z-transform of the restriction $x|_{[0,\infty)}$ and by $\hat{u}(s)$ the Z-transform of $u|_{[0,\infty)}$. For each $t = 0, 1, \ldots$ we multiply both sides of (4.2.5) by s^{-t}. Summation with respect to t from zero to infinity yields

$$s\sum_{t=0}^{\infty}s^{-(t+1)}x(t+1) = \sum_{i=0}^{N}A_i s^{-h_i}\sum_{t=0}^{\infty}s^{-(t-h_i)}x(t-h_i) +$$
$$+ \sum_{i=0}^{N}B_i s^{-h_i}\sum_{t=0}^{\infty}s^{-(t-h_i)}u(t-h_i) + \sum_{t=0}^{\infty}s^{-t}f(t),$$

and

$$s\hat{x}(s) - sx(0) = \sum_{i=1}^{N}\sum_{t=-h_i}^{-1}[A_i s^{-(t+h_i)}\varphi(t) + B_i s^{-(t+h_i)}v(t)] +$$
$$+ \sum_{i=0}^{N}A_i s^{-h_i}\hat{x}(s) + \sum_{i=0}^{N}B_i s^{-h_i}\hat{u}(s) + \hat{f}(s).$$

Hence

$$\hat{x}(s) = \Delta_1(s)^{-1}[P(s) + H(s)\hat{u}(s) + \hat{f}(s)] \qquad (4.2.12)$$

where

$$\Delta_1(s) = sI - \sum_{i=0}^{N}A_i s^{-h_i} \qquad (4.2.13)$$

$$P(s) = \sum_{i=1}^{N}\sum_{t=-h_i}^{-1}s^{-(t+h_i)}[A_i\varphi(t) + B_i v(t)] + s\varphi(0)$$
$$H(s) = \sum_{i=0}^{N}B_i s^{-h_i}. \qquad (4.2.14)$$

Comparing (4.2.13) with (4.2.8) we see that, provided zero is not a characteristic root, all poles of $\Delta_1(s)^{-1}$ are equal to the characteristic roots of the system. The original of the solution we obtain from (4.2.11):

$$x(t) = \frac{1}{2\pi i}\int_C \Delta_1(s)^{-1}[P(s) + H(s)\hat{u}(s) + \hat{f}(s)]s^{t-1}ds. \qquad (4.2.15)$$

4.3 State and Output Equations; Transfer Functions

We shall here present a state space approach to discrete-time systems (4.1.1) and (4.2.1). The state space H will be identical with the space of initial conditions

$$H = R^n \times H_1 \times H_2 \tag{4.3.1}$$

where H_1 is the space of all functions $\{-h_N, -h_N+1, \ldots, -1\} \to R^n$ and H_2 is the space of all functions $\{-h_N, -h_N+1, \ldots, -1\} \to R^m$. If there are no delays in control, $H = R^n \times H_1$, and if there are no delays in trajectory, $H = R^n \times H_2$. In the general case the state of system (4.1.1) at moment t is a triplet

$$\tilde{x}(t) = \operatorname{col}\bigl(\tilde{x}(t)^0, \tilde{x}(t)^1, \tilde{x}(t)^2\bigr) \tag{4.3.2}$$

where

$$\begin{aligned}
&\tilde{x}(t)^0 = x(t) \in R^n \\
&\tilde{x}(t)^1 \in H_1 \\
&\tilde{x}(t)^1(\tau) = x(t+\tau) \quad \tau = -h_N, -h_N+1, \ldots, -1 \\
&\tilde{x}(t)^2 \in H_2 \\
&\tilde{x}(t)^2(\tau) = u(t+\tau) \quad \tau = -h_N, -h_N+1, \ldots, -1.
\end{aligned} \tag{4.3.3}$$

We introduce a mapping $\tilde{f} \colon H \times R^m \times \{t_0, t_0+1, \ldots\} \to H$,

$$\begin{aligned}
&\tilde{f} = \operatorname{col}(\tilde{f}^0, \tilde{f}^1, \tilde{f}^2) \\
&\tilde{f}^0 \colon H \times R^m \times \{t_0, t_0+1, \ldots\} \mapsto R^n \\
&\tilde{f}^1 \colon H \times R^m \times \{t_0, t_0+1, \ldots\} \mapsto H_1 \\
&\tilde{f}^2 \colon H \times R^m \times \{t_0, t_0+1, \ldots\} \mapsto H_2 \\
&\tilde{f}^0(v, w, t) = f(v^0, v^1(-h_1), \ldots, v^1(-h_N), w, v^2(-h_1), \ldots, v^2(-h_N), t) \\
&[\tilde{f}^1(v, w, t)](-1) = v^0 \quad [\tilde{f}^1(v, w, t)](\tau) = v^1(\tau+1) \quad \tau \neq -1 \\
&[\tilde{f}^2(v, w, t)](-1) = w \quad [\tilde{f}^2(v, w, t)](\tau) = v^2(\tau+1) \quad \tau \neq -1.
\end{aligned} \tag{4.3.4}$$

The state equation for system (4.1.1) then has the form

$$\tilde{x}(t+1) = \tilde{f}\bigl(\tilde{x}(t), u(t), t\bigr) \quad t = t_0, t_0+1, \ldots \tag{4.3.5}$$

$$\tilde{x}(t_0) = \operatorname{col}\bigl(\varphi(0), \varphi|_{[-h_N, -1]}, \nu\bigr). \tag{4.3.6}$$

It is worthwhile noting that the transformation of the system equation (4.1.1) to its state form is at least in principle equivalent to the procedure described in Section 4.1 which allows us to replace the system equation by a set of first-order difference equations.

4.3 State and Output Equations; Transfer Functions

To complete the description of a dynamical system we have to add an output equation to the state equation. The output equation has the form
$$y(t) = g(\tilde{x}(t), u(t), t) \tag{4.3.7}$$
where the output value $y(t)$ is an element of R^p.

Let us note that the system description (4.3.5), (4.3.7) is local in the sense of Section 3.3.1, and all the requirements of the state space theory are fulfilled.

Now we shall apply this approach to the linear system (4.2.1). We define mappings $\tilde{A}(t)\colon H \to H$, $\tilde{B}(t)\colon R^m \to H$, and $\tilde{f}$,
$$\tilde{A}(t)v = \mathrm{col}\Big(A_0(t)v^0 + \sum_{i=1}^{N}[A_i(t)v^1(-h_i) + B_i(t)v^2(-h_i)], w^1, w^2\Big) \tag{4.3.8}$$
where
$$w^1 \in H_1, w^1(-1) = v^0 \quad w^1(t) = v^1(t+1) \quad t = -h_N, \ldots, -2$$
$$w^2 \in H_2, w^2(-1) = 0 \quad w^2(t) = v^2(t+1) \quad t = -h_N, \ldots, -2$$
$$\tilde{B}(t)w = \mathrm{col}(B_0(t)w, 0, z) \quad z(-1) = w \quad z(t) = 0 \quad t \ne -1$$
$$\tilde{f}(t) = \mathrm{col}(f(t), 0, 0). \tag{4.3.9}$$

The state equation equivalent to (4.2.1) is then
$$\tilde{x}(t+1) = \tilde{A}(t)\tilde{x}(t) + \tilde{B}(t)u(t) + \tilde{f}(t) \quad t = t_0, t_0+1, \ldots \tag{4.3.10}$$
with initial condition
$$\tilde{x}(t_0) = \mathrm{col}(\varphi(0), \varphi|_{[-h_N, -1]}, \nu). \tag{4.3.11}$$

A variation-of-constants formula analogous to that in the continuous-time case (3.3.25) is valid for the solution of the state equation (4.3.9):
$$\tilde{x}(t) = \tilde{\Phi}(t, t_0)\tilde{x}(t_0) + \sum_{k=t_0}^{t-1} \tilde{\Phi}(t, k)[\tilde{B}(k)u(k) + \tilde{f}(k)]. \tag{4.3.12}$$

The fundamental solution $\tilde{\Phi}(t, t_0)\colon H \to H$ is determined by equation
$$\tilde{\Phi}(t+1, t_0) = \tilde{A}(t)\tilde{\Phi}(t, t_0) \quad t = t_0, t_0+1, \ldots \tag{4.3.13}$$
$$\tilde{\Phi}(t_0, t_0) = I.$$

The output equation for the linear discrete time system (4.2.1) is
$$y(t) = \sum_{i=0}^{N}[C_i(t)x(t-h_i) + D_i(t)u(t-h_i)]. \tag{4.3.14}$$

The reader should verify that by an appropriate definition of $\tilde{C}(t)\colon H \to R^p$, $\tilde{D}(t)\colon R^m \to R^p$, equation (4.3.13) can be transformed into
$$y(t) = \tilde{C}(t)\tilde{x}(t) + \tilde{D}(t)u(t). \tag{4.3.15}$$

Using the Z-transform, the analogy with the continuous-time case can be extended by introducing transfer functions. Consider the stationary system

$$x(t+1) = \sum_{i=0}^{N} [A_i x(t-h_i) + B_i u(t-h_i)] + f(t) \quad t = 0, 1, \ldots \quad (4.3.16)$$

$$y(t) = \sum_{i=0}^{N} [C_i x(t-h_i) + D_i u(t-h_i)]. \quad (4.3.17)$$

Assuming zero initial conditions, we obtain from (4.2.12)

$$\hat{x}(s) = \Delta_1(s)^{-1} [H(s)\hat{u}(s) + \hat{f}(s)]. \quad (4.3.18)$$

Application of the Z-transform to (4.3.17) yields

$$\hat{y}(s) = \sum_{i=0}^{N} s^{-h_i} [C_i \hat{x}(s) + D_i \hat{u}(s)] \quad (4.3.19)$$

and by virtue of (4.3.18)

$$\hat{y}(s) = G_u(s)\hat{u}(s) + G_f(s)\hat{f}(s) \quad (4.3.20)$$

$$G_u(s) = \sum_{i=0}^{N} s^{-h_i} C_i \Delta_1(s)^{-1} H(s) + \sum_{i=0}^{N} s^{-h_i} D_i \quad (4.3.21)$$

$$G_f(s) = \sum_{i=0}^{N} s^{-h_i} C_i \Delta_1(s)^{-1}. \quad (4.3.22)$$

4.4 Approximation of Continuous-time Systems by Discrete-time Systems

4.4.1 Linear systems

Broadly speaking, most methods of difference or discrete approximation used for ordinary differential systems can also be applied to functional-differential systems. The main difference lies in the fact that smoothness of the right-hand side of the FDE does not ensure smoothness of its solutions. This may be a real inconvenience if the points at which the derivatives of the solution are discontinuous cannot be determined *a priori*. This is why the difference approximation methods of higher order need special care and why, on the other hand, first-order methods are more popular.

We shall begin with a short presentation of Delfour's first-order approximation scheme for the linear retarded system (Delfour, 1977a):

$$\dot{x}(t) + \sum_{i=0}^{N} B_i(t) x(t-h_i) + \int_{-h}^{0} B(t,\tau) x(t+\tau) d\tau + C(t) u(t) = f(t) \quad t \in [0, T] \quad (4.4.1)$$

4.4 Approximation of Continuous-time Systems

with assumptions as in (3.2.5). The initial condition is

$$x(t) = \varphi^1(t) \quad t \in [-h, 0)$$
$$x(0) = \varphi^0. \quad (4.4.2)$$

A characteristic feature of this approach is the averaging on discretization steps.

We assume that there are integers $M, L, L_0, \ldots, L_N$ and a discretization step $\delta > 0$ such that $T = M\delta$, $h = L\delta$, $h_i = L_i\delta$, $i = 0, \ldots, N$. If it is impossible to find such numbers, a variable discretization step may be used and everything remains true after obvious changes. By $\{i, j; k\}$ we denote the space of all functions $\{i, \ldots, j-1, j\} \to R^k$. The state space H is approximated by a finite-dimensional space $H^\delta = \{-L, 0; n\}$ endowed with inner product

$$(\xi|\zeta)_{H^\delta} = \sum_{i=-L}^{0} (\xi(i)|\zeta(i)). \quad (4.4.3)$$

We define functions $r_H: H \to H^\delta$

$$\varphi = (\varphi^0, \varphi^1) \mapsto r_H(\varphi) = \bar{\varphi} \quad (4.4.4)$$

where

$$\bar{\varphi}(i) = \int_{i\delta}^{(i+1)\delta} \varphi^1(t)\,dt \quad i = -L, \ldots, -1, \quad \bar{\varphi}(0) = \varphi^0 \quad (4.4.5)$$

and

$$q_H: H^\delta \to H$$

$$q_H(y) = \left(y(0), \sum_{i=-L}^{-1} y(i)\chi_i\right). \quad (4.4.6)$$

χ_i denotes the characteristic function of the interval $[i\delta, (i+1)\delta)$. Let us note that

$$\|q_H(r_H(\varphi))\|^2 = \|\varphi^0\|^2 + \delta \sum_{i=-L}^{-1} \left[\frac{1}{\delta} \int_{i\delta}^{(i+1)\delta} \varphi^1(t)\,dt\right]^2 \leq \|\varphi\|_H^2. \quad (4.4.7)$$

We proceed to the approximation of FDE (4.4.1). We associate functions $\bar{B}_i: \{0, \ldots, M-1\} \to L(R^n)$ with the matrix functions B_i,

$$\bar{B}_i(j) = \frac{1}{\delta} \int_{j\delta}^{(j+1)\delta} B_i(t)\,dt \quad j = 0, \ldots, M-1 \quad i = 0, \ldots, N. \quad (4.4.8)$$

With B we associate $\bar{B}$: $\{0, \ldots, M-1\} \times \{-L, \ldots, 0\} \to L(R^n)$,

$$\bar{B}(j, 0) = \delta^{-2} \int_{j\delta}^{(j+1)\delta} \int_{j\delta - t}^{0} B(t, \tau) d\tau dt \quad j = 0, \ldots, M-1$$

$$\bar{B}(j, -L) = \delta^{-2} \int_{j\delta}^{(j+1)\delta} \int_{-h}^{(j+1)\delta - t - h} B(t, \tau) d\tau dt$$

$$\bar{B}(j, i) = \delta^{-2} \int_{j\delta}^{(j+1)\delta} \int_{(j+i)\delta - t}^{(j+i+1)\delta - t} B(t, \tau) d\tau dt \quad i = -L+1, \ldots, -1 \quad (4.4.9)$$

The space $L^2(0, T; R^n)$ is approximated by $\{0, M-1; n\}$. We define r_f: $L^2(0, T; R^n) \to \{0, M-1; n\}$, $r_f(f) = \bar{f}$ where

$$\bar{f}(i) = \frac{1}{\delta} \int_{i\delta}^{(i+1)\delta} f(t) dt \quad i = 0, \ldots, M-1. \quad (4.4.10)$$

The finite-dimensional approximation of the control space $L^2(0, T; R^m)$ is $\{0, M-1; m\}$. We introduce the maps r_u: $L^2(0, T; R^m) \to \{0, M-1; m\}$, $r_u(u) = \bar{u}$, where

$$\bar{u}(i) = \frac{1}{\delta} \int_{i\delta}^{(i+1)\delta} u(t) dt \quad i = 0, \ldots, M-1, \quad (4.4.11)$$

and q_u: $\{0, M-1; m\} \to L^2(0, T; R^m)$

$$q_u(v) = \sum_{i=0}^{M-1} v(i) \chi_i. \quad (4.4.12)$$

Let us note that

$$\|q_u(r_u(u))\|_{L^2} \leq \|u\|_{L^2}. \quad (4.4.13)$$

Finally, we associate with C a function $\bar{C}$: $\{0, \ldots, M-1\} \to L(R^m, R^n)$,

$$\bar{C}(i) = \frac{1}{\delta} \int_{i\delta}^{(i+1)\delta} C(t) dt \quad i = 0, \ldots, M-1. \quad (4.4.14)$$

Thus, the continuous-time problem (4.4.1) is approximated by the discrete problem

$$z(i+1) - z(i) + \delta \left[\sum_{j=0}^{N} \bar{B}_j(i) z(i - L_j) + \delta \sum_{j=-L}^{0} \bar{B}(i, j) z(i+j) + \right.$$
$$\left. + \bar{C}(i) \bar{u}(i) - \bar{f}(i) \right] = 0 \quad i = 0, \ldots, M-1 \quad (4.4.15)$$
$$z(i) = \bar{\varphi}(i) \quad i = -L, \ldots, 0 \quad (4.4.16)$$

where $\bar{\varphi}$ is a given element of H^δ.

4.4 Approximation of Continuous-time Systems

Theorem 4.4.1 (Delfour 1977a) (i) (Stability) *Denote by z the solution of* (4.4.15) *corresponding to the initial condition* $\bar{\varphi} = r_H(\varphi)$, *control* $\bar{u} = r_u(u)$, $\bar{f} = r_f(f)$ *for some* $\varphi \in H$, $u \in L^2(0, T; R^m)$, $f \in L^2(0, T; R^n)$. *We define functions*

$$z^\delta(t) = \sum_{i=0}^{M-1} z(i)\chi_i(t) \quad t \in [0, T) \quad z^\delta(T) = z(M) \quad (4.4.17)$$

$$Dz^\delta(t) = \sum_{i=0}^{M-1} \frac{z(i+1) - z(i)}{\delta} \chi_i(t) \quad t \in [0, T]. \quad (4.4.18)$$

There is a constant $c > 0$ (independent of φ, u, f, and δ) such that

$$\max\{\|z(i)\|, i = 0, \ldots, M\} + \|z^\delta\|_{L^\infty} + \|Dz^\delta\|_{L^2} \leq c(\|\varphi\| + \|u\| + \|f\|). \quad (4.4.19)$$

(ii) (Convergence) *Fix $\varphi \in H$, $f \in L^2(0, T; R^n)$, $u \in L^2(0, T; R^m)$. As $\delta \to 0$ with $M\delta = T$, $L\delta = h$, $L_i\delta = h_i$, $i = 0, \ldots, N$,*

$$\max\{\|z(i) - x(i\delta)\|, i = 0, \ldots, M\} + \|z^\delta - x\|_{L^2} + \|Dz^\delta - \dot{x}\|_{L^2} \quad (4.4.20)$$

converges to zero, where x is the solution in $W^{1,2}(0, T; R^n)$ of (4.4.1).

Theorem 4.4.2 (Delfour, 1977a) *Let the notation be as in Theorem 4.4.1. Assume that $B_1, \ldots, B_N$ and C are constant matrices and $B = 0$. Then there exists a constant $c > 0$ (independent of φ, δ, u, and f) such that*

$$\max\{\|z(i) - x(i\delta)\|, i = 0, \ldots, M\} + \|z^\delta - x\|_{L^2} \leq c\delta\|\dot{x}\|_{L^2}. \quad (4.4.21)$$

Using a similar approach, Delfour (1977a) obtains a set of difference equations to approximate the state differential equation. A function of two discrete variables $z: \{0, \ldots, M\} \times \{-L, \ldots, 0\} \to R^n$ satisfies the following finite difference scheme:

$$z(i+1, 0) - z(i, 0) + \delta\left[\sum_{k=0}^{N} \bar{B}_k(i)z(i, -L_k) + \delta\sum_{k=-L}^{0} \bar{B}(i, k)z(i, k) + \right.$$
$$\left. + \bar{C}(i)\bar{u}(i) - \bar{f}(i)\right] = 0 \quad i = 0, \ldots, M-1 \quad (4.4.22)$$
$$z(0, j) = \bar{\varphi}(j) \quad j = -L, \ldots, 0$$
$$z(i+1, j) = z(i, j+1) \quad i = 0, \ldots, M-1 \quad j = -L, \ldots, -1.$$

To rewrite this set of equations in a more compact form, we introduce $\tilde{B}(i) \in L(H^\delta)$ and $\tilde{C}(i) \in L(R^m, H^\delta)$ where

$$[\tilde{B}(i)v](0) = \sum_{j=0}^{N} \bar{B}_j(i)v(-L_j) + \delta\sum_{j=-L}^{0} \bar{B}(i, j)v(j) \quad (4.4.23)$$

$$[\tilde{B}(i)v](j) = \frac{v(j)-v(j+1)}{\delta} \quad j = -L, \ldots, -1 \quad (4.4.24)$$

$$[\tilde{C}(i)w](0) = \bar{C}(i)w \quad [\tilde{C}(i)w](j) = 0 \quad j = -L, \ldots, -1, \quad (4.4.25)$$

and vector functions $\tilde{z}: \{0, \ldots, M\} \to H^\delta, \tilde{f}: \{0, \ldots, M\} \to H^\delta$

$$\begin{aligned}[\tilde{z}(i)](j) &= z(i,j) \quad j = -L, \ldots, 0 \\ [\tilde{f}(i)](0) &= \bar{f}(i) \quad [\tilde{f}(i)](j) = 0 \quad j = -L, \ldots, -1.\end{aligned} \quad (4.4.26)$$

We then have by virtue of (4.4.22)

$$\begin{aligned}\tilde{z}(i+1) - \tilde{z}(i) + \delta[\tilde{B}(i)\tilde{z}(i) + \tilde{C}(i)\bar{u}(i) - \tilde{f}(i)] &= 0 \quad i = 0, \ldots, M-1 \\ \tilde{z}(0) &= \bar{\varphi}.\end{aligned} \quad (4.4.27)$$

It is easy to see that this scheme has a unique solution.

Theorem 4.4.3 (Delfour, 1977a) Let $\tilde{z}$ be the solution of (4.4.27), with $\bar{\varphi} = r_H(\varphi), \bar{u} = r_u(u), \bar{f} = r_f(f)$ for some $\varphi \in H, u \in L^2(0, T; R^m), f \in L^2(0, T; R^n)$. We introduce the function $\tilde{z}^\delta: [0, T] \to H$

$$[\tilde{z}^\delta(t)]^0 = \sum_{i=0}^{M-1} z(i, 0)\chi_i(t) \quad t \in [0, T) \quad (4.4.28)$$

$$[\tilde{z}^\delta(t)]^1(\tau) = \sum_{i=0}^{M-1} \sum_{j=-L}^{-1} [z(i,j)\chi^1_{ij}(t, \tau) + z(i, j+1)\chi^2_{ij}(t, \tau)]$$

$$(t, \tau) \in [0, T) \times (-h, 0] \quad (4.4.29)$$

$$\tilde{z}^\delta(T) = q_H(\tilde{z}(M)) \quad (4.4.30)$$

where χ^1_{ij} is the characteristic function of

$$\{(t, \tau) \in [i\delta, (i+1)\delta) \times [j\delta, (j+1)\delta): t+\tau < (i+j+1)\delta\}$$

and χ^2_{ij} is the characteristic function of

$$\{(t, \tau) \in [i\delta, (i+1)\delta) \times [j\delta, (j+1)\delta): t+\tau \geq (i+j+1)\delta\}.$$

(i) (Stability) *There is a constant $c > 0$, independent of $\delta, \varphi, u,$ and f, such that for all $\varphi \in H, u \in L^2(0, T; R^m), f \in L^2(0, T; R^n)$*

$$\max\{\|q_H(\tilde{z}(i))\|, i = 0, \ldots, M\} + \|\tilde{z}^\delta\|_{L^\infty} \leq c(\|\varphi\| + \|f\| + \|u\|). \quad (4.4.31)$$

(ii) (Convergence) *Fix φ, f and u. As $\delta \to 0$ with $M\delta = T, L\delta = h, L_i\delta = h_i, i = 0, \ldots, N$,*

$$\max\{\|q_H(\tilde{z}(i)) - \Lambda\tilde{x}(i\delta)\|, i = 0, \ldots, M\} \to 0 \quad (4.4.32)$$

and $\tilde{z}^\delta$ converges to $\Lambda\tilde{x}$ in $L^\infty(0, T; H)$ where $\tilde{x}$ denotes the solution of equation (3.3.20).

4.4 Approximation of Continuous-time Systems

4.4.2 Nonlinear systems

We now consider the system

$$\dot{x}(t) = f(x(t), x(t-h(t)), t) \quad h(t) \geq 0, t \in [a, b] \quad (4.4.33)$$

in which dependence on control is not expressed explicitly. We denote $g = \inf(t-h(t)), t \in [a, b]$, and assume the initial condition

$$x(t) = \varphi(t) \quad t \in [g, a]. \quad (4.4.34)$$

We shall describe a family of discretization methods due to Feldstein and Goodman (1973). Let δ be a (positive) discretization step and $t_i = a + i\delta$, $i = 0, \pm 1, \pm 2, \ldots$

$$q(i) = \left[i - \frac{h(t_i)}{\delta} \right] \quad (4.4.35)$$

where $[\cdot]$ denotes the integer part, and

$$r(i) = i - \frac{h(t_i)}{\delta} - q(i). \quad (4.4.36)$$

Observe that $t_i - h(t_i) \geq t_j$ if and only if $q(i) \geq j$. For some finite nonnegative integers k_1 and k_2 we define

$$\mu_i(t) = \sum_{j=-k_1}^{k_2} a_{ji} t^j \quad (i = 0, 1, 2, \ldots). \quad (4.4.37)$$

It is assumed that each sequence $\{a_{ji}\}$ is uniformly bounded in i and $\mu_i(1) = 1$. Further let $O(k)$ denote a term bounded by a constant times k and E be the shift operator, $Ex(i) = x(i+1)$.

The general algorithm is as follows. Define functions z and y: $\{0, \pm 1, \pm 2, \ldots\} \to R^n$,

$$z(i) = \varphi(t_i) \quad \text{for} \quad g \leq t_i \leq a \quad (4.4.38)$$

$$y(i) = \begin{cases} \varphi(t_i - h(t_i)) + O(\delta) & \text{for} \quad t_i - h(t_i) \leq a \\ \mu_i(E) z(q(i)) + O(\delta) & \text{for} \quad t_i - h(t_i) > a \end{cases} \quad (4.4.39)$$

$$\bar{f}(i) = f(z(i), y(i), t_i) \quad (4.4.40)$$

$$z(i+1) = z(i) + \delta \bar{f}(i) + O(\delta^2). \quad (4.4.41)$$

The restriction is made that $a_{ji} = 0$ when either $q(i) + j > i$ or $a + \delta(q(i) + j) < g$. All algorithms of this family reduce to Euler's method if $h = 0$. Two simple algorithms are obtained if $O(\delta)$ and $O(\delta^2)$ are equal to zero and (1) $\mu_i = 1$ or (2) $\mu_i(t) = r(i)t + 1 - r(i)$. The methods presented here can easily be extended to systems with multiple delays h_j, $j = 1, \ldots$; one has only to introduce a function $y^{(j)}$ for each delay in the obvious way. The functions μ_i are usually chosen so that $r(i) = 0$ implies that $\mu_i = 1$, that is if $t_i - h(t_i)$

$= t_{q(i)}$, then $y(i) = z(q(i)) + O(\delta)$. This means that when $t_i - h(t_i)$ is the mesh point $t_{q(i)}$ then $z(q(i))$ is the approximation to $x(t_i - h(t_i))$.

To formulate a theorem on convergence and error estimate we define function F, $F(t) = f(x(t), x(t-h(t)), t)$, where x is the solution of (4.4.33), (4.4.34).

Theorem 4.4.4 (Feldstein and Goodman, 1973) *Let f satisfy a Lipschitz condition in its last two arguments. Let z be determined by one of the algorithms (4.4.38)–(4.4.41).*

(i) *If F is bounded and Riemann integrable, then*

$$\lim_{\substack{i\to\infty \\ \delta\to 0}} (z(i) - x(t_i)) = 0. \qquad (4.4.42)$$

(ii) *If also F is of bounded variation, then $z(i) - x(t_i) = O(\delta)$. Furthermore $y(i) - x(t_i - h(t_i)) = O(\delta)$.*

(iii) *Let $0 < d \leqslant 1$. If $[a, b]$ may be split into finitely many subintervals on each of which*

$$\|F(t) - F(\bar{t})\| \leqslant c|t - \bar{t}|^d \quad \text{for some } c > 0, \qquad (4.4.43)$$

then

$$z(i) - x(t_i) = O(\delta^d). \qquad (4.4.44)$$

For proof see Feldstein and Goodman (1973).

If the points at which the derivatives of the solution are discontinuous can be determined *a priori*, one can modify a k-th order method and maintain $O(\delta^k)$ convergence, taking account of the jumps. This is the approach of Zverkina (1964) who gives modified Adams formulas. Let us assume for simplicity that the delay h in (4.4.33) is constant and that $h = M\delta$, where M is an integer and δ is the discretization step. Let also function f be smooth enough so that the solution can have discontinuities in its derivatives only at mesh points $t_i = a + i\delta$. The second-order Adams difference scheme then takes the form

$$z(i+1) = z(i) + q(i) + \tfrac{1}{2}\Delta q(i-1) - \tfrac{1}{2}\delta S_i^1 + \tfrac{1}{2}\delta^2 S_i^2 \qquad (4.4.45)$$

where

$$q(i) = \delta \bar{f}(i) \quad \bar{f}(i) = f(z(i), z(i-M), t_i)$$
$$\Delta q(i-1) = q(i) - q(i-1) \quad S_i^k = F^{(k)}(t_i+) - F^{(k)}(t_i-).$$

The third-order scheme is given by the formula

$$z(i+1) = z(i) + q(i) + \tfrac{1}{2}\Delta q(i-1) + \tfrac{5}{12}\Delta^2 q(i-2) - \tfrac{11}{12}\delta S_i^1 + \tfrac{5}{12}\delta S_{i-1}^1 +$$
$$+ \tfrac{1}{2}\delta^2 S_i^2 - \tfrac{5}{12}\delta^2 S_{i-1}^2 + \tfrac{1}{3}\delta^3 S_i^3 + \tfrac{5}{24}\delta^3 S_{i-1}^3 \qquad (4.4.46)$$

where

$$\Delta^2 q(i) = \Delta q(i) - \Delta q(i-1).$$

5

Identification of Linear Systems

5.1 Problem Statement

Generally speaking, an identification problem involves finding unknown parameters of a model on the basis of measurement data. Many different identification problems can be formulated in the following common way. Let the map

$$S: U \times P \to Y \qquad (5.1.1)$$

be given, where U, P, Y are real Hilbert spaces or their subsets. Generally U, P, Y may be infinite-dimensional. Scalar products in these spaces are denoted by $(\cdot|\cdot)_U$, $(\cdot|\cdot)_P$ and $(\cdot|\cdot)_Y$, respectively. The space U is called the control space, Y the output space and P the parameter space. The map S represents a model of the real system under consideration adjusted so as to describe the measurement needs and possibilities.

The identification problem is

given S, U, P, Y,

given $y_0 \in Y$, $u_0 \in U$ (e.g. by measurement)

find $p \in P$ such that

$$y_0 = S(u_0, p). \qquad (5.1.2)$$

In many cases we have many measurements $\{u_i\}_{i=1}^n$, $\{y_i\}_{i=1}^n$. This situation can be reduced to the previous one by redefining S, U, Y. For instance we may put $U_n = U^n$, $Y_n = Y^n$, $S_n: (u_1, u_2, \ldots, u_n, p) \mapsto (S(u_1, p), \ldots, S(u_n, p)) \in Y^n$. Thus the identification problem with many measurements may be stated as

given $u = (u_1, \ldots, u_n) \in U_n$, $y = (y_1, \ldots, y_n) \in Y_n$
find $p \in P$ such that $y = S_n(u, p)$.

Now we give examples illustrating a few of the possible situations. We consider the following FDE system:

$$\dot{x}(t) = \int_{-h}^{0} dA(s)x(t+s) + Bu(t) \quad t \geq 0 \quad x(t) \in R^n \quad u(t) \in R^m$$

$$y(t) = cx(t) \quad y(t) \in R^k \quad (5.1.3)$$

$$x(s) = f(s) \quad \text{for} \quad s \in [-h, 0]$$

with the variation-of-constants formula

$$x(t) = H(t)x(0) + \int_0^h H(t-\tau) \int_{-h}^{-\tau} dA(s)x(s+\tau)d\tau + \int_0^t H(t-\tau)Bu(\tau)d\tau \quad (5.1.4)$$

where $H(t)$, $t \geq -h$ is the fundamental matrix of solutions.

Example 5.1.1 Known $h, h \geq 0$. Measured $u \in U = L^2(0, T; R^m)$, $y \in Y = L^2(0, T; R^k)$. Find $A(s)$, $f(s)$, $s \in [-h, 0]$, c, B. The space of parameters is $P = L^1(-h, 0; R^{n \times n}) \times L^1(-h, 0; R^n) \times R^{kn} \times R^{nm}$.

Example 5.1.2 Known f, B, c, h, $h > 0$. Measured u, y. Find $A(s)$ for $s \in [-h, 0]$.

Example 5.1.3 Measured y, u. Known $f, f = 0$. Find $H(t)$, $t \geq 0$.

Example 5.1.4 Measured y, u. Known c, B. Find f, H.

It appears that the most appropriate form of the map S for identification purposes is given by formula (5.1.4), because all operations on u and x in (5.1.4) are continuous as functions defined on spaces of L^2-type. Moreover it suffices to identify two functions of time $cH(t)$, $cH(t)B$, $t \geq 0$ to be able to forecast the future behaviour of the dynamic system. The examples given above may be complicated by additional information; for instance, $A(s)$ may be of multistep character, or the output may be of the form $y(t) = \int_{-h}^{0} c(s)x(t+s)ds$, and so on.

From the practical point of view there are two different situations. The first is when one can choose the control in an arbitrary way. In this case it is possible to use the most appropriate controls for identification, for instance signals of great variation, sinusoidal signals, unit step functions, some approximations of Dirac distributions, signals which form some basis in $L^2(0, T)$ or $L^1(0, T)$ spaces, or exponential signals. Application of such signals makes the identification simpler. If we use an approximation of $\delta(t)$ (Dirac distribution) as the control u, then the output y in (5.1.3), (5.1.4) immediately gives the function $cH(t)B$, $t \geq 0$ provided $f(s) = 0$ for $s \in]-h, 0]$.

The second situation is when we have no possibility of choice of the control signals because the system is in operation and one cannot change it. In this case the identification must be based on measurements of those controls and outputs which appear in operation time. This makes the identification much more difficult, especially if the system works in a level stabilization regime and u and y are almost constant as functions of time. Then the knowledge of their values contains little information about the dynamic properties of the system. Moreover, the measurements of u and y are strongly randomized. This kind of identification is called on-line identification.

5.2 Mathematical Methods in Identification Problems

As was stated in the previous section an identification problem can be formulated as the problem of solution of the equation

$$y_0 = S(u_0, p) \quad y_0 \in Y \quad u_0 \in U \quad p \in P \quad (5.2.1)$$

with respect to the variable p. We assume that S, u_0 and y_0 are given.

There are some difficulties involved in solving (5.2.1). The first is that the measurements of u_0 and y_0 are randomized, e.g. due to inaccuracy of measurements or the presence of noise. It may happen that there is no solution to (5.2.1) in the set P for given u_0, y_0. The second difficulty appears when there are many solutions to equation (5.2.1). This is possible when the data, measured or given *a priori*, are insufficient for full identification or the proposed model S is not adequate (not correctly built) for the real system under identification. The third difficulty appears when the spaces Y, U, P are infinite-dimensional. In this case appropriate approximations of elements of these spaces and of the map S are required which would enable us to use the methods of solution for equations in finite-dimensional spaces. For this reason it will be helpful to reformulate the problem of identification (5.2.1).

Let $\{\bar{u}_i\}_{i=1}^{\infty}$, $\{\bar{y}_i\}_{i=1}^{\infty}$ and $\{\bar{p}_i\}_{i=1}^{\infty}$ denote bases in the spaces U, Y, P, respectively. Let $\bar{U}_m \subset U$, $\bar{Y}_k \subset Y$ and $\bar{P}_r \subset P$ denote the finite-dimensional subspaces of dimensions m, k and r, spanned by elements $\{\bar{u}_i\}_{i=1}^{m}$, $\{\bar{y}_i\}_{i=1}^{k}$ and $\{\bar{p}_i\}_{i=1}^{r}$, respectively. Thus

$$\bar{U}_m = \left\{u: u = \sum_{i=1}^{m} a_i \bar{u}_i, a_i \in R^1\right\}$$

$$\bar{Y}_k = \left\{y: y = \sum_{i=1}^{k} a_i \bar{y}_i, a_i \in R^1\right\} \quad (5.2.2)$$

$$\bar{P}_r = \left\{p: p = \sum_{i=1}^{r} a_i \bar{p}_i, a_i \in R^1\right\}.$$

For elements $u \in U$ and $y \in Y$ we denote by $\bar{u}$ and $\bar{y}$ their best approximations in subspaces $\bar{U}_m$ and $\bar{Y}_k$, respectively,

$$\|u-\bar{u}\| = \min_{x \in \bar{U}_m} \|u-x\|, \quad \|y-\bar{y}\| = \min_{x \in \bar{Y}_k} \|y-x\|. \quad (5.2.3)$$

Similarly we obtain the approximation $\bar{p}$ for an element $p \in P$. In our case, when U, Y, P are Hilbert spaces one can give explicit expressions for $\bar{u}$, $\bar{y}$ and $\bar{p}$. Introducing the notation

$$G_m = \begin{bmatrix} (\bar{u}_1|\bar{u}_1) & \ldots & (\bar{u}_1|\bar{u}_m) \\ (\bar{u}_m|\bar{u}_1) & \ldots & (\bar{u}_m|\bar{u}_m) \end{bmatrix}_{m \times m}$$

$$G_k = \begin{bmatrix} (\bar{y}_1|\bar{y}_1) & \ldots & (\bar{y}_1|\bar{y}_k) \\ (\bar{y}_k|\bar{y}_1) & \ldots & (\bar{y}_k|\bar{y}_k) \end{bmatrix}_{k \times k} \quad (5.2.4)$$

$$G_r = \begin{bmatrix} (\bar{p}_1|\bar{p}_1) & \ldots & (\bar{p}_1|\bar{p}_r) \\ (\bar{p}_r|\bar{p}_1) & \ldots & (\bar{p}_r|\bar{p}_r) \end{bmatrix}_{r \times r}$$

$$w_m = \begin{bmatrix} \bar{u}_1 \\ \vdots \\ \bar{u}_m \end{bmatrix} \quad w_k = \begin{bmatrix} \bar{y}_1 \\ \vdots \\ \bar{y}_k \end{bmatrix} \quad w_r = \begin{bmatrix} \bar{p}_1 \\ \vdots \\ \bar{p}_r \end{bmatrix} \quad (5.2.5)$$

$$w_m \cdot u = \begin{bmatrix} (\bar{u}_1|u) \\ \vdots \\ (\bar{u}_m|u) \end{bmatrix} \in R^m \quad \text{for any } u \in U$$

$$w_k \cdot y = \begin{bmatrix} (\bar{y}_1|y) \\ \vdots \\ (\bar{y}_k|y) \end{bmatrix} \in R^k \quad \text{for any } y \in Y \quad (5.2.6)$$

$$w_r \cdot p = \begin{bmatrix} (\bar{p}_1|p) \\ \vdots \\ (\bar{p}_r|p) \end{bmatrix} \in R^p \quad \text{for any } p \in P$$

$$w_m^T \cdot a = \sum_{i=1}^{m} a_i \bar{u}_i \in \bar{U}_m \subset U \quad \text{for any } a \in R^m \quad a = \begin{bmatrix} a_1 \\ \vdots \\ a_m \end{bmatrix} \quad (5.2.7)_1$$

5.2 Mathematical Methods in Identification Problems

$$w_k^T \cdot a = \sum_{i=1}^{k} a_i \bar{y}_i \in Y_k \subset Y \quad \text{for any } a \in R^k \quad (5.2.7)_2$$

$$w_r^T \cdot a = \sum_{i=1}^{r} a_i \bar{p}_i \in \bar{P}_r \subset P \quad \text{for any } a \in R^r$$

we have the following formulas for the approximations $\bar{u}, \bar{y}, \bar{p}$ of elements $u \in U$, $y \in Y$, $p \in P$, respectively:

$$\begin{aligned}\bar{u} &= w_m^T[G_m^{-1}(w_m u)] \\ \bar{y} &= w_k^T[G_k^{-1}(w_k y)] \\ \bar{p} &= w_r^T[G_r^{-1}(w_r p)].\end{aligned} \quad (5.2.8)$$

One obtains an extremely simple form of (5.2.8) if the chosen bases $\{\bar{u}_i\}$, $\{\bar{y}_i\}$, $\{\bar{p}_i\}$ are orthonormal. G_m, G_k and G_r are then identity matrices and

$$\begin{aligned}\bar{u} &= \sum_{i=0}^{m} (\bar{u}_i | u) \bar{u}_i = w_m^T(w_m u) \\ \bar{y} &= \sum_{i=1}^{k} (\bar{y}_i | y) \bar{y}_i = w_k^T(w_k y) \\ \bar{p} &= \sum_{i=1}^{r} (\bar{p}_i | p) \bar{p}_i = w_r^T(w_r p).\end{aligned} \quad (5.2.9)$$

Let us denote by $\overline{S}(u_0, p) \in Y_k$ the best approximation of the element $S(u_0, p)$, for any u_0, p. We shall give a list of problems which are more specific, and therefore more applicable in practice, than the general problem (5.2.1).

Problem 5.1 For given u_0, y_0 minimize

$$f(a) = \left\| y_0 - S\left(u_0, \sum_{i=1}^{r} a_i \bar{p}_i\right) \right\|_Y^2 \quad (5.2.10)$$

with respect to $a = [a_1, \ldots a_r]^T \in R^r$.

Problem 5.1A For given u_0, y_0 minimize

$$f(a) = \left\| y_0 - S\left(u_0, \sum_{i=1}^{r} a_i \bar{p}_i\right) \right\|_Y^2 + b\|a\|_{R^r}^2 \quad (b \geq 0) \quad (5.2.11)$$

with respect to $a = [a_1 \ldots a_r]^T \in R^r$. This formulation sometimes enables us to avoid the possible non-uniqueness of solution to Problem 5.1. The coefficient b plays the same role as the penalty function coefficient in optimization techniques.

Problem 5.2 For given u_0, y_0 minimize

$$f(a) = \left\|y_0 - S\left(\bar{u}_0, \sum_{i=1}^{r} a_i \bar{p}_i\right)\right\|_Y^2 \quad (5.2.12)$$

with respect to $a = [a_1 \ldots a_r]^T \in R^r$.

Problem 5.2A For given u_0, y_0 minimize

$$f(a) = \left\|y_0 - S\left(\bar{u}_0, \sum_{i=1}^{r} a_i \bar{p}_i\right)\right\|_Y^2 + b\|a\|_{R^r}^2 \quad b \geq 0 \quad (5.2.13)$$

with respect to $a = [a_1 \ldots a_r]^T \in R^r$.

Problem 5.3 For given u_0, y_0 minimize

$$f(a) = \left\|y_0 - \bar{S}\left(\bar{u}_0, \sum_{i=1}^{r} a_i \bar{p}_i\right)\right\|_Y^2 \quad (5.2.14)$$

with respect to $a = [a_1 \ldots a_r]^T \in R^r$.

Problem 5.3A For given u_0, y_0 minimize

$$f(a) = \left\|\bar{y}_0 - \bar{S}\left(\bar{u}_0, \sum_{i=1}^{r} a_i \bar{p}_i\right)\right\|_Y^2 + b\|a\|_{R^r}^2 \quad b \geq 0 \quad (5.2.15)$$

with respect to $a = [a_1 \ldots a_r]^T \in R^r$.

The most convenient from the practical point of view are Problems 5.3 and 5.3A. They are frequently referred to as the momentum method. Let us observe that the solution of (5.2.1) may be obtained in the following way:

$$y_0 = S(u_0, p) \Leftrightarrow (f_i|y_0) = (f_i|S(u_0, p)) \quad \forall i \geq 1$$

where $\{f_i\}_{i=1}^{\infty}$ is a complete family of functionals on Y, for instance, $\{f_i\}$ is a basis in Y. But this method is the same as in Problems 5.3 and 5.3A with increasing $k = \dim \bar{Y}_k$.

A relatively simple but important problem arises when S is an affine-bilinear map, i.e. $\forall p \in P$, the map $u \mapsto S(u, p)$ is affine and $\forall u \in U$ the map $p \mapsto S(u, p)$ is also affine. This problem is called the linear regression analysis problem. Let H be a continuous linear map

$$H: P \ni p \mapsto H(p) \in B(U, Y) \quad (5.2.16)$$

where $B(U, Y)$ is the linear space of continuous linear mappings from U into Y. Consider the map S,

$$S: U \times P \ni (u, p) \mapsto H(p)u + d \in Y \quad (5.2.17)$$

where d is some element of Y. This map is affine and bilinear.

5.2 Mathematical Methods in Identification Problems

The following problem is a particular case of Problem 5.1: given u_0, y_0, minimize

$$f(a) = \left\| y_0 - S\left(u_0, \sum_{i=1}^{r} a_i \bar{p}_i\right) \right\|_Y^2$$

with respect to $a = [a_1 \ldots a_r]^T \in R^r$. We have

$$f(a) = \left(y_0 - d - \sum_{i=1}^{r} a_i H(\bar{p}_i) u_0 \mid y_0 - d - \sum_{i=1}^{r} a_i H(\bar{p}_i) u_0\right)_Y \quad (5.2.18)$$

and the minimum is obtained at a point $p_0 \in \bar{P}_r \subset P$, $a_0 \in R^r$

$$p_0 = w_r^T a_0 \quad a_0 = (G_r^H)^{-1} v \quad (5.2.19)$$

where w_r is as in (5.2.6); we assume that

$$G_r^H = \begin{bmatrix} (H(\bar{p}_1)u_0 \mid H(\bar{p}_1)u_0) \ldots (H(\bar{p}_1)u_0 \mid H(\bar{p}_r)u_0) \\ \ldots \ldots \ldots \ldots \ldots \ldots \ldots \ldots \\ (H(\bar{p}_r)u_0 \mid H(\bar{p}_1)u_0) \ldots (H(\bar{p}_r)u_0 \mid H(\bar{p}_r)u_0) \end{bmatrix} \quad (5.2.20)$$

has an inverse and

$$v = \begin{bmatrix} (H(\bar{p}_1)u_0 \mid y_0 - d) \\ \vdots \\ (H(\bar{p}_r)u_0 \mid y_0 - d) \end{bmatrix}.$$

The minimal value of f is

$$f(a_0) = (y_0 - d \mid y_0 - d)_Y - v^T (G_r^H)^{-1} v. \quad (5.2.21)$$

The following problem is a particular case of Problem 5.1A: given u_0, y_0, minimize

$$f(a) = \left\| y_0 - \left(H\left(\sum_{i=1}^{r} a_i \bar{p}_i\right) + d\right) \right\|_Y^2 + b \|a\|_{R^r} \quad b \geq 0$$

with respect to $a = [a_1 \ldots a_r]^T \in R^r$.

The solution $p_0 \in \bar{P}_r \subset P$, $a_0 \in R^r$ to this problem is

$$p_0 = w_r^T a_0 \quad a_0 = (G_r^H(b))^{-1} v \quad (5.2.22)$$

where w_r is as in (5.2.6), v is as in (5.2.20),

$$G_r^H(b) = G_r^H + bI,$$

G_r^H is as in (5.2.20), and the minimal value of f is equal to

$$f(a_0) = (y_0 - d \mid y_0 - d) - v^T (G_r^H(b))^{-1} v. \quad (5.2.23)$$

It may easily be observed that for $b > 0$, $G_r^H(b)$ has an inverse because G_r^H is nonnegatively definite (as a Gram matrix) and bI is positively definite. The value of $f(a_0)$ is a measure of accuracy of approximation.

Let us consider Problems 5.3 and 5.3A for the affine-bilinear map S. Instead of the original formulation we give a more convenient one. For given u_0, y_0 and a system $\{\bar{y}_i\}_{i=1}^{k}$ of functionals of the space Y, find $a_0 \in R^r$ (and $p_0 \in \bar{P}_r$) which minimizes

$$f(a) = \sum_{j=1}^{k} \left(\bar{y}_j | y_0 - d - H\left(\sum_{i=1}^{r} a_i \bar{p}_i\right) u_0\right)^2. \quad (5.2.24)$$

The solution to this problem is

$$\begin{aligned} a_0 &= (F^T F)^{-1} F^T h \\ p_0 &= w_r^T a_0 \end{aligned} \quad (5.2.25)$$

where

$$F = \begin{bmatrix} (y_1|H(\bar{p}_1)u_0) & \cdots & (y_1|H(\bar{p}_r)u_0) \\ \cdots & \cdots & \cdots \\ (y_k|H(\bar{p}_1)u_0) & \cdots & (y_k|H(\bar{p}_r)u_0) \end{bmatrix}_{k \times r}$$

$$h = \begin{bmatrix} (y_1|y_0 - d) \\ \vdots \\ (y_k|y_0 - d) \end{bmatrix}. \quad (5.2.26)$$

The minimal value $f(a_0)$ is given by

$$f(a_0) = h^T h - h^T F(F^T F)^{-1} F^T h. \quad (5.2.27)$$

If for this solution we have $f(a_0) = 0$, then $h = Fa_0$.

Similarly, for Problem 5.3A we have the minimal point p_0, a_0 and the minimal value of f

$$\begin{aligned} a_0 &= (F^T F + bI)^{-1} F^T h \\ p_0 &= w_r^T a_0 \\ f(a_0) &= h^T h - h^T (F^T F + bI)^{-1} F^T h. \end{aligned} \quad (5.2.28)$$

Some simplifications are of course possible if the identification is not performed on-line. Then one can plan the control u_0 in such a manner that, for instance, the matrices G_r^H or H take a diagonal or other form convenient for calculations.

In nonlinear identification problems one can offer many numerical methods for numerical minimization or the numerical solving of equations. The most popular among them are the gradient and conjugate gradient methods, Newton's method, Rosenbrock's method and Powell's methods. We refer the reader to the vast literature related to this area.

5.3 Some Particular Identification Problems and Methods

In this section we focus our attention on the problem of autonomous model identification. We will treat this problem as an illustration of the methods presented in the previous section.

Let there be given the dynamic system

$$\dot{x}(t) = \int_{-h}^{0} dA_1(z)x(t+z) + \int_{-h}^{0} dA_2(z)\dot{x}(t+z) +$$
$$+ \int_{-h}^{0} dB_1(z)u(t+z) + \int_{-h}^{0} dB_2(z)\dot{u}(t+z)$$
$$y(t) = \int_{-h}^{0} dC_1(z)x(t+z) + \int_{-h}^{0} dC_2(z)\dot{x}(t+z) +$$
$$+ \int_{-h}^{0} dD_1(z)u(t+z) + \int_{-h}^{0} dD_2(z)\dot{u}(t+z)$$

$$u(t) \in R^m \quad x(t) \in R^n \quad (yt) \in R^k \quad t \geq 0. \quad (5.3.1)$$

Let us assume that $y(t)$ and $u(t)$ are known for $t \in [0, T]$. We assume that we have no other information about the system (5.3.1) except the knowledge of the dimensions k and m of the output and control vectors and of the value of h, $h \geq 0$. Thus, we do not know the dimension of $x(t)$. We only know from the causality principle that a state space exists.

The problem is to obtain a model of the given system which will produce the same output trajectories with respect to the same controls as the original system (5.3.1). There are several ways to do this. Firstly, we may try to obtain some equivalent FDE model with the same instantaneous state trajectories. Secondly, we may try to obtain a model based on convolution, i.e. a transient response model. Thirdly, we may try to obtain the transmittance describing the control–output relation.

Now we present a few results which guarantee the correctness of the above mentioned approaches.

Theorem 5.3.1 *For any system of the form* (5.3.1) *there exists an equivalent system*

$$\sum_{i=0}^{2n} \int_{-h_1}^{0} d\bar{A}_i(z)f^{(i)}(t+z) + \sum_{i=0}^{2n+1} \int_{-h_1}^{0} d\bar{B}_i(z)u^{(i)}(t+z) = 0 \quad t \geq 0$$
$$f(t) = g(t) \quad t \in [-h_1, 0] \quad (5.3.2)$$

where $\bar{A}_i$, $i = 0, \ldots, 2n$ are scalar functions, $\bar{B}_i$, $i = 0, \ldots, 2n+1$ are matrix functions and $h_1 > 0$. The equivalence is understood in the following sense: For every initial condition $x(t)$, $t \in [-h, 0]$ in (5.3.1) there is a function g such that for every control u, (5.3.2) has a unique solution (with respect to f) satisfying

$$y(t) = f(t) \quad t \geq 0. \tag{5.3.3}$$

Moreover, if in (5.3.1) $x(t) = 0$, $t \in [-h, 0]$, then for every control u the solution of (5.3.2) satisfies (5.3.3) with $g = 0$.

Proof Let us take the Laplace transform of (5.3.1) with initial conditions equal to zero:

$$\Delta(s)\hat{x}(s) = B(s)\hat{u}(s) \tag{5.3.4}$$

$$\hat{y}(s) = C(s)\hat{x}(s) + D(s)\hat{u}(s) \tag{5.3.5}$$

where

$$\Delta(s) = sI - \int_{-h}^{0} [dA_1(z) + s\,dA_2(z)]e^{sz}$$

$$B(s) = \int_{-h}^{0} [dB_1(z) + s\,dB_2(z)]e^{sz}$$

$$C(s) = \int_{-h}^{0} [dC_1(z) + s\,dC_2(z)]e^{sz}$$

$$D(s) = \int_{-h}^{0} [dD_1(z) + s\,dD_2(z)]e^{sz}. \tag{5.3.6}$$

From (5.3.5) we have

$$C(s)\hat{x}(s) = \hat{y}(s) - D(s)\hat{u}(s). \tag{5.3.7}$$

From (5.3.4) we have

$$\det\Delta(s)\hat{x}(s) = \operatorname{adj}\Delta(s)B(s)\hat{u}(s)$$
$$\det\Delta(s)C(s)\hat{x}(s) = C(s)\operatorname{adj}\Delta(s)B(s)\hat{u}(s) \tag{5.3.8}$$

where $\operatorname{adj}\Delta(s)$ is the adjoint matrix to $\Delta(s)$. Substitution of (5.3.7) into (5.3.8) gives

$$\det\Delta(s)\hat{y}(s) = [C(s)\operatorname{adj}\Delta(s)B(s) + \det\Delta(s)D(s)]\hat{u}(s).$$

$\det\Delta(s)$ is a function of a complex variable. Its elements are of the form

$$s^i \int_{-h}^{0} da_1(s)e^{sz} \int_{-h}^{0} da_2(s)e^{sz} \ldots \int_{-h}^{0} da_n(s)e^{sz}$$

5.3 Some Particular Identification Problems and Methods 97

where $i \leq 2n$ and the product of integrals contains maximally n terms. Each element of this form represents the i-th derivative of the convolution of n functions with bounded support in the time domain. The support of this convolution will be at most equal to $[-(2n+1)h, 0]$, hence $h_1 \geq (2n+1)h$. Finally, in the time domain we have

$$\sum_{i=0}^{2n} \int_{-h_1}^{0} d\bar{A}_i(z) y^{(i)}(t+z) = \sum_{i=0}^{2n+1} \int_{-h_1}^{0} d\bar{B}_i(z) u^{(i)}(t+z)$$

with $\bar{A}_i, \bar{B}_i$ are scalar functions, which gives the theorem. □

Generally, the set of solutions to (5.3.2) is richer than the set of all possible outputs of system (5.3.1).

Lemma 5.3.1 *Any output of an autonomous linear causal system with the continuous map*

$$L^2(0, T; R^m) \ni u \mapsto y \in L^2(0, T; R^k)$$

can be represented in the form

$$y(t) = g(t) + \int_0^t dH(\tau) u(t-\tau) \qquad (5.3.9)$$

where the $(k \times m)$-matrix H is of bounded variation, and g is some element from $L^2(0, T; R^k)$. Moreover, if the integral term is absolutely continuous in t for every control $u \in L^2(0, T; R^m)$, then relation (5.3.9) can be rewritten in the form

$$y(t) = g(t) + \int_0^t \bar{H}(t-\tau) u(\tau) d\tau \qquad (5.3.10)$$

where the $(k \times m)$-matrix function $\bar{H}$ is from $L^2(0, T; \mathscr{L}(R^m, R^k))$. □

In view of Theorem 5.3.1 the identification of an almost unknown system can be reduced to the identification of a system of type (5.3.2).

Let us consider different approaches to the identification of the system described by model (5.3.10). We confine ourselves only to the case $y(t)$, $u(t) \in R^1$, because in view of the methods presented in Section 5.2 there is no essential difference between the multidimensional and the scalar case. There are two different situations: system (5.3.10) is asymptotically stable and the converse. In the first case one can assume that after a sufficiently long time, say T_1, $T_1 > 0$, $T_1 < T$, the values of $g(t)$ which is the response to the initial conditions are equal to zero, $g(t) = 0$ for $t \geq T_1$. Hence the output y of the system is

$$y(t) = \int_0^t H(t-\tau)u(\tau)\,d\tau \quad \text{for} \quad t \in [T_1, T], \tag{5.3.11}$$

and this information together with the knowledge of u, y in the interval $[0, T]$ is the basis for any identification procedure.

In the case of unstable systems it is always assumed that $g(t) = 0$. If this cannot be done, the following is possible. First we construct a stabilizing feedback, then we identify the closed-loop system. In the sequel we create a model of the open-loop system by means of the identified model of the closed-loop system.

Now we consider the relation (5.3.11). Let us introduce a basis in $P = L^2(0, T; R^1)$. The basis can be chosen from some well-known set of functions, for instance

$$\{\bar{p}_i\}_{i=1}^r = \{e^{i\lambda t}\}_{i=1}^r \quad t \in [0, T]$$

$$\{\bar{p}_i\}_{i=1}^r = \{\sin\omega_i t, \cos\omega_i t\}_{i=1}^r \quad t \in [0, T]$$

$$\{\bar{p}_i\}_{i=1}^r = \{t^i\}_{i=1}^r \quad t \in [0, T]$$

or in the form of Lagrange polynomials

$$\{\bar{p}_i\}_{i=1}^r = \frac{\sqrt{\alpha}}{i!} e^{\alpha t} \frac{d^i}{d(\alpha t)^i}[(\alpha t)^i e^{-\alpha t}]$$

or as the characteristic function basis

$$\{\bar{p}_i\}_{i=1}^r = \{\varkappa_i(t)\}_{i=1}^r \quad \varkappa_i(t) = \begin{cases} 1 & \text{for} \quad t \in \left[\frac{T}{r}(i-1), \frac{T}{r}i\right] \\ 0 & \text{otherwise.} \end{cases}$$

We formulate the following problem: given $u_0(t)$, $t \in [0, T]$, $y_0(t)$, $t \in [T_1, T]$, find $p \in P_r \subset P = L^2(0, T)$ which minimizes

$$f(a) = \left\| y_0 - \int_0^t \sum_{i=1}^r \bar{p}_i(t-\tau)a_i u_0(\tau)\,d\tau \right\|_{L^2(T_1, T_2)}^2$$

with respect to $a = (a_1 \ldots a_r)^T \in R^r$. The solution is given by formulas (5.2.19) with

$$w_r = \begin{bmatrix} \bar{p}_1 \\ \vdots \\ \bar{p}_r \end{bmatrix}$$

5.3 Some Particular Identification Problems and Methods

$$G_r^H = \int_{T_1}^{T} \begin{bmatrix} \Gamma_1(t)^2 & \ldots & \Gamma_1(t)\Gamma_r(t) \\ \cdots & \cdots & \cdots \\ \Gamma_r(t)\Gamma_1(t) & \ldots & \Gamma_r(t)^2 \end{bmatrix} dt$$

$$v = \int_{T_1}^{T} \begin{bmatrix} \Gamma_1(t) \\ \vdots \\ \Gamma_r(t) \end{bmatrix} y_0(t) \, dt$$

where

$$\Gamma_k(t) = \int_0^t \bar{p}_k(t-\tau) u_0(\tau) \, d\tau \quad (k=1,\ldots,r).$$

Following a procedure similar to that in Problem 5.3, we obtain analogous formulas.

PART III
CONTROL PROBLEMS

6

Stability

6.1 Lyapunov's Second Method for FDE in the Space C

Let $h \in (0, +\infty)$, $C = C([-h, 0], R^n)$—a Banach space of continuous functions defined on an interval $[-h, 0]$ taking values in R^n with a norm $\|\phi\| \stackrel{df}{=} \max_{\theta \in [-h, 0]} |\phi(\theta)|$, $|\cdot|$ an arbitrary norm in R^n (e.g. euclidean). The system of equations

$$\dot{x}(t) = f(x_t) \quad t \geq 0 \quad (6.1.1a)$$
$$x_0 = \phi \quad \phi \in C \quad (6.1.1b)$$

where

$$x_t \colon [-h, 0] \ni \theta \mapsto x_t(\theta) \stackrel{df}{=} x(t+\theta) \in R^n \text{ for fixed } t \geq 0, \ x_t \in C, f \colon C \ni \psi \mapsto$$
$$\mapsto f(\psi) \in R^n \quad (6.1.2)$$

is called the initial-value problem for functional-differential equation (FDE) in the space C.

A continuous function $x(\phi)(\cdot) \colon [-h, T) \ni t \mapsto x(\phi)(t) \in R^n$, such that
(i) $\dot{x}(\phi)(t)$ exists for $t \geq 0$ (the right-hand derivative at 0)
(ii) $\begin{cases} \dot{x}(\phi)(t) = f[x_t(\theta)] & \forall t \geq 0 \\ x(\phi)(\theta) = x_0(\theta) = \phi(\theta) & \theta \in [-h, 0] \end{cases}$

where

$$x_t(\phi)(\theta) = x(\phi)(t+\theta) \quad (6.1.3)$$

for $\theta \in [-h, 0]$ and fixed $t \geq 0$ is said to be the solution of problem (6.1.1) on an interval $[-h, T)$, $0 < T \leq +\infty$. We call $x_t(\phi)$ the current state, evolving from the initial state ϕ. The space C is the state space.

6.1 Lyapunov's Second Method for FDE in Space C

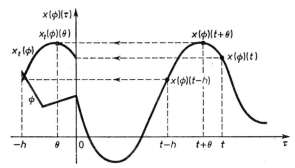

Figure 6.1.1 An explanation of a segmentation principle for the solution $x(\phi)(\cdot)$ according to (6.1.2). The segmentation enables us to pass from solution $x(\phi)(\cdot)$ to a family of continuous functions $\{x_t(\phi)\}_{t \geq 0}$.

In the sequel we make the following general assumptions concerning the regularity of the function f:

f is completely continuous (i.e. f is continuous and maps bounded sets in C into bounded sets in R^n) and is regular such that for any initial state $\phi \in C$ there exists a unique solution $x(\phi)(\cdot)$ of system (6.1.1) with a right-maximal interval of existence $[0, \omega(\phi))$, $\omega(\phi) \in (0, +\infty]$. (6.1.4)

Under these assumptions, $x_t(\phi)$ is continuous with respect to ϕ at fixed t and is continuous with respect to $t \in [0, \omega(\phi))$ at fixed ϕ (Hale, 1977).

A solution $x(\phi)(\cdot)$ of system (6.1.1a) is said to be stable in Lyapunov's sense if

$$\forall \varepsilon > 0 \quad \exists \delta > 0 : \|\phi_1 - \phi\| < \delta \Rightarrow \|x_t(\phi_1) - x_t(\phi)\| < \varepsilon \quad \forall t \geq 0.$$

A constant function $\phi \in C$ is called an equilibrium point, if $x_t(\phi) = \phi$ $\forall t \geq 0$ or, equivalently, $f(\phi) = 0$. An equilibrium point $\phi \in C$ is called stable if the solution $x(\phi)(\cdot)$ is stable. We say that an equilibrium or a solution is unstable if it is not stable. The stability analysis of any equilibrium ϕ may be reduced to the analysis of the zero equilibrium by the substitution $z(t) = x(\phi)(t) - \phi(0)$; z then is a solution of the initial value problem

$$\dot{z}(t) = f(z_t + \phi) \quad z(\theta) = 0 \quad \forall \theta \in [-h, 0]. \qquad (6.1.1a')$$

The set $A(D) = \{\phi \in C : |x_t(\phi), D| \underset{t \to \infty}{\to} 0\}$, where $|\psi, D|$ is the distance of element $\psi \in C$ from the set $D \subset C$, is called the domain of attraction of the set D. We say that an equilibrium point $0 \in C$ is asymptotically stable if 0 is stable and an open ball with centre at 0 can be inscribed in $A(\{0\})$. The point $0 \in C$ is globally asymptotically stable if it is stable and $A(\{0\}) = C$.

A set $M \subset C$ is called invariant if the function $x_t(\cdot)$ maps M onto M for every $t \geqslant 0$, that is, $x_t(M) = M \ \forall t \geqslant 0$.

Every equilibrium point is an invariant set.

The set $\gamma^+(\phi) = \bigcup_{t \geqslant 0} x_t(\phi)$ is called the trajectory of system (6.1.1a) through the point $\phi \in C$.

The set $\Lambda(\phi) = \{\psi \in C : \exists \{t_n\}_{n \in \mathbb{N}}, t_n \underset{n \to \infty}{\to} \infty, x_{t_n}(\phi) \underset{n \to \infty}{\to} \psi\}$ is called the limit set of the trajectory starting from $\phi \in C$.

Lemma 6.1.1 (La Salle–Hale's invariance principle) *Suppose that* $\exists \phi \in C$, $\exists H > 0$: $\|x_t(\phi)\| \leqslant H \ \forall t \geqslant 0$. *Then*

  (i) $\gamma^+(\phi)$ *is a relatively compact (precompact) set in* C,

  (ii) *The limit set* $\Lambda(\phi)$ *is nonempty, compact, connected, invariant, and* $\phi \in A(\Lambda(\phi))$.

We say that $V: C \to R$ is a Lyapunov functional on the set $G \subset C$ if

  (i) V is continuous on $\overline{G}$, the closure of G,

  (ii) $\dot{V}(\phi) \leqslant 0 \ \forall \phi \in G$, where

$$\dot{V}(\phi) \stackrel{\text{df}}{=} \overline{\lim_{h \to 0+}} \frac{1}{h} \{V[x_h(\phi)] - V(\phi)\}. \tag{6.1.5}$$

$\dot{V}(\phi)$ is called the derivative of $V(\phi)$ along solutions of system (6.1.1).

Theorem 6.1.1 *Assumptions*:

  (i) V *is a Lyapunov functional on* G,

  (ii) $x_t(G) \subset G \ \forall t \geqslant 0$ *and* $\forall \phi \in G \ \exists H_1 > 0$: $\|x_t(\phi)\| \leqslant H_1 \ \forall t \geqslant 0$,

  (iii) *Let* M *be the largest invariant set, in the sense of inclusion, included in the set* $E \stackrel{\text{df}}{=} \{\phi \in \overline{G} : \dot{V}(\phi) = 0\}$.

Thesis: $G \subset A(M)$.

Proof Let $\phi \in G$. According to (ii) and Lemma 6.1.1, $\overline{\gamma^+(\phi)}$ is compact. From the continuity of V on $\overline{\gamma^+(\phi)}$ implied by (i), and from Weierstrass' theorem we conclude that V is bounded below on $\overline{\gamma^+(\phi)}$. (i), (ii) imply that the function $R^* \ni t \mapsto V(x_t(\phi)) \in R$ is well-defined, continuous and weakly decreasing. This, together with the boundedness of V from below on $\overline{\gamma^+(\phi)}$, leads to the conclusion that there exists $c \in R$ such that $\lim_{t \to \infty} V[x_t(\phi)] = c$. Since V is continuous on $\overline{G}$ and since from Lemma 6.1.1 we have $|x_t(\phi), \Lambda(\phi)| \underset{t \to \infty}{\to} 0$, it follows that V is constant on $\Lambda(\phi)$. But we know from Lemma 6.1.1 that $\Lambda(\phi)$ is an invariant set, and so $\Lambda(\phi) \subset M \subset E$. Now, as $|x_t(\phi),

6.1 Lyapunov's Second Method for FDE in Space C

$\Lambda(\phi)| \underset{t\to\infty}{\to} 0$ we obtain $|x_t(\phi), M| \underset{t\to\infty}{\to} 0$. Thus we have proved the implication $\phi \in M \Rightarrow |x_t(\phi), M| \underset{t\to\infty}{\to} 0$ from which the inclusion $G \subset A(M)$ follows. □

Theorem 6.1.2 *Let V be a continuous functional on C. Consider a set $\Omega_l^* = \{\phi \in C: V(\phi) \leq l\}$. Let Ω_l be a component of the set Ω_l^* such that V is a Lyapunov functional on Ω_l and there exists a constant $K = K(l) > 0$: $\phi \in \Omega_l \Rightarrow |\phi(0)| < K$. Then $\Omega_l \subset A(M)$, where M is the largest, in the sense of the relation of inclusion of sets, invariant set, included in the set $E = \{\phi \in \Omega_l: V(\phi) = 0\}$.*

Proof $\phi \in \Omega_l \Rightarrow x_t(\phi) \in \Omega_l \; \forall t \geq 0$ which follows from the fact that V is a Lyapunov functional on Ω_l. Moreover, from the assumptions we have $|x_t(\phi)(0)| = |x(\phi)(t)| < K$. Thus the solution $x(\phi)(\cdot)$ is bounded which implies that there exists $H_1 > 0$: $||x_t(\phi)|| \leq H_1$. Now we may put $G = \Omega_l = \overline{\Omega}_l$ in Theorem 6.1.1 which yields the result. □

Theorem 6.1.3 (A generalization of Theorem 5.2.1, Hale (1977)) *Assumptions*: $C \supset \Omega$—*an open set, $0 \in \Omega$. There exist continuous functionals $V: \Omega \to R$ and $u: R^n \to R$ such that*:
(i) $V(\phi) \geq u[\phi(0)] > 0 \quad \forall \phi \in \Omega \setminus \{\psi \in C: \psi(0) = 0\} \quad u(0) = 0$
$V(0) = 0$,
(ii) $\dot{V}(\phi) \leq 0 \quad \forall \phi \in \Omega$.
Thesis: *The equilibrium point 0 is stable.*

Proof Take any $\varepsilon > 0$. The set $\Omega_0 \overset{df}{=} \{s \in R^n: s = \phi(0), \phi \in \Omega\}$ is open, moreover $0 \in \Omega_0$. From the Bhatia–Szegö lemma (Bhatia and Szegö, 1970) it follows that $\exists l_0 > 0: \forall l \in (0, l_0)$ the set $\mathscr{P}_l^* = \{s \in \Omega_0: u(s) \leq l\}$ has a compact component $\mathscr{P}_l$ of $0 \in R^n$ and thus $l = l(\varepsilon) > 0$ may be chosen so that $\mathscr{P}_l \subset K(0, \varepsilon) \subset R^n$, where $K(0, \varepsilon)$ is an open ball with centre at 0 and radius ε. V is continuous on Ω, $V(0) = 0$, then $\forall l > 0 \; \{\phi \in C: V(\phi) < l\}$ is an open set in C containing 0. This implies that there exists $\delta_1 > 0$ such that $K(0, \delta_1) \subset \{\phi \in C: V(\phi) < l\}$. $0 \in \text{Int}\,\mathscr{P}_l$, the interior of $\mathscr{P}_l$, and there exists $\delta_2 = \delta_2(\varepsilon) > 0$ (because $l = l(\varepsilon)$): $K(0, \delta_2) \subset \mathscr{P}_l \subset R^n$. Now take $\delta = \min\{\delta_1, \delta_2\} = \delta(\varepsilon) > 0$, and let $\phi \in K(0, \delta) \subset C$. From (ii) it follows that $V[x_t(\phi)] < l \; \forall t \in [0, \omega(\phi))$ but according to (i) one obtains $u[x(\phi)(t)] < l, \forall t \in [0, \omega(\phi))$. Since $\phi(0) \in \text{Int}\,\mathscr{P}_l$, then $x(\phi)(t) \in \mathscr{P}_l \subset K(0, \varepsilon)$ for all $t \in [0, \omega(\phi))$. But $\mathscr{P}_l$ is bounded, so $x(\phi)(\cdot)$ can be extended to all $t > 0$, thus $x(\phi)(t) \in \mathscr{P}_l \subset K(0, \varepsilon) \; \forall t \geq 0$. It immediately follows that $||x_t(\phi)|| < \varepsilon \; \forall t \geq 0$, and thus we have proved that $\forall \varepsilon > 0 \; \exists \delta = \delta(\varepsilon) > 0: ||\phi|| < \delta \Rightarrow ||x_t(\phi)|| < \varepsilon \; \forall t \geq 0$. Hence 0 is a stable equilibrium point. □

Taking Theorems 6.1.2 and 6.1.3 together, we obtain sufficient conditions for the asymptotic stability (AS) of the equilibrium point 0.

Example 1

Consider the system described by the equations:

$$\dot{x}_1(t) = x_2(t)$$

$$\dot{x}_2(t) = -\frac{a}{h} x_2(t) - \frac{b}{h} \sin x_1(t) + \frac{b}{h} \int_{-h}^{0} x_2(t+\theta) \cos x_1(t+\theta) d\theta$$

$$a, b, h > 0. \tag{6.1.6}$$

For $t \geq h$ the solution of the above system satisfies the differential equation

$$\ddot{x}_1(t) + \frac{a}{h} \dot{x}_1(t) + \frac{b}{h} \sin x_1(t-h) = 0. \tag{6.1.7}$$

Thus the asymptotic behaviour of (6.1.6) and (6.1.7) is identical. The last equation arises in modelling of circumnutations of plants ('sunflower equation') (Israelson and Johnson, 1968, 1969).

Let

$$\phi(\theta) = \begin{bmatrix} \phi_1(\theta) \\ \phi_2(\theta) \end{bmatrix} \in C([-h, 0], R^2).$$

System (6.1.6) is of the form (6.1.1), with

$$f(\phi) = \begin{bmatrix} \phi_2(0) \\ -\frac{a}{h} \phi_2(0) - \frac{b}{h} \sin \phi_1(0) + \frac{b}{h} \int_{-h}^{0} \phi_2(\theta) \cos \phi_1(\theta) d\theta \end{bmatrix}$$

For the function f assumptions (6.1.4) hold. Constant functions of the form $\begin{bmatrix} k\pi \\ 0 \end{bmatrix}$, k an integer, are equilibria. Now we investigate the stability of the point $\begin{bmatrix} 0 \\ 0 \end{bmatrix}$. Consider a functional

$$V(\phi) = \frac{h}{2} \phi_2^2(0) + b[1 - \cos \phi_1(0)] + \frac{a}{2h} \int_{-h}^{0} \int_{\theta}^{0} \phi_2^2(\sigma) d\sigma d\theta. \tag{6.1.8}$$

Making use of (6.1.6) we easily find

$$\dot{V}(\phi) = h\phi_2(0) \left[-\frac{a}{h} \phi_2(0) - \frac{b}{h} \sin \phi_1(0) + \frac{b}{h} \int_{-h}^{0} \phi_2(\theta) \cos \phi_1(\theta) d\theta \right]$$

6.1 Lyapunov's Second Method for FDE in Space C

$$+ b\phi_2(0)\sin\phi_1(0) + \frac{a}{2h}\int_{-h}^{0}[\phi_2^2(0) - \phi_2^2(\theta)]d\theta$$

$$= -\int_{-h}^{0}\left[\frac{a\phi_2^2(0)}{2h} - b\phi_2(0)\phi_2(\theta)\cdot\cos\phi_1(\theta) + \frac{a}{2h}\phi_2^2(\theta)\right]d\theta$$

$$\leq -\int_{-h}^{0}[|\phi_2(0)|,|\phi_2(\theta)|]\begin{bmatrix}\dfrac{a}{2h} & -\dfrac{b}{2} \\ -\dfrac{b}{2} & \dfrac{a}{2h}\end{bmatrix}\begin{bmatrix}|\phi_2(0)| \\ |\phi_2(\theta)|\end{bmatrix}d\theta \stackrel{df}{=} -W(\phi)$$

(6.1.9)

$\forall \phi \in D = \{\psi \in C: \psi' \in C, \psi'(0) = f(\psi)\}.$

Note that
(i) $\overline{D} = C$ (D is dense in C (Webb, 1974)),
(ii) $\phi \in D \Rightarrow x_t(\phi) \in D \quad \forall t \in [0, \omega(\phi))$,
(iii) $W: C \to R$ is a continuous functional;
$h \leq a/b \Rightarrow W(\phi) \geq 0 \ \forall \phi \in C$ and V is a Lyapunov functional on D.

Let $\{\phi_n\}_{n \in N} \subset D$, $\phi_n \underset{n \to \infty}{\to} \phi \in C$. As the function $[0, \omega(\phi_n)) \ni t \mapsto x_t(\phi_n) \in C$ is continuous, V is the Lyapunov functional on D and (ii) holds, we have that $[0, \omega(\phi_n)) \ni t \mapsto V[x_t(\phi_n)]$ is a continuous and non-increasing function. Since $[0, \omega(\phi)) \ni t \mapsto x_t(\phi) \in C$ is continuous, V is continuous functional on C (as a Lyapunov one on D) we obtain the fact that $[0, \omega(\phi)) \ni t \mapsto V[(x_t(\phi)$ is a continuous function. By the continuity of $x_t(\phi)$ with respect to ϕ (on C) and by the continuity of V on C (as V is a Lyapunov functional on D) we have $\lim_{n \to \infty} V[x_t(\phi_n)] = \Delta(x_n(\phi)]$. Thus the function $[0, \omega(\phi)) \ni t \mapsto V[x_t(\phi)]$ is non-increasing. $\dot{V}(\phi)$ then exists everywhere on C, and moreover $\dot{V}(\phi) \leq 0$. We conclude that V is a Lyapunov functional on C. Furthermore

$$-V[x_t(\phi_n)] + V[\phi_n] = \int_0^t (-\dot{V}[x_s(\phi_n)])ds \quad \forall t \in [0, \omega(\phi_n)).$$

Taking the limits $\liminf_{n \to \infty}$ of both sides of the above inequality one obtains by Fatou's lemma (Rudin, 1964)

$$-V[x_t(\phi)] + V(\phi) = \liminf_{n \to \infty}\int_0^t(-\dot{V}(x_s(\phi_n)))ds \geq \int_0^t \liminf_{n \to \infty} W[x_s(\phi_n)]ds$$

$$= \int_0^t W[x_s(\phi)]ds \quad \forall t \in [0, \omega(\phi)).$$

Applying the integral version of the mean-value theorem to the right-hand side we get

$$\dot{V}(\phi) \leq -W(\phi) \leq 0 \quad \forall \phi \in C \qquad (6.1.10)$$

Thus for $h < a/b$

$$E \stackrel{df}{=} \{\phi \in C: \dot{V}(\phi) = 0\} \subset \{\phi \in C: W(\phi) = 0\} = \{\phi \in C: \phi_2 = 0\}.$$

It is easy to see now that by (6.1.6), the largest invariant set contained in E is the set of equilibria of system (6.1.6).

Now it follows from Theorem 6.1.1 that every bounded solution of system (6.1.6) tends to an equilibrium point.

Notice that

$$V(\phi) \geq \frac{h}{2}\phi_2^2(0) + b[1 - \cos\phi_1(0)] = u[\phi(0)] \quad \forall \phi \in C. \qquad (6.1.11)$$

Let $\Omega \stackrel{df}{=} C \setminus \{\phi \in C: \phi_1(0) = 2n\pi$, for all nonzero integers $n\}$. As the set in { } is closed in C, the set Ω is open, and moreover from (6.1.10) and (6.1.11) it follows that the assumptions of Theorem 6.1.3 hold for V. By virtue of this theorem we say that $0 \in C$ is stable. Two problems now arise—the evaluation of the domain of attraction $A(\{0\})$ and the problem of the asymptotic stability of the point $0 \in C$.

In what follows some topological properties of the level-sets of the functional $u: R^2 \to R$ are of great importance. In connection with this the level sets of u are shown in Figure 6.1.2.

Let $\Omega_l^* = \{\phi \in C: V(\phi) \leq l < 2b\}$. Denote by Ω_l the component of Ω_l^* which contains the point 0. Notice that $\Omega_l = Z \stackrel{df}{=} \{\phi \in C: V(\phi) \leq l < 2b,$

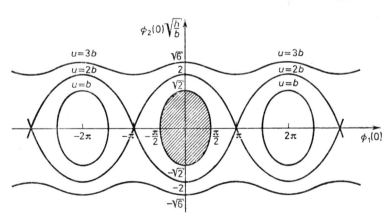

Figure 6.1.2 The level sets of functional u, defined by (6.1.11)

6.1 Lyapunov's Second Method for FDE in Space C

$|\phi_1(0)| \leq \pi\}$ since $\phi \in Z \Rightarrow \lambda\phi \in Z \; \forall \lambda \in [0, 1]$ and $|\phi_1(0)| = \pi \Rightarrow \phi \notin \Omega_l^*$. Further, if $\phi \in \Omega_l$, then

$$\frac{h}{2}\phi_2^2(0) + b(1 - \cos\phi_1(0)) \leq l < 2b \quad \text{and} \quad |\phi_1(0)| < \pi$$

hence $|\phi(0)| < K$ where K is the radius of a ball containing the set depicted in Figure 6.1.2. The largest invariant set contained in the set

$$E = \{\phi \in \Omega_l : \dot{V}(\phi) = 0\} = \{\phi \in \Omega_l : \phi_2 = 0\}$$

is $M = \{0\} \subset C$.

Now, Theorem 6.1.2 implies that for all $l \in (0, 2b)$ the set Ω_l is a domain of attraction of the stable equilibrium 0. By continuity of V, an open ball centred at $0 \in C$ can be inscribed in the set Ω_l, thus 0 is an asymptotically stable equilibrium point.

Finally, we come to an estimation of the domain of attraction in the form

$$\{\phi \in C : V(\phi) < 2b, |\phi_1(0)| \leq \pi\} \quad \text{for} \quad h < \frac{a}{b}.$$

We have thus refined and improved Hale's result (Hale, 1977).

Sinha (1972a, 1972b, 1973) and Gayshun (1972) give a tutorial survey of methods for the construction of Lyapunov functionals for first to fourth-order delay-differential systems.

Example 2 Stability analysis of a class of control systems with delay
Consider the system

$$\dot{x}(t) = Ax(t) + bF[c^T x(t-h)] \quad t \geq 0 \tag{6.1.12}$$

where A is a real $(n \times n)$-matrix, $b, c \in R^n$, $x(t) \in R^n$, $F: R \to R$ is locally Lipschitzian, $F(0) = 0$, $0 < h < \infty$.

In particular, system (6.1.12) describes the dynamics of the control system shown in Figure 6.1.3.

Equation (6.1.12) has the form (6.1.1a) with $f(\phi) = A\phi(0) + bF[c^T\phi(-h)]$. As F is a locally Lipschitzian function, assumptions (6.1.4) hold for f. $0 \in C$

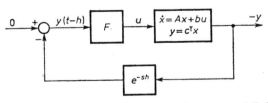

Figure 6.1.3 Block diagram of control system modelled by (6.1.12)

is an equilibrium point, because $F(0) = 0$. We seek a Lyapunov functional for system (6.1.12) in the following form:

$$V(\phi) = \phi^T(0) H \phi(0) + \int_{-h}^{0} \phi^T(\theta) c c^T \phi(\theta) d\theta \qquad (6.1.13)$$

where $H = H^T$ is a real $(n \times n)$-matrix.

It is easy to calculate the derivative $\dot{V}(\phi)$ defined in (6.1.5) for $\phi \in \{\psi \in C: \psi' \in C, \psi'(0) = f(\psi)\}$:

$$\dot{V}(\phi) = [\phi^T(0) \quad F[c^T \phi(-h)]] \begin{bmatrix} A^T H + HA + cc^T & Hb \\ b^T H & 0 \end{bmatrix} \begin{bmatrix} \phi(0) \\ F[c^T \phi(-h)] \end{bmatrix} - $$
$$- \phi^T(-h) cc^T \phi(-h)$$

$$= [\phi^T(0) \quad F[c^T \phi(-h)]] \begin{bmatrix} A^T H + HA + cc^T & Hb \\ b^T H & -\dfrac{1}{k} \end{bmatrix} \begin{bmatrix} \phi(0) \\ F[c^T \phi(-h)] \end{bmatrix} -$$
$$- \dfrac{1}{k} \cdot \phi^T(-h) cc^T \phi(-h) \left[k - \left\{ \dfrac{F[c^T \phi(-h)]}{c^T \phi(-h)} \right\}^2 \right] \qquad (6.1.14)$$

where $k > 0$.

Assume that the system of equations (known as Lur'e's system of resolving equations)

$$A^T H + HA + cc^T = -gg^T \qquad (6.1.15a)$$

$$Hb = -\sqrt{\dfrac{1}{k}}\, g \qquad (6.1.15b)$$

has a solution with respect to (H, g), $H = H^T$, $g \in R^n$, and that the nonlinearity F satisfies the sector condition

$$-\sqrt{k} < \dfrac{F(\sigma)}{\sigma} < \sqrt{k} \quad \forall \sigma \neq 0 \quad F(0) = 0 \qquad (6.1.16)$$

The geometric interpretation of (6.1.16) is given in Figure 6.1.4. By a similar argument as in the previous example and by (6.1.14), we can write the following inequality:

$$\dot{V}(\phi) \leq -\left\{ g^T \phi(0) + \sqrt{\dfrac{1}{k}} F[c^T \phi(-h)] \right\}^2 -$$
$$- \dfrac{1}{k} \phi^T(-h) cc^T \phi(-h) \left[k - \left\{ \dfrac{F[c^T \phi(-h)]}{c^T \Phi(-h)} \right\}^2 \right]$$
$$= -W(\phi) \leq 0 \quad \forall \phi \in C. \qquad (6.1.17)$$

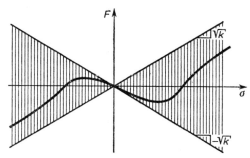

Figure 6.1.4 The geometric interpretation of the sector condition (6.1.16): a nonlinearity F lies in the cone constrained by the lines with slopes $\pm\sqrt{k}$

Thus, under the assumptions made so far V is Lyapunov functional on C. For this functional we have

$$E = \{\phi \in C: \dot{V}(\phi) = 0\} \subset \{\phi \in C: W(\phi) = 0\}$$
$$= \{\phi \in C: c^T\phi(-h) = 0, g^T\phi(0) = 0\}.$$

Any solution x of (6.1.12) such that $W(x_t) = 0 \; \forall t \geq 0$ must be the solution of a linear system without delay $\dot{x} = Ax$, then $c^T e^{At} x(0) = 0, \; \forall t \geq 0$. Assume that the pair (A, c^T) is observable. Then $x(0) = 0$, and the only solution of (6.1.12) which lies in E is the null solution. The largest invariant set M, in the sense of inclusion, contained in E is thus $M = \{0\} \subset C$.

Notice that system (6.1.15) can be reduced to only one matrix Riccati equation with respect to $H = H^T$:

$$kHbb^TH + A^TH + HA + cc^T = 0 \qquad (6.1.15')$$

Assume now that the pair (A, b) is controllable. It is well known (Popov, 1973; Willems, 1971) that a necessary and sufficient condition for (6.1.15) (equivalently (6.1.15')) to have a solution is

$$\frac{1}{k} - |G(i\omega)|^2 \geq 0 \qquad \forall \omega \in R \quad i\omega \notin \lambda(A) \qquad (6.1.18)$$

where $\lambda(A)$ denotes the spectrum of A, and

$$G(s) = c^T(A - sI)^{-1}b \qquad (6.1.19)$$

is the transmittance of the linear part of system (6.1.12). The geometric interpretation of condition (6.1.18) is given in Figure 6.1.5.

If inequality (6.1.18) holds, then the matrix A has no eigenvalues on the imaginary axis. Indeed, let $i\tilde{\omega}$ be such an eigenvalue. Then for $\omega \to \tilde{\omega}$ we have $|G(i\omega)| \to \infty$, contrary to (6.1.18).

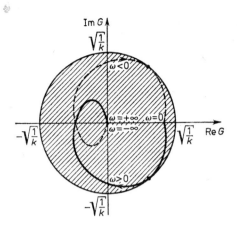

Figure 6.1.5 The geometric interpretation of the frequency-domain inequality (6.1.18): the plot $G(i\omega)$ for $\omega \geq 0$ lies inside the closed circle $K\left(0, -\dfrac{1}{\sqrt{k}}\right)$

We further assume that

$$\operatorname{Re} \lambda(A) < 0 \quad (A \text{ is Hurwitzian}) \qquad (6.1.20)$$

By the observability of the pair (A, c^T) and the well-known properties of Lyapunov matrix equation (Anderson and Vongpanitlerd, 1973) we must have $H = H^T > 0$. This means that $\lambda_{\min}(H), \lambda_{\max}(H) > 0$, where the symbols $\lambda_{\min}(Z), \lambda_{\max}(Z)$ denote the smallest and the greatest eigenvalue of a matrix $Z = Z^T$, respectively. Now we have

$$V(\phi) \geq \phi^T(0) H \phi(0) = u[\phi(0)] > 0 \quad \forall \phi \in C \setminus \{\psi \in C: \psi(0) = 0\}$$
$$V(0) = 0 \quad u(0) = 0. \qquad (6.1.21)$$

By (6.1.17), (6.1.20) and Theorem 6.1.3 the equilibrium point 0 is stable.

As V is a convex functional on C then any level-set $\Omega_l^* = \{\phi \in C: V(\phi) \leq l\}$ has exactly one component, thus $\Omega_l = \Omega_l^*$ in Theorem 6.1.2. On the other hand

$$\phi \in \Omega_l \Rightarrow u[\phi(0)] \leq V(\phi) \leq l \Rightarrow \lambda_{\min}(H) |\phi(0)|^2 \leq l$$
$$\Rightarrow |\phi(0)| \leq \sqrt{\dfrac{1}{\lambda_{\min}(H)}} = K(l)$$

in Theorem 6.1.2. From the preceding results we know also that $M = \{0\}$. By Theorem 6.1.2 we have $\forall l > 0$: $\{\phi \in C: V(\phi) \leq l\} \subset A(\{0\})$, where $A(\{0\})$ denotes the domain of attraction of the equilibrium 0. Then $A(\{0\}) = C$

6.1 Lyapunov's Second Method for FDE in Space C

The results obtained so far may now be summarized in the form of the following lemma.

Lemma 6.1.2 *Assumptions*:
 (i) *the pair (A, b) is controllable, (A, c^T) is observable, and (6.1.20) holds;*
 (ii) *there exists $k > 0$ such that conditions (6.1.18) and (6.1.16) hold.*
Thesis: *For every $h \in (0, +\infty)$ the equilibrium point 0 is globally asymptotically stable, i.e. it is stable, and its domain of attraction covers the whole space C.*

System (6.1.15) can be solved using the method given in Walker and McClamroch (1967), Grabowski (1980).

1. We rewrite the left-hand side of inequality (6.1.18) in a factorized form

$$\pi(\omega) = \frac{1}{k} - |G(i\omega)|^2 = \frac{\psi(i\omega)}{\det(A - i\omega I)} \frac{\psi(-i\omega)}{\det(A + i\omega I)} \qquad (6.1.21a)$$

where $\psi(s)$ is a polynomial of degree n with leading coefficient $(-1)^n \sqrt{1/k}$, i.e.

$$\psi(s) = (-1)^n \sqrt{\frac{1}{k}} s^n + \ldots \qquad (6.1.21b)$$

2. We determine the vector g from the identity

$$\psi(s) \equiv \sqrt{\frac{1}{k}} \det(A - sI) - g^T \operatorname{adj}(A - sI) b \qquad (6.1.22)$$

3. We put the vector g into (6.1.15), calculated in 2, and hence determine the matrix $H = H^T$.

Numerical examples

(a) $n = 1$, $c = 1$, $A < 0$

By (6.1.19) we have $G(s) = \dfrac{b}{A-s}$, $\pi(\omega) = \dfrac{1}{k} - |G(i\omega)|^2 = \dfrac{\omega^2 + A^2 - b^2 k}{k(A^2 + \omega^2)}$.

The maximal number k such that (6.1.18) holds is $k = A^2/b^2 \left(\sqrt{k} = -A/|b|\right)$.

The equilibrium point 0 is globally asymptotically stable independently of the delay $h \in (0, +\infty)$ if the nonlinearity F satisfies condition (6.1.16), which now takes the form

$$\frac{A}{|b|} < \frac{F(\sigma)}{\sigma} < -\frac{A}{|b|} \quad \forall \sigma \neq 0 \quad F(0) = 0 \qquad (6.1.23)$$

We easily find (directly from (6.1.15) or through factorization) that

$$H = \frac{1}{-A} > 0 \quad g = -\operatorname{sign} b.$$

(b) $n = 2$, $A = \begin{bmatrix} -\alpha & 1 \\ -1 & 0 \end{bmatrix}$, $\alpha > 0$, $b = \begin{bmatrix} -1 \\ 0 \end{bmatrix}$, $c = \begin{bmatrix} 1 \\ 0 \end{bmatrix}$.

These data relate, in particular, to a model of an electronic oscillator (see Rubanik, 1969). (A, b) is controllable, (A, c^T) is observable, and $\operatorname{Re} \lambda(A) < 0$. From (6.1.19)

$$G(s) = \frac{s}{s^2 + \alpha s + 1}.$$

The maximal number k such that (6.1.18) holds is $k = \alpha^2 (\sqrt{k} = \alpha)$. The equilibrium point 0 is globally asymptotically stable independently of the delay $h \in (0, +\infty)$ if the nonlinearity F satisfies the following sector condition:

$$-\alpha < \frac{F(\sigma)}{\sigma} < \alpha \quad \forall \sigma \neq 0 \quad F(0) = 0 \quad (6.1.24)$$

For $k = \alpha^2$ we have $\pi(\omega) = \dfrac{(\omega^2 - 1)^2}{\alpha^2 [(1 - \omega^2)^2 + \alpha^2 \omega^2)]}$ and by (6.1.21) one obtains

$$\psi(s) = \frac{1}{\alpha} s^2 + \frac{1}{\alpha}.$$

From (6.1.22) and (6.1.15) we obtain $g^T = [1, 0]$, $H = \dfrac{1}{\alpha} I$.

(c) $n = 2$, $A = \begin{bmatrix} 0 & 1 \\ -ab & -a-b \end{bmatrix}$, $a, b > 0$, $b = \begin{bmatrix} 0 \\ -1 \end{bmatrix}$, $c = \begin{bmatrix} 1 \\ 0 \end{bmatrix}$.

This case was considered by Heiden (1979). The pair (A, b) is controllable and the pair (A, c^T) observable. A is a Hurwitzian matrix. As

$$G(s) = \frac{1}{s^2 + (a+b)s + ab}$$

the maximal number k for which (6.1.18) holds is $k = a^2 b^2$ ($\sqrt{k} = ab$). The equilibrium point 0 is globally asymptotically stable independently of the delay $h \in (0, +\infty)$ if the nonlinearity F satisfies (6.1.16) which now takes the form

$$-ab < \frac{F(\sigma)}{\sigma} < ab \quad \forall \sigma \neq 0 \quad F(0) = 0. \quad (6.1.25)$$

For $k = a^2 b^2$ we have

$$\pi(\omega) = \frac{\omega^2 (\omega^2 + a^2 + b^2)}{a^2 b^2 [(ab - \omega^2)^2 + (a+b)^2 \omega^2]}$$

and our procedure yields

$$\psi(s) = \frac{s^2}{ab} + \frac{s \sqrt{a^2 + b^2}}{ab}.$$

6.1 Lyapunov's Second Method for FDE in Space C

$$g^T = \begin{bmatrix} 1 & \dfrac{a+b-\sqrt{a^2+b^2}}{ab} \end{bmatrix}$$

$$H = \begin{bmatrix} \dfrac{a+b}{ab} & \dfrac{1}{ab} \\ \dfrac{1}{ab} & \dfrac{a+b-\sqrt{a^2+b^2}}{a^2b^2} \end{bmatrix} = H^T > 0.$$

Discussion of the results

1. The situation frequently arises where the nonlinearity F appearing in (6.1.12) does not satisfy (6.1.16) but only the weaker condition

$$\exists \sigma_1 < 0 \quad \exists \sigma_2 > 0: \quad -\sqrt{k} < \dfrac{F(\sigma)}{\sigma} < \sqrt{k} \quad \forall \sigma \in (\sigma_1, \sigma_2) \setminus \{0\} \quad (6.1.26)$$

$$F(0) = 0$$

The geometric interpretation of the last condition is given in Figure 6.1.6.

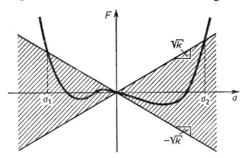

Figure 6.1.6 The geometric interpretation of the related sector condition (6.1.26)

The Lasota–Ważewska model of red blood cell production is a practical example of this situation. This model has the form (Chow, 1974)

$$\dot{x}(t) = -ax(t) - ag[1 - e^{-x(t-h)}] \quad a > 0 \quad (6.1.27)$$

where g is a real solution of the equation $e^{-g} = ag$. Note that $g \in (0, 1)$ for $a > 1/e$.

Equation (6.1.27) has the form discussed in Example 1, with $A = -a$, $b = -ag$, $F(\sigma) = 1 - e^{-\sigma}$, $c = 1$. The nonlinearity does not satisfy condition (6.1.23) for any $a > 0$. However, if $a > 1/e$, then condition (6.1.26) is fulfilled with $\sigma_1 < 0$ such that $\sigma_1 = g(1 - e^{-\sigma_1})$, and $\sigma_2 = +\infty$. For any $a \in (0, 1/e]$ even (6.1.26) does not hold.

By methods given later (Theorem 6.3.1) it may be proved that the origin is locally asymptotically stable, independently of delay h, for every $a \geq 1/e$. If $a < 1/e$, the origin is unstable for sufficiently large h.

In Chow (1974), Kaplan and Yorke (1977) it has been proved that (6.1.27) has nontrivial periodic solutions for small $a > 0$. This means that condition (6.1.23) is so close to a necessary one that a small perturbation may lead to the loss of global asymptotic stability.

For systems without delay, condition (6.1.26) is the starting point for an evaluation of the domain of attraction of an equilibrium that is not globally asymptotically stable (Walker and McClamroch, 1967; Grabowski, 1980). If we try to apply this approach to system (6.1.12) we obtain results of little value. This is because in the family of level sets $\{\phi \in C: V(\phi) \leqslant l\}_{l \geqslant 0}$ of the functional V, only the level set corresponding to $l = 0$ is contained in the set $\{\phi \in C: \sigma_1 \leqslant c^T\phi(-h) \leqslant \sigma_2\}$ on which V is now a Lyapunov functional. Therefore Theorem 6.1.2 yields only

$$\{\phi \in C: V(\phi) = 0\} = \{\phi \in C: \phi(0) = 0, c^T\phi(\theta) = 0$$
$$\forall \theta \in [-h, 0]\} \subset A(\{0\}).$$

However every initial condition taken from the above set gives a null solution.

2. If we set $F(\sigma) = \mu\sigma$ then we obtain a linear system with delay,

$$\dot{x}(t) = Ax + \mu bc^T x(t-h). \tag{6.1.28}$$

To compare stability conditions for this system with those for the nonlinear system of Lemma 6.1.2 we attempt to discover for which $\mu \in R$ system (6.1.28) has the equilibrium 0 asymptotically stable independently of delay $h \in (0, +\infty)$.

A solution to this problem can be obtained by employing the following result obtained by Yoshizawa.

Lemma 6.1.3 (Yoshizawa, 1975, see also Brierley et al. 1982; Kamen, 1980; Jury and Mansour, 1982 for further results) *The necessary and sufficient conditions for the equilibrium point 0 of system (6.1.12), in which*

$$f(\phi) = A\phi(0) + \sum_{k=1}^{m} B_k\phi(-h_k) \quad h = \max\{h_k, k = 1, 2, ..., m\} \tag{6.1.29}$$

to be asymptotically stable independently of delays $h_k \in [0, +\infty), k = 1, 2, 3, ..., m$, are

$$\text{Re}\,\lambda\left(A + \sum_{k=1}^{m} B_k\right) < 0 \tag{6.1.30}$$

and

$$\forall \{\Omega_k\} \subset [0, 2\pi] \text{ the matrix } A + \sum_{k=1}^{m} e^{i\Omega_k} B_k \tag{6.1.31}$$

has no nonzero imaginary eigenvalue.

6.1 Lyapunov's Second Method for FDE in Space C

To solve the posed problem we first have to determine the set of such $\mu \in R$ that

$$\operatorname{Re} \lambda(A + \mu bc^T) < 0 \qquad (6.1.32)$$

and to check that

$$\forall \Omega \in [0, 2\pi] \text{ the matrix } A + e^{i\Omega}\mu bc^T \text{ has no nonzero imaginary eigenvalue.} \qquad (6.1.33)$$

By (6.1.33), the plot $G(i\omega)$ does not possess any common points with a circle $|z| = 1/|\mu|$ in the complex domain ($\operatorname{Re} G, \operatorname{Im} G$). Since $G(i\omega) \underset{\omega \to \infty}{\to} 0$, which follows from (6.1.19), the plot $G(i\omega)$ must lie in the circle $K(0, 1/|\mu|)$. Thus

$$\frac{1}{\mu^2} - |G(i\omega)|^2 > 0 \qquad \forall \omega \in R. \qquad (6.1.34)$$

This condition excludes the possibility of the matrix A having any eigenvalue on the imaginary axis. It also excludes the possibility of encirclement of the point $(-1/|\mu|, 0)$ by the plot $G(i\omega)$. Lastly, (6.1.34) implies that the plot $G(i\omega)$ lies on the right side of the half-fline $\{s \in C: \operatorname{Im} s = 0, \operatorname{Re} s \leq -1/|\mu|\}$. In virtue of the Nyquist criterion for the system without delay, (6.1.32) can hold only when $\operatorname{Re} \lambda(A) < 0$, i.e. when the condition (6.1.20) is satisfied. Conversely, if (6.1.20) holds then (6.1.31) is true by (6.1.34) and the Nyquist criterion. Taking the upper bound of the numbers $|\mu|$ satisfying (6.1.34) we have conditions (6.1.20) and (6.1.18). In this manner we have proved that (6.1.20) and (6.1.18) are necessary conditions for system (6.1.28) to be asymptotically stable for every $h > 0$. Thus assumption (ii) of Lemma 6.1.2 cannot be weakened in Lemma 6.1.2.

Example 3 *Analysis of a class of control systems with delay introduced by Gromova and Pelevina (1977)*

In this example we consider the system

$$\dot{x}(t) = Ax(t) + Bx(t-h) + bF[c^T x(t)] \qquad t \geq 0 \qquad (6.1.35)$$

where A, B are real, $(n \times n)$-matrices; $b, c, x(t) \in R^n$; $F: R \to R$ locally Lipschitzian, $F(0) = 0$.

Of course, system (6.1.35) has the form (6.1.1a) with $f(\phi) = A\phi(0) + B\phi(-h) + bF[c^T\phi(0)]$. As F is locally Lipschitzian, so is f and the assumptions (6.1.4) hold for f. $0 \in C$ is an equilibrium point, because $F(0) = 0$. We seek a Lyapunov functional for system (6.1.35) in the form

$$V(\phi) = \phi^T(0)H\phi(0) + \int_{-h}^{0} \phi^T(\theta)Q\phi(\theta)d\theta + q \int_{0}^{c^T\phi(0)} F(\sigma)d\sigma \qquad (6.1.36)$$

where $H = H^T$, $Q = Q^T$ are real, $(n \times n)$-matrices, $q \in R$. The derivative $\dot{V}(\phi)$, defined by (6.1.5) can be readily calculated for $\phi \in D \stackrel{\text{df}}{=} \{\psi \in C: \psi' \in C, \psi'(0) = f(\psi)\}$:

$$\dot{V}(\phi) = [\phi^T(0), \phi^T(-h), F] \begin{bmatrix} A^T H + HA + Q & HB & Hb + \frac{q}{2} A^T c \\ B^T H & -Q & \frac{q}{2} B^T c \\ b^T H + \frac{q}{2} c^T A & \frac{q}{2} c^T B & qc^T b \end{bmatrix} \begin{bmatrix} \phi(0) \\ \phi(-h) \\ F \end{bmatrix}.$$

(6.1.37)

Adding and subtracting the term $\{F[c^T\phi(0)] - k_1 c^T\phi(0)\}c^T\phi(0)$ one can rewrite $\dot{V}(\phi)$ in the form

$$\dot{V}(\phi) = [\phi^T(0), \phi^T(-h), F] G \begin{bmatrix} \phi(0) \\ \phi(-h) \\ F \end{bmatrix} - \{F[c^T\phi(0)] -$$

$$- k_1 c^T\phi(0)\} c^T\phi(0) \quad \forall \phi \in D \quad (6.1.38)$$

where

$$G = \begin{bmatrix} A^T H + HA + Q - k_1 cc^T & HB & Hb + \frac{q}{2} A^T c + \frac{1}{2} c \\ B^T H & -Q & \frac{q}{2} B^T c \\ b^T H + \frac{q}{2} c^T A + \frac{1}{2} c^T & \frac{q}{2} c^T B & qc^T b \end{bmatrix}.$$

We now introduce an auxiliary functional

$$W(\phi) = [\phi^T(0) \quad \phi^T(-h) \quad F] G \begin{bmatrix} \phi(0) \\ \phi(-h) \\ F \end{bmatrix} +$$

$$+ \{F[c^T\phi(0)] - k_1 c^T\phi(0)\} c^T\phi(0) \quad \phi \in C. \quad (6.1.39)$$

W is continuous on C. Furthermore, if

$$-G \geqslant 0_{(2n+1) \times (2n+1)} \quad (6.1.40)$$

$$k_1 < \frac{F(\sigma)}{\sigma} \quad \forall \sigma \neq 0 \quad F(0) = 0 \quad (6.1.41)$$

then W is nonnegative on C.

Applying the standard procedure for the estimation of $\dot{V}(\phi)$ for $\phi \in C$ as described in Example 1 we get

$$\dot{V}(\phi) \leqslant -W(\phi) \leqslant 0 \quad \forall \phi \in C \quad (6.1.42)$$

6.1 Lyapunov's Second Method for FDE in Space C

hence V is a Lyapunov functional on C for system (6.1.35). If the equilibrium point $0 \in C$ is the largest, invariant set in the sense of inclusion, contained in $\{\phi \in C: W(\phi) = 0\}$ then by virtue of Theorem 6.1.1 it follows that every bounded solution of (6.1.35) tends to 0 as t tends to infinity.

Note that (6.1.40) implies the inequality $Q \geqslant 0$, hence by (6.1.36)

$$V(\phi) \geqslant \phi^T(0) H \phi(0) + q \int_0^{c^T \phi(0)} F(\sigma) d\sigma \quad \forall \phi \in C. \qquad (6.1.43)$$

Under the additional conditions

$$z^T H z + q \int_0^{c^T z} F(\sigma) d\sigma > 0 \quad \forall z \in R^n \setminus \{0\} \qquad (6.1.44)$$

$$z^T H z + q \int_0^{c^T z} F(\sigma) d\sigma \underset{|z| \to \infty}{\to} \infty \qquad (6.1.45)$$

it follows from Theorems 6.1.2 and 6.1.3 that $0 \in C$ is globally asymptotically stable.

Similar sufficient conditions for the global asymptotic stability of 0 may be obtained by adding and subtracting the term $\{F[c^T \phi(0)] - k_1 c^T \phi(0)\} \times \{k_2 c^T \phi(0) - F[c^T \phi(0)]\}$ in (6.1.37) instead of the one before. In this case the sufficient conditions for global asymptotic stability are

$$\begin{bmatrix} -A^T H - HA - Q + k_1 k_2 cc^T & -HB & -Hb - \dfrac{q}{2} A^T c - \dfrac{k_1+k_2}{2} c \\ -B^T H & Q & -\dfrac{q}{2} B^T c \\ -b^T H - \dfrac{q}{2} c^T A - \dfrac{k_1+k_2}{2} c^T & -\dfrac{q}{2} c^T B & 1 - qc^T b \end{bmatrix}$$
$$\geqslant 0_{(2n+1) \times (2n+1)} \qquad (6.1.46)$$

$$k_1 < \frac{F(\sigma)}{\sigma} < k_2 \quad \forall \sigma \neq 0 \quad F(0) = 0. \qquad (6.1.47)$$

$\{0\}$ is the largest invariant set contained in $\{\phi \in C: W(\phi) = 0\}$, where $W(\phi)$ has the form (6.1.39), with the $(2n+1)$-dimensional quadratic matrix replaced by matrix (6.1.46) $\}$ (6.1.48)

$$z^T H z + q \int_0^{c^T z} F(\sigma) d\sigma > 0 \quad \forall z \in R^n \setminus \{0\} \qquad (6.1.49)$$

$$z^T H z + q \int_0^{c^T z} F(\sigma) d\sigma \underset{|z| \to \infty}{\to} \infty. \qquad (6.1.50)$$

Our results represent a generalization of those of Gromova and Pelevina.

In the numerical examples discussed below inequality (6.1.40) is satisfied due to the assumption

$$Q = \frac{q}{-4c^Tb} B^T cc^T B \qquad qc^Tb < 0 \tag{6.1.51}$$

and the requirement that there exist $k, q \in R$ such that the system

$$A^T H + HA - \frac{q}{4c^Tb} B^T cc^T B - k_1 cc^T = -gg^T \tag{6.1.52a}$$

$$H[B\ b] + \left[0\quad \tfrac{q}{2} A^T c + \tfrac{1}{2} c\right] = -g\left[-\frac{q}{2\sqrt{-qc^Tb}} c^T B \quad \sqrt{-qc^Tb}\right] \tag{6.1.52b}$$

has a solution with respect to H, $H = H^T$, $g \in R^n$. In this case (6.1.38) can be rewritten in the form

$$\dot{V}(\phi) = -\left[g^T\phi(0) - \frac{q}{2\sqrt{-qc^Tb}} c^T B\phi(-h) + \sqrt{-qc^Tb}\ F\right]^2 -$$
$$- \{F[c^T\phi(0)] - k_1 c^T\phi(0)\}c^T\phi(0) \qquad \forall \phi \in D. \tag{6.1.52c}$$

Numerical examples

(a) $n = 1$, $b = [-1]$, $c = [1]$ in (6.1.35). This is a system considered by Gruber (1969). According to (6.1.51) we assume $Q = qB^2/4$, $q > 0$ and system (6.1.52) then takes the form

$$-2AH + \frac{qB^2}{4} - k_1 = -g^2$$

$$\left[-HB \quad -H + \frac{Aq}{2} + \frac{1}{2}\right] = -g\left[\frac{\sqrt{q}\ B}{2} \quad \sqrt{q}\right].$$

The minimal number $k_1 \in R$ for which there exists $q > 0$, such that the above system has a solution with respect to (H, g) is $k_1 = |B| + A$. Next, we obtain

$$q = \frac{2}{|B|} \qquad H = \frac{-A}{|B|} - \frac{1}{2} \qquad g = \frac{-2A - |B|}{\sqrt{2|B|}} \qquad Q = \frac{|B|}{2}$$

$$V(\phi) = \left(\frac{-A}{|B|} - \frac{1}{2}\right)\phi^2(0) + \frac{|B|}{2}\int_{-h}^{0}\phi^2(\theta)d\theta + \frac{2}{|B|}\int_{0}^{\phi(0)} F(\sigma)d\sigma$$

$$\dot{V}(\phi) = -\frac{1}{2|B|}[(2A+|B|)\phi(0) + B\phi(-h) - 2F]^2 - [F - (|B|+A)\phi(0)]\phi(0).$$

Condition (6.1.41) now takes the form

$$|B| + A < \frac{F(\sigma)}{\sigma} \qquad \forall \sigma \neq 0 \qquad F(0) = 0. \tag{6.1.5}$$

6.1 Lyapunov's Second Method for FDE in Space C

It may be easily noticed that the equilibrium point $\{0\}$ is the only invariant set contained in the set $\{\phi \in C: W(\phi) = 0\}$ where W is defined by (6.1.39).

Conditions (6.1.44) and (6.1.45) hold, since $q > 0$ and by virtue of (6.1.43) and (6.1.53)

$$V(\phi) \geq \left(\frac{-A}{|B|} - \frac{1}{2}\right)\phi^2(0) + \frac{2}{|B|} \int_0^{\phi(0)} F(\sigma)\,d\sigma$$

$$\geq \left[\frac{-A}{|B|} - \frac{1}{2} + \frac{2}{|B|}\frac{k_1}{2}\right]\phi^2(0) = \frac{1}{2}\phi^2(0).$$

Finally, (6.1.53) is a sufficient condition for the global asymptotic stability of the equilibrium.

The results we have obtained are much stronger than those of Gromova and Pelevina or of Gruber. For instance, using a quite different Lyapunov functional, Gruber proved that 0 is globally asymptotically stable if $0 < F(\sigma)/\sigma$ $\forall \sigma \neq 0$, $F(0) = 0$, $A < 0$, $|B| < -A$.

It turns out that condition (6.1.53) is the best result obtainable, because the linear system $\dot{x}(t) = (A-\mu)x(t) + Bx(t-h)$ obtained from the nonlinear system by substituting $F(\sigma) = \mu\sigma$ is stable independently of delay $h \in [0, +\infty)$ when $\mu > |B| + A$, by virtue of Lemma 6.1.3. Thus condition (6.1.53) cannot be weakened.

(b) $n = 2$, $A = \begin{bmatrix} 0 & 0 \\ 0 & -1 \end{bmatrix}$, $B = \begin{bmatrix} 0 & -a \\ 0 & 0 \end{bmatrix}$, $b = \begin{bmatrix} -p \\ 1 \end{bmatrix}$, $p > 0$, $c = \begin{bmatrix} 1 \\ 0 \end{bmatrix}$.

This system is considered by Goryachenko (1971) in connection with a stability analysis for a simple nuclear reactor model. According to (6.1.51) we assume

$$Q = \begin{bmatrix} 0 & 0 \\ 0 & \dfrac{a^2 q}{4p} \end{bmatrix} \qquad q > 0.$$

System (6.1.52) can be written in the form

$$\begin{bmatrix} 0 & 0 \\ 0 & -1 \end{bmatrix}\begin{bmatrix} h_1 & h_2 \\ h_2 & h_3 \end{bmatrix} + \begin{bmatrix} h_1 & h_2 \\ h_2 & h_3 \end{bmatrix}\begin{bmatrix} 0 & 0 \\ 0 & -1 \end{bmatrix} + \begin{bmatrix} 0 & 0 \\ 0 & \dfrac{a^2 q}{4p} \end{bmatrix} + \begin{bmatrix} -k_1 & 0 \\ 0 & 0 \end{bmatrix}$$

$$= \begin{bmatrix} -g_1^2 & -g_1 g_2 \\ -g_1 g_2 & -g_2^2 \end{bmatrix}$$

$$\begin{bmatrix} h_1 & h_2 \\ h_2 & h_3 \end{bmatrix}\begin{bmatrix} 0 & -a & -p \\ 0 & 0 & 1 \end{bmatrix} + \begin{bmatrix} 0 & 0 & \tfrac{1}{2} \\ 0 & 0 & 0 \end{bmatrix} = \begin{bmatrix} -g_1 \\ -g_2 \end{bmatrix}\begin{bmatrix} 0 & \dfrac{a\sqrt{q}}{2\sqrt{p}} & \sqrt{qp} \end{bmatrix}$$

and has a solution with respect to (H, g) if and only if $p^2 > a^2$, $q = 2p/\sqrt{p^2-a^2}$ and $k_1 = 1/2\sqrt{p^2-a^2}$. This solution is of the form

$$H = \begin{bmatrix} k_1 & -\tfrac{1}{2}-pk_1 \\ -\tfrac{1}{2}-pk_1 & p^2k_1+\tfrac{1}{2}p \end{bmatrix} \quad g = \begin{bmatrix} \sqrt{k_1} \\ -\dfrac{1}{2\sqrt{k_1}} - p\sqrt{k_1} \end{bmatrix}.$$

Furthermore,

$$Q = \begin{bmatrix} 0 & 0 \\ 0 & a^2k_1 \end{bmatrix}.$$

The functional V corresponding to this solution is determined by the formula

$$V(\phi) = k_1\phi_1^2(0) - \phi_1(0)\phi_2(0)(1+2pk_1) + (\tfrac{1}{2}p+p^2k_1)\phi_2^2(0) +$$
$$+ a^2k_1 \int_{-h}^{0} \phi_2^2(\theta)\,d\theta + 4pk_1 \int_{0}^{\phi_1(0)} F(\sigma)\,d\sigma,$$

for which

$$\dot{V}(\phi) = -\left\{\sqrt{k_1}\phi_1(0) - \left(\dfrac{1}{2\sqrt{k_1}} + p\sqrt{k_1}\right)\phi_2(0) + \right.$$
$$\left. + a\sqrt{k_1}\phi_2(-h) + 2p\sqrt{k_1}F[\phi_1(0)]\right\}^2 - \{F[\phi_1(0)] - k_1\phi_1(0)\}\phi_1(0).$$

The only invariant set contained in $\{\phi \in C: \phi(0) = 0\}$ is the equilibrium point $\{0\} \subset C$, hence $\{0\}$ is the only invariant set contained in $\{\phi \in C: W(\phi) = 0\}$ where W is defined in (6.1.39).

As $q > 0$ and (6.1.41) holds,

$$V(\phi) \geqslant \phi^T(0)H\phi(0) + q \int_{0}^{c^T\phi(0)} F(\sigma)\,d\sigma \geqslant \phi^T(0)[H+\tfrac{1}{2}qk_1cc^T]\phi(0)$$
$$= [\phi_1^T(0) \; \phi_2(0)] \begin{bmatrix} k_1+2pk_1^2 & -\tfrac{1}{2}pk_1-\tfrac{1}{2} \\ -\tfrac{1}{2}pk_1-\tfrac{1}{2} & p^2k_1+\tfrac{1}{2}p \end{bmatrix} \begin{bmatrix} \phi_1(0) \\ \phi_2(0) \end{bmatrix}.$$

The quadratic matrix appearing in the right-hand side is positive-definite (Sylvester criterion) thus conditions (6.1.44), (6.1.45) are fulfilled.

We conclude that

$$\dfrac{1}{2\sqrt{p^2-a^2}} < \dfrac{F(\sigma)}{\sigma} \quad \forall \sigma \neq 0 \quad F(0) = 0 \quad p^2 > a^2 \quad p > 0 \quad (6.1.54)$$

is a sufficient condition for the equilibrium $0 \in C$ to be globally asymptotically stable.

6.1 Lyapunov's Second Method for FDE in Space C

Discussion of the result

If $a = 0$, $p > 0$, it is possible to obtain a sufficient condition of global symptotic stability in the form $0 < F(\sigma)/\sigma \ \forall \sigma \neq 0$, $F(0) = 0$, therefore (6.1.54) is not the best possible stability condition for the system under consideration. An improved version of the stability condition (6.1.54) would be necessary, for example, in the stability analysis of nuclear reactor model considered by Goryachenko (1971). Since in this case $F(\sigma) = e^\sigma - 1$, it is important to check whether the condition

$$0 < \frac{F(\sigma)}{\sigma} \quad \forall \sigma \neq 0 \quad F(0) = 0 \quad p^2 > a^2 \quad p > 0 \quad (6.1.55)$$

is a sufficient one for global asymptotic stability of the equilibrium $0 \in C$.

However, we shall show that this cannot be done by means of the Lyapunov functional of the general form (6.1.36). Indeed, even if we confine ourselves to linear functions F,

$$F(\sigma) = \mu\sigma, \mu > 0$$

this particular Lyapunov approach does not yield the necessary and sufficient conditions of asymptotic stability for all delays $h \geq 0$. System (6.1.35) then takes the form

$$\begin{aligned} \dot{x}_1(t) &= -ax_2(t-h) - p\mu x_1(t) \\ \dot{x}_2(t) &= -x_2(t) + \mu x_1(t) \quad p^2 > a^2 \quad p > 0. \end{aligned} \quad (6.1.56)$$

The reader may check with the use of Lemma 6.1.3 that system (6.1.56) is asymptotically stable independently of the delay $h \geq 0$, for every $\mu > 0$. These shortcomings of the Lyapunov functional (6.1.36) were first observed by Calvarho et al. (1980) (see section 4 of their work). We shall follow their argument.

Lemma 6.1.4 (Calvarho et al., 1980) *Given $(n \times n)$-complex matrices M, N, P with M and N Hermitian, the following two statements are equivalent*:
  (i) $M + N + e^{i\omega}P + e^{-i\omega}P^* \leq 0 \quad \forall \omega \in [0, 2\pi]$,
  (ii) *there exists a Hermitian matrix Q such that*

$$\begin{bmatrix} M+Q & P \\ P^* & N-Q \end{bmatrix} \leq 0. \quad (6.1.57)$$

In the case of system (6.1.56) we have

$$\dot{V}(\phi) = [\phi^T(0) \quad \phi^T(-h)] \begin{bmatrix} F_1 & F_2 \\ F_2^T & -Q \end{bmatrix} \begin{bmatrix} \phi(0) \\ \phi(-h) \end{bmatrix} \quad \forall \phi \in D \quad (6.1.58)$$

where
$$F_1 = (A+\mu bc^T)^T(H+\tfrac{1}{2}\mu qcc^T)+(H+\tfrac{1}{2}\mu qcc^T)(A+\mu bc^T)+Q$$
$$F_2 = (H+\tfrac{1}{2}\mu qcc^T)B.$$

Now, the necessary and sufficient condition for a V of the form (6.1.58) to be a Lyapunov functional for (6.1.56) on D and also on C, is the existence of real $(n\times n)$-matrices H, Q, $H = H^T$, $Q = Q^T$ and $q \in R$ such that

$$\begin{bmatrix} F_1 & F_2 \\ F_2^T & -Q \end{bmatrix} \leqslant 0. \qquad (6.1.59)$$

We assume H, μ, q to be fixed in our example. Without loss of generality, one may take

$$H = \begin{bmatrix} 1 & \beta \\ \beta & \delta \end{bmatrix}. \qquad (6.1.60)$$

Substituting

$$M = (A+\mu bc^T)^T(H+\tfrac{1}{2}\mu qcc^T)+(H+\tfrac{1}{2}\mu qcc^T)(A+\mu bc^T)$$
$$N = 0 \quad P = (H+\tfrac{1}{2}\mu qcc^T)B \qquad (6.1.61)$$

in Lemma 6.1.4, we find due to this lemma that a necessary condition for the existence) of a real matrix $Q, Q = Q^T$ is

$$(A^T+\mu cb^T+e^{-i\omega}B^T)(H+\tfrac{1}{2}\mu qcc^T)+(H+\tfrac{1}{2}\mu qcc^T)(A+\mu bc^T+e^{i\omega}B) \leqslant 0$$
$$\forall \omega \in [0, 2\pi]. \qquad (6.1.62)$$

Taking account of (6.1.60) and the form of A, B, c, b in (6.1.62), we obtain an equivalent form of inequality (6.1.62),

$$\begin{bmatrix} 2p\mu+pq\mu^2-2\mu\beta & p\mu\beta-\mu\delta+\beta+ae^{i\omega}+\tfrac{1}{2}\mu aqe^{i\omega} \\ p\mu\beta-\mu\delta+\beta+ae^{-i\omega}+\tfrac{1}{2}\mu aqe^{-i\omega} & 2\delta+2a\beta\cos\omega \end{bmatrix} \geqslant 0_{2\times 2}$$
$$\forall \omega \in [0, 2\pi] \qquad (6.1.62')$$

A necessary condition for (6.1.62') to hold is the nonnegativity of the determinant of the matrix in the left-hand side of the inequality above, which in turn yields the inequality

$$\mu^2\{[aq\delta+ap\beta q]\cos\omega+[2\delta pq-\tfrac{1}{4}a^2q^2-(p\beta-\delta)^2]\}+$$
$$+\mu[a(2\delta-4\beta^2-q\beta+2p\beta)\cos\omega+4p\delta-2\delta\beta-2p\beta^2-a^2q]-$$
$$-a^2-\beta^2-2a\beta\cos\omega \geqslant 0 \quad \forall\omega \in [0, 2\pi] \qquad (6.1.63)$$

Treating the left-hand side of (6.1.63) as a trinomial with respect to μ we notice that the free term of this trinomial is negative for ω different from $0, \pi, 2\pi$. Hence (6.1.63) cannot occur for all positive μ.

Let us summarize our considerations. By any choice of the matrices H and Q and the parameter q in the Lyapunov functional (6.1.36), we cannot

6.2 Stability Theory of Abstract Linear Systems

obtain the necessary and sufficient conditions of asymptotic stability independently of delay of the linear system (6.1.56). Further, such conditions cannot be obtained for all nonlinearities satisfying (6.1.55). In conclusion, a different functional is needed.

Similar results can be derived by frequency-domain techniques (Răsvan, 1975, 1983). The results presented here, obtained with the use of Lyapunov functionals, require less restrictive assumptions about the nonlinearity in the differential equation (6.1.12).

6.2 Stability Theory of Abstract Linear Systems and its Applications to Systems Involving Delay

6.2.1 Some basic results in abstract linear systems theory

Definition 6.2.1 *Let $\mathscr{X}$ be a real Banach space with norm $\|\cdot\|$. A family $\{S(t)\}_{t \geqslant 0}$ of linear bounded operators $S(\cdot): \mathscr{X} \to \mathscr{X}$ is called a linear C_0-semigroup if*
(i) $S(t+\tau) = S(t)S(\tau) \quad \forall t, \tau \geqslant 0$,
(ii) $S(0) = I$, where I is an identity on $\mathscr{X}$,
(iii) $S(\cdot)u: R^* \to \mathscr{X}$ is continuous (right-continuous at $t = 0$), $\forall u \in \mathscr{X}$, $R^* = [0, +\infty)$.

An operator $\mathscr{A}: (\mathscr{D}(\mathscr{A}) \subset \mathscr{X}) \to \mathscr{X}$ defined as follows

$$\mathscr{A}u = \lim_{t \to 0+} \frac{1}{t}[S(t)u - u] \quad u \in \mathscr{D}(\mathscr{A}) \Leftrightarrow \exists \lim_{t \to 0+} \frac{1}{t}[S(t)u - u]$$

is called an infinitesimal generator of the linear C_0-semigroup $\{S(t)\}_{t \geqslant 0}$.

Definition 6.2.2 *Let $\mathscr{X}$ be a real Banach space with norm $\|\cdot\|$. Denote its dual by $\mathscr{X}^*$. An operator $\mathscr{F}: (\mathscr{D}(\mathscr{F}) \subset \mathscr{X}) \to \mathscr{X}$ is called dissipative, if*

$$\forall \{u_1, u_2\} \subset \mathscr{D}(\mathscr{F}) \exists j \in \mathscr{X}^*: j(u_1 - u_2) = \|u_1 - u_2\|^2,$$
$$\|j\| = \|u_1 - u_2\| \text{ and } j(\mathscr{F}u_1 - \mathscr{F}u_2) \leqslant 0.$$

If $\mathscr{X}$ is a real Hilbert space with the inner product $\langle \cdot, \cdot \rangle$ then $\mathscr{F}: (\mathscr{D}(\mathscr{F}) \subset \mathscr{X}) \to \mathscr{X}$ is a dissipative operator if and only if

$$\langle u_1 - u_2, \mathscr{F}u_1 - \mathscr{F}u_2 \rangle \leqslant 0 \quad \forall u_1, u_2 \in \mathscr{D}(\mathscr{F}). \tag{6.2.1}$$

For a linear operator $\mathscr{F}$ condition (6.2.1) reduces to

$$\langle u, \mathscr{F}u \rangle \leqslant 0 \quad \forall u \in \mathscr{D}(\mathscr{F}). \tag{6.2.1'}$$

Definition 6.2.3 *Let $\mathscr{X}$ be a Banach space with a norm $\|\cdot\|$. Consider the abstract initial-value problem (AIVP) for the differential equation with an operator $\mathscr{A}: (\mathscr{D}(\mathscr{A}) \subset \mathscr{X}) \to \mathscr{X}$:*

(a) $\dot{u}(t) = \mathscr{A}u(t)$ $t > 0$
(b) $u(0) = u_0$.

A function $u: [0, T] \ni t \mapsto u(t) \in \mathscr{X}$ where $T \in (0, +\infty]$, is called the strong solution to AIVP on interval $[0, T)$ if

(i) u is locally Lipschitz-continuous on $[0, T)$,
(ii) u is strongly differentiable almost everywhere on $(0, T)$,
(iii) $u(t) \in \mathscr{D}(\mathscr{A})$ for almost all $t \in (0, T)$,
(iv) u satisfies (AIVPb), and satisfies (AIVPa) almost everywhere on $(0, T)$.

The following theorem which is J. A. Walker's version of the well-known Hille–Philips–Yosida theorem is a fundamental result.

Theorem 6.2.1 (Walker, 1976, 1980). *Let $\mathscr{X}$ be a real Banach space with norm $\|\cdot\|$. The necessary and sufficient conditions for a linear operator $\mathscr{A}: (\mathscr{D}(A) \subset \mathscr{X}) \to \mathscr{X}$ to be the infinitesimal generator of a linear C_0-semigroup $\{S(t)\}_{t \geq 0}$, satisfying the inequality*

$$\|S(t)\|_{\mathscr{L}(\mathscr{X})} \leq Me^{\omega t} \quad \forall t \geq 0 \tag{6.2.2}$$

for some $\omega \in R$ and $M \geq 1$, are

(i) $\overline{\mathscr{D}(\mathscr{A})} = \mathscr{X}$ (the domain of $\mathscr{A}$ is dense in $\mathscr{X}$), \hfill (6.2.3)

(ii) $\exists \lambda_0 > 0: \forall \lambda \in (0, \lambda_0): \mathscr{R}(I - \lambda \mathscr{A}) = \mathscr{X}$, \hfill (6.2.4)

(iii) *There exists a norm $\|\cdot\|_e$, equivalent to $\|\cdot\|$ in which the operator $\mathscr{A} - \omega I$ is dissipative.* \hfill (6.2.5)

Furthermore, if conditions (6.2.3)–(6.2.5) are fulfilled then

$$S(t)u = \lim_{n \to \infty} \left(I - \frac{t}{n}\mathscr{A}\right)^{-n} u \quad \forall u \in \mathscr{X} \tag{6.2.6}$$

$$\|S(t)u\|_e \leq \|u\|_e e^{\omega t} \quad \forall u \in \mathscr{X} \quad \forall t \geq 0 \tag{6.2.7}$$

$\forall u_0 \in \mathscr{D}(\mathscr{A}): S(\cdot)u_0$ *is the unique strong solution to* (AIVP) *on* $[0, +\infty)$, *moreover conditions* (ii), (iii), (iv) *in Definition 6.2.3 hold everywhere on* $(0, +\infty)$.

Lemma 6.2.1 (Walker, 1980) *Let $\mathscr{X}$ be a real Hilbert space. Assume that $\mathscr{A}: (\mathscr{D}(\mathscr{A}) \subset \mathscr{X}) \to \mathscr{X}$ is a linear operator which satisfies condition (6.2.4), and $\mathscr{A} - \omega I$ is dissipative for some $\omega \in R$ in the norm generated by the inner product. Then (6.2.3) is fulfilled.*

The significance of this lemma for the verification of the hypotheses of Theorem 6.2.1 is obvious.

6.2 Stability Theory of Abstract Linear Systems

Remark 6.2.1 (Grabowski, 1979) Condition (6.2.5) can also be written in another equivalent form: There exists a norm $\|\cdot\|_e$ equivalent to $\|\cdot\|$, such that

$$\|(I-\lambda\mathscr{A})u_1 - (I-\lambda\mathscr{A})u_2\|_e \geqslant (1-\lambda\omega)\|u_1 - u_2\|_e$$

$$\forall u_1, u_2 \in \mathscr{D}(\mathscr{A}) \quad \forall \lambda > 0 \quad \lambda\omega < 1 \tag{6.2.5'}$$

6.2.2 Stability of abstract linear systems

Definition 6.2.4 Let $\mathscr{X}$ be a real Banach space with a norm $\|\cdot\|$. We say that a linear C_0-semigroup $\{S(t)\}_{t\geqslant 0}$ is *exponentially stable* (EXS) if there exist $M \geqslant 1$ and $\Omega < 0$ such that

$$\|S(t)\|_{\mathscr{L}(\mathscr{X})} \leqslant Me^{\Omega t} \quad \forall t \geqslant 0 \tag{6.2.8}$$

and it is L^2-*stable if*

$$\int_0^\infty \|S(t)u\|^2 dt < \infty \quad \forall u \in \mathscr{X}. \tag{6.2.9}$$

The quantity

$$\omega_0 = \inf_{t>0} \frac{\ln\|S(t)\|}{t} \tag{6.2.10}$$

is called the *type of the linear* C_0-*semigroup* $\{S(t)\}_{t\geqslant 0}$.

We use here the convention $\ln 0 = -\infty$.

Lemma 6.2.2 (Yosida, 1980) *In the notation given above*

$$\omega_0 = \lim_{t\to\infty} \frac{\ln\|S(t)\|}{t}. \tag{6.2.11}$$

Lemma 6.2.3 (Foiaş, 1973) *Let* $\sigma(\mathscr{A})$ *denote the spectrum of an operator* $\mathscr{A}$, *the infinitesimal generator of a linear* C_0-*semigroup* $\{S(t)\}_{t\geqslant 0}$. *Then*

$$\sup\{\operatorname{Re}\lambda: \lambda \in \sigma(\mathscr{A})\} \leqslant \omega_0 \quad \text{if} \quad \sigma(\mathscr{A}) \neq \phi. \tag{6.2.12}$$

Unfortunately, the full characterization of the case in which the equality takes place in (6.2.12) is not known. However, it is known that this equality holds when

$$C\sigma[S(1)] \subset \{e^\lambda: \lambda \in \sigma(\mathscr{A})\} \cup \{0\} \tag{6.2.13}$$

where $C\sigma[S(1)]$ is the continuous part of the spectrum of the operator $S(1)$ $\in \mathscr{L}(\mathscr{X})$.

In Hilbert space a more precise characterization is possible.

Lemma 6.2.4 (Prüss, 1984) *Let $\mathscr{X}$ be a Hilbert space and $\{S(t)\}_{t\geq 0}$ a C_0-semigroup on $\mathscr{X}$, generated by $\mathscr{A}$. Then $\{S(t)\}_{t\geq 0}$ is EXS if and only if*
 (i) *The spectrum $\sigma(\mathscr{A})$ lies in the open left half-plane,*
 (ii) $\exists M \geq 1$: $\|(\lambda I - \mathscr{A})^{-1}\| \leq M$ *for every λ such that* $\operatorname{Re}\lambda \geq 0$.

Theorem 6.2.2 *Let $\mathscr{X}$ be a real Banach space with norm $\|\cdot\|$, let $\{S(t)\}_{t\geq 0}$ be a linear C_0-semigroup on $\mathscr{X}$, and let ω_0 be its type. Then $\{S(t)\}_{t\geq 0}$ is EXS if and only if $\omega_0 < 0$.*

Proof. Necessity If $\{S(t)\}_{t\geq 0}$ is EXS then there exist $M \geq 1$ and $\Omega < 0$ such that $\|S(t)\| \leq Me^{\Omega t}$ $\forall t \geq 0$. Hence $\dfrac{\ln\|S(t)\|}{t} \leq \dfrac{\Omega t + \ln M}{t}$ $\forall t > 0$.
After taking the limits of both sides as t tends to infinity we obtain $\omega_0 \leq \Omega < 0$.

Sufficiency If $\omega > \omega_0$, then by (6.2.11) there exists $t_0 > 0$ such that $\|S(t)\| \leq e^{\omega t}$ $\forall t \geq t_0$. On the other hand $\|S(t)\| \leq M_0$ $\forall t \in [0, t_0]$, for some $M_0 \geq 1$. Thus $\|S(t)\| \leq Me^{\omega t}$ where $M = M_0$ if $\omega \geq 0$ and $M = M_0 e^{-\omega t_0}$ if $\omega < 0$.

Now, if $\omega_0 < 0$ then taking an arbitrary $\omega \in (\omega_0, 0)$ we see that $\{S(t)\}_{t\geq 0}$ is EXS. □

In Hilbert spaces a more detailed characterization of the exponential stability of linear C_0-semigroup is available.

Theorem 6.2.3 *Let $\mathscr{X}$ be a real Hilbert space with an inner product $\langle\cdot,\cdot\rangle$. Assume that $\mathscr{A}: (\mathscr{D}(\mathscr{A}) \subset \mathscr{X}) \to \mathscr{X}$ is the infinitesimal generator of a linear C_0-semigroup $\{S(t)\}_{t\geq 0}$ on $\mathscr{X}$. Let also $Q \in \mathscr{L}(\mathscr{X})$, $Q = Q^*$, $\langle u, Qu\rangle \geq q\|u\|^2 > 0$ $\forall u \neq 0$. Then $\{S(t)\}_{t\geq 0}$ is EXS if and only if there exists an $\mathscr{H} \in \mathscr{L}(\mathscr{X})$, $\mathscr{H} = \mathscr{H}^*$, $\langle u, \mathscr{H}u\rangle \geq 0$ $\forall u \in \mathscr{X}$ such that*

$$\langle \mathscr{A}u, \mathscr{H}u\rangle + \langle u, \mathscr{H}\mathscr{A}u\rangle = -\langle u, Qu\rangle \quad \forall u \in \mathscr{D}(\mathscr{A}). \quad (6.2.14)$$

Note that a particular version of this theorem has been proved by Datko (1970). Our version follows Delfour (1974) and Wexler (1980) who gave this theorem without a proof.

Proof Necessity will follow from the lemma below.

Lemma 6.2.5 *Let $\mathscr{X}$ and $\mathscr{A}$ be as in Theorem 6.2.3. Assume that the semigroup $\{S(t)\}_{t\geq 0}$ generated by $\mathscr{A}$ is EXS. Then for every $\mathscr{P} \in \mathscr{L}(\mathscr{X})$, $\mathscr{P} = \mathscr{P}^*$ the operator $\mathscr{K} \in \mathscr{L}(\mathscr{X})$ defined by*

$$\mathscr{K}u = \int_0^\infty S^*(\tau)\mathscr{P}S(\tau)u\,d\tau \quad u \in \mathscr{X} \quad (6.2.15)$$

6.2 Stability Theory of Abstract Linear Systems

is the unique, selfadjoint operator in $\mathscr{L}(\mathscr{X})$ satisfying the operator Lyapunov equation

$$\langle \mathscr{A}u, \mathscr{K}u\rangle + \langle u, \mathscr{K}\mathscr{A}u\rangle = -\langle u, \mathscr{P}u\rangle \quad \forall u \in \mathscr{D}(\mathscr{A}). \tag{6.2.16}$$

Moreover, if $\langle u, \mathscr{P}u\rangle \geq 0$ $\forall u \in \mathscr{X}$ then also $\langle u, \mathscr{K}u\rangle \geq 0$ $\forall u \in \mathscr{X}$.

Proof of Lemma 6.2.5 Our proof follows that of Wexler (1980). EXS implies that (6.2.15) really defines an operator belonging to $\mathscr{L}(\mathscr{X})$. The selfadjointness of the so-defined operator $\mathscr{K}$ is obvious. Now we check that $\mathscr{K}$ satisfies (6.2.16). Let $u \in \mathscr{D}(\mathscr{A})$. By virtue of Theorem 6.2.1 we have on the one hand

$$\lim_{h\to 0+} \frac{1}{h}[\langle S(h)u, \mathscr{K}S(h)u\rangle - \langle u, \mathscr{K}u\rangle]$$
$$= \lim_{h\to 0+} \frac{1}{h}[\langle u+h\mathscr{A}u, \mathscr{K}u+h\mathscr{K}\mathscr{A}u\rangle - \langle u, \mathscr{K}u\rangle]$$
$$= \langle \mathscr{A}u, \mathscr{K}u\rangle + \langle u, \mathscr{K}\mathscr{A}u\rangle,$$

and on the other hand by Definition 6.2.1 and due to the integral version of the mean-value theorem

$$\lim_{h\to 0+} \frac{1}{h}[\langle S(h)u, \mathscr{K}S(h)u\rangle - \langle u, \mathscr{K}u\rangle]$$
$$= \lim_{h\to 0+} \frac{1}{h}\left[\int_0^\infty \langle S(t+h)u, \mathscr{P}S(h+t)u\rangle\,\mathrm{d}t - \int_0^\infty \langle S(t)u, \mathscr{P}S(t)u\rangle\,\mathrm{d}t\right]$$
$$= -\lim_{h\to 0+} \frac{1}{h}\left[\int_0^h \langle S(z)u, \mathscr{P}S(z)u\rangle\,\mathrm{d}z\right] = -\langle u, \mathscr{P}u\rangle.$$

Thus we have proved that (6.2.16) holds. Next,

$$\langle u, \mathscr{P}u\rangle \geq 0 \quad \forall u \in \mathscr{X}$$
$$\Rightarrow \int_0^\infty \langle S(t)u, \mathscr{P}S(t)u\rangle\,\mathrm{d}t = \langle u, \mathscr{K}u\rangle \geq 0 \quad \forall u \in \mathscr{X},$$

and it now only remains to show that $\mathscr{K}$ is a unique linear bounded and selfadjoint operator satisfying (6.2.16). To prove this assume that there exists another operator $\hat{\mathscr{K}} \in \mathscr{L}(\mathscr{X})$, $\hat{\mathscr{K}} = \hat{\mathscr{K}}^*$, which satisfies (6.2.16). Let $\hat{\mathscr{K}} = \mathscr{K} + \Delta$. By virtue of (6.2.16)

$$\langle \mathscr{A}S(t)u, \Delta S(t)u\rangle + \langle S(t)u, \Delta\mathscr{A}S(t)u\rangle = 0 \quad \forall t \geq 0.$$

This relation can be rewritten as

$$\frac{\mathrm{d}}{\mathrm{d}t}\langle S(t)u, \Delta S(t)u\rangle = 0 \quad \forall t \geq 0,$$

but EXS implies $\lim_{t\to\infty} \langle S(t)u, \Delta S(t)u \rangle = 0$, hence

$$\langle u, \Delta u \rangle = 0 \quad \forall u \in \mathscr{D}(\mathscr{A}).$$

As $\mathscr{D}(\mathscr{A})$ is dense in $\mathscr{X}$ and due to the selfadjointness of Δ this yields $\Delta = 0$, uniqueness is proved. □

To complete the proof of necessity in Theorem 6.2.3 it is now sufficient to replace $\mathscr{P}$ by Q in Lemma 6.2.5.

Sufficiency (Theorem 6.2.3) Recall the definition of Lyapunov functional given in Section 6.1. If we there replace the space C by $\mathscr{X}$, ϕ by u and $x_h(\phi)$ by $S(h)u$ we obtain a definition of Lyapunov function for the semigroup $\{S(t)\}_{t \geq 0}$ on a subset of $\mathscr{X}$.

Notice that $V(u) = \langle u, \mathscr{H}u \rangle$ is a Fréchet-differentiable Lyapunov functional on $\mathscr{X}$ for AIVP. Moreover

$$\dot{V}(u) = -\langle u, Qu \rangle \leq -q\|u\|^2 \quad \forall u \in \mathscr{D}(\mathscr{A}).$$

One of Walker's results (1976, Theorem 3.9) yields

$$\dot{V}[S(t)u] - V(u) \leq -q \int_0^t \|S(\tau)u\|^2 d\tau \quad \forall t \geq 0 \quad \forall u \in \mathscr{X}.$$

Hence we easily obtain the estimate

$$\int_0^\infty \|S(t)u\|^2 dt \leq \frac{1}{q} \|\mathscr{H}\| \|u\|^2 \quad \forall u \in \mathscr{X}.$$

Now the proof can be completed by applying the following lemma.

Lemma 6.2.6 *Let $\mathscr{X}$ and $\mathscr{A}$ be as in Theorem 6.2.2. Assume that for the semigroup $\{S(t)\}_{t \geq 0}$ generated by $\mathscr{A}$ the following condition holds*:

$$\exists K > 0: \int_0^\infty \|S(t)u\|^2 dt \leq K\|u\|^2 \quad \forall u \in \mathscr{X}. \tag{6.2.17}$$

Then the semigroup is EXS.

Proof of Lemma 6.2.6 By the assumptions there are $\omega \geq 0$, $M \geq 1$ such that

$$\|S(t)\| \leq Me^{\omega t} \quad \forall t \geq 0. \tag{6.2.18}$$

Therefore, due to Definition 6.2.1

6.2 Stability Theory of Abstract Linear Systems

$$\int_0^\infty \|S(t)u\|^2 dt \geq \int_0^t \|S(t-\xi)u\|^2 d\xi \geq \int_0^t \frac{\|S(\xi)\|^2}{M^2 e^{2\omega\xi}} \|S(t-\xi)u\|^2 d\xi$$

$$\geq \frac{\|S(t)u\|^2}{M^2} \int_0^t e^{-2\omega\xi} d\xi = \frac{\|S(t)u\|^2}{M^2} \frac{1-e^{-2\omega t}}{2\omega}. \quad (6.2.19)$$

From this and (6.2.17)

$$\|S(t)u\|^2 \leq \frac{2\omega K M^2}{1-e^{-2\omega t}} \|u\|^2 \quad \forall t > 0$$

and so

$$\|S(t)\|^2 \leq \frac{2\omega K M^2}{1-e^{-2\omega t}} \quad \forall t > 0. \quad (6.2.20)$$

Comparison of (6.2.18) with (6.2.20) gives

$$\|S(t)\|^2 \leq M^2(1+2\omega K) \quad \forall t \geq 0. \quad (6.2.21)$$

Now we repeat the estimation as in (6.2.19), but using (6.2.21) instead of (6.2.18). Thus

$$\int_0^\infty \|S(t)u\|^2 dt \geq \int_0^t \frac{\|S(\xi)\|^2}{M^2(1+2\omega K)} \|S(t-\xi)u\|^2 d\xi \geq \frac{\|S(t)u\|^2}{M^2(1+2\omega K)} t$$

and by virtue of (6.2.17)

$$\|S(t)\|^2 \leq \frac{1}{t} K M^2(1+2\omega K) \quad \forall t > 0.$$

On the other hand

$$\|S(t)\|^2 \geq e^{2\omega_0 t} \quad \forall t \geq 0$$

where ω_0 is the type of the semigroup $\{S(t)\}_{t \geq 0}$ defined by (6.2.10). But the inequality $\omega_0 < 0$ is a necessary and sufficient condition of EXS (see Theorem 6.2.2). □

Remark 6.2.2 In Theorem 6.2.3 the statement '$\{S(t)\}_{t \geq 0}$ is EXS' may be replaced by '$\{S(t)\}_{t \geq 0}$ is L^2-stable' or by '$\|S(t)\| \underset{t \to \infty}{\to} 0$'.

The proof involves only slight modifications of the proofs of Lemmas 6.2.5 and 6.2.6.

A general scheme for the application of this theory
1. Write the system as an abstract initial-value problem in a Banach or Hilbert space.

2. Check the assumptions of Theorem 6.1.1 to know whether a linear C_0-semigroup is generated.

3. Apply Theorems 6.2.2 or 6.2.3 to obtain the required stability conditions. If one applies Theorem 6.2.2 then one ought to establish whether the properties of the generator spectrum imply exponential stability. This involves a verification of whether the equality in (6.2.12) holds. If one applies Theorem 6.2.3, one ought to solve the operator Lyapunov equation (6.2.14). As a rule the general form of the solution to (6.2.14) can be established by employing the necessary conditions for stability, especially (6.2.15) in a transformed form

$$\langle u, \mathcal{K} u \rangle = \int_0^\infty \langle S(t)u, \mathcal{P}S(t)u \rangle \, dt \quad \forall u \in \mathcal{X}. \qquad (6.2.15')$$

A survey of applications of the above scheme to systems with delays is given in the next section.

6.2.3 Application of the abstract theory to systems with delay

1. *Stability analysis of linear* FDE *in the space* $C = C([-h, 0], R^n)$

We now consider the class of systems described by the following initial-value problem:

$$\dot{x}(t) = L(x_t) \quad t \geq 0$$
$$x_0 = \phi \qquad (6.2.22)$$

where $x_t(\theta) = x(t+\theta)$ for fixed $t \geq 0$, $\theta \in [-h, 0]$; $L: C \to R^n$ is a linear continuous operator. Due to the Riesz theorem on the general form of linear continuous functional defined on C (Balakrishnan, 1976), we may assume with no loss of generality that $L(\phi) = \int_{-h}^{0} d\eta(\theta)\phi(\theta)$, where $\eta(\theta)$ is a real $(n \times n)$-matrix whose elements are functions of bounded variation. The operator L has the norm $\|L\| = \underset{-h}{\text{Var}} \eta$. Introducing the function $\psi(\theta, t) = x(t+\theta)$ we may formally rewrite (6.2.22) as

$$\frac{\partial \psi(\theta, t)}{\partial t} = \frac{\partial \psi(\theta, t)}{\partial \theta}$$

$$\psi(\theta, 0) = \phi(\theta) \quad \text{(initial condition)}$$

$$\frac{\partial \psi(0, t)}{\partial t} = L[\psi(\cdot, t)] \quad \text{(boundary condition)},$$

which can, in turn, be interpreted as an abstract initial-value problem of the form

6.2 Stability Theory of Abstract Linear Systems

$$\dot{\psi} = \mathscr{A}\psi$$
$$\psi(0) = \phi \qquad (6.2.23)$$
$$\mathscr{A}\psi = \psi' \quad \mathscr{D}(\mathscr{A}) = \{\psi \in C : \psi' \in C, \psi'(0) = L(\psi)\}.$$

The operator $\mathscr{A}$ has the following properties

(a) $\overline{\mathscr{D}(\mathscr{A})} = C$,

(b) $\mathscr{R}(I - \lambda\mathscr{A}) = C \quad \forall \lambda \in \left(0, \dfrac{1}{||L||}\right)$,

(c) $\mathscr{A} - ||L||I$ is dissipative in the norm $||\cdot||$.

The proof is left to the reader as an exercise. It was originally given by Webb (1974). It should be noted that the proof of (b) is crucial, and for the proof of (c) (6.2.5') is especially useful as the condition of dissipativity.

One can now establish by Theorem 6.1.1 that operator $\mathscr{A}$ generates a linear C_0-semigroup $\{S(t)\}_{t \geq 0}$ on C, which provides a strong solution to (6.2.23). For the initial conditions taken from $\mathscr{D}(\mathscr{A})$ this strong solution for every t is identical with the segment x_t of the solution of (6.2.22). According to the results of Flaschka and Leitmann (1975) this is also true for the initial conditions from the whole space C. Turning now to the problem of the EXS of the semigroup $\{S(t)\}_{t \geq 0}$, we first observe that by the results of Hale (1977), $S(t)$ is a compact linear operator for $t \geq h$. As the continuous spectrum of a compact operator is empty or $\{0\}$ (Balakrishnan, 1976) we establish that condition (6.2.13) holds, after rescaling of time, thus in turn we have the equality in (6.2.12), more precisely $\omega_0 = \sup\{\operatorname{Re}\lambda : \lambda \in \sigma(\mathscr{A})\}$ where ω_0 is the type of the semigroup $\{S(t)\}_{t \geq 0}$.

By Theorem 6.2.2 the necessary and sufficient condition of the EXS of the semigroup $\{S(t)\}_{t \geq 0}$ is $\sup\{\operatorname{Re}\lambda : \lambda \in \sigma(\mathscr{A})\} < 0$. This condition can be simplified because of the following two facts.

1. The operator $\mathscr{A}$ only has a nonempty point spectrum of the form

$$\sigma(\mathscr{A}) = \left\{\lambda \in C : \det\Delta(\lambda) = 0, \Delta(\lambda) = \lambda I - \int_{-h}^{0} e^{\lambda\theta} d\eta(\theta)\right\}.$$

2. If $\det\Delta(\lambda)$ has an infinite number of roots λ_k then

$$\operatorname{Re}\lambda_k \to -\infty, \quad k \to \infty.$$

1. is obvious and for the proof of 2 observe that by the Gershgorin theorem (Lancaster, 1969) for an arbitrary real matrix $\{\alpha_{ij}\}_{i,j=1,2,\ldots,n}$ we have the following implications:

$$|\alpha_{ii}| > \sum_{j \neq i} |\alpha_{ij}| \quad (i = 1, 2, \ldots, n) \Rightarrow \det\{\alpha_{ij}\} \neq 0.$$

Hence one obtains $\det \Delta(\lambda) \neq 0$ when $|\lambda| > \max\{e^{\operatorname{Re}\lambda}, 1\} \cdot \max_i \sum_j V^o_{-h} \operatorname{ar} \eta_{ij}$, which in turn implies 2. Thus we have' established the following important result.

Corollary 6.2.1 (Hale, 1977) *A necessary and sufficient condition for the semigroup $\{S(t)\}_{t \geq 0}$ as above to be EXS is*

$$\left\{\lambda \in C: \det\Delta(\lambda) = 0, \quad \Delta(\lambda) = \lambda I - \int_{-h}^{0} e^{\lambda \theta} d\eta(\theta)\right\} \subset \{\lambda \in C : \operatorname{Re}\lambda < 0\}. \quad (6.2.24)$$

It turns out that there are some degenerate cases in which $\det\Delta(\lambda)$ only has a finite number of zeros, e.g., when

$$\eta(\theta) = \begin{bmatrix} \begin{cases} 0 & -h \leq \theta < 0 \\ 1 & \theta = 0 \end{cases} & \begin{cases} 0 & \theta = -h \\ 1 & -h < \theta \leq 0 \end{cases} \\ 0 \quad -h \leq \theta \leq 0 & \begin{cases} 0 & -h \leq \theta < 0 \\ 1 & \theta = 0 \end{cases} \end{bmatrix}$$

In this case $\det\Delta(\lambda) = (\lambda - 1)^2$.

The relations between the EXS of the semigroup $\{S(t)\}_{t \geq 0}$ and the properties of the solutions of (6.2.22) are well explained by the following lemma.

Lemma 6.2.7 *Let $\{S(t)\}_{t \geq 0} = \{x_t\}_{t \geq 0}$ be the semigroup described above and let $x(\phi)(\cdot)$ be a solution of (6.2.22). Then a necessary and sufficient condition for $\{S(t)\}_{t \geq 0}$ to be EXS is*

$$\exists M \geq 1, \exists \Omega < 0: |x(\phi)(t)| \leq M e^{\Omega t} \|\phi\| \quad \forall \phi \in C \quad \forall t \geq 0. \quad (6.2.25)$$

Proof. Necessity $|x(\phi)(t)| = |(S(t)\phi)(0)| \leq \max_{\theta \in [-h, 0]} |(S(t)\phi)(\theta)| = \|S(t)\phi\|$
$\leq M e^{\Omega t} \|\phi\| \quad \forall \phi \in C, \forall t \geq 0$, when $M \geq 1$, $\Omega < 0$.

Sufficiency For $t \geq h$ we have by virtue of (6.2.25)

$$\|S(t)\phi\| \leq \max_{\theta \in [-h, 0]} M e^{\Omega(t+\theta)} \|\phi\| = M e^{\Omega(t-h)} \|\phi\|.$$

For $t \in [0, h]$ we have

$$\|S(t)\phi\| \leq \max\{\|\phi\|, \max_{\theta \in [-h, 0]} M e^{\Omega(t+\theta)} \|\phi\|\} = M\|\phi\|.$$

Thus $\|S(t)\| \leq M e^{\Omega(t-h)}$, $\forall t \geq 0$ which ends the proof. $\square$

Some of the results given above can be generalized to $C([-h, 0], X)$ where X is a Banach space (see Dickerson and Gibson, 1976).

6.2 Stability Theory of Abstract Linear Systems

2. Stability analysis of linear functional equations in the space C

As in the previous section $C = C([-h, 0], R^n)$ denotes a Banach space of continuous functions with a norm $\|\phi\| = \max_{\theta \in [-h, 0]} |\phi(\theta)|$. Together with this norm we shall use another equivalent norm $\|\phi\|_\omega = \max_{\theta \in [-h, 0]} |e^{-\omega\theta}\phi(\theta)|$ where $\omega \in R$. This norm will be called Plant's weighted norm (Plant, 1977). Consider the class of systems described by the following initial-value problem

$$\dot{x}(t) = L(x_t) \quad t \geq 0$$
$$x_0 = \phi \in C \quad (6.2.26)$$

where

$$x_t(\theta) = x(t+\theta) \quad \text{for any fixed } t \geq 0 \text{ and } \theta \in [-h, 0]$$

$$L(\phi) = \int_{-h}^{0} d\mu(\theta)\phi(\theta)$$

with

$$\operatorname*{Var}_{-h}^{0} \mu < \infty, \operatorname*{Var}_{-s}^{0} \mu \underset{s \to 0+}{\to} 0. \quad (6.2.27)$$

Condition (6.2.27) indicates that the influence of $\phi(0)$ on the value of $L(\phi)$ is not too strong, e.g., it excludes the case in which $L(\phi) = M\phi(0)$, M is a real $(n \times n)$-matrix. Introducing the function $\psi(\theta, t) = x(t+\theta)$ we may formally rewrite (6.2.26) in the form

$$\frac{\partial \psi(\theta, t)}{\partial t} = \frac{\partial \psi(\theta, t)}{\partial \theta} \quad t \geq 0 \quad \theta \in [-h, 0]$$

$$\psi(\theta, 0) = \phi(\theta) \quad \theta \in [-h, 0] \text{ (initial condition)}$$

$$\psi(0, t) = L[\psi(\cdot, t)] \quad t \geq 0 \text{ (boundary condition)}.$$

We introduce also a Banach space $(C_D, \|\cdot\|)$ (or $(C_D, \|\cdot\|_\omega)$), where $C_D = \{\psi \in C: \psi(0) = L(\psi)\}$. Now we can rewrite the last system as an abstract initial-value problem

$$\dot{\psi} = \mathscr{A}\psi \quad t > 0$$
$$\psi(0) = \phi \in \mathscr{D}(\mathscr{A}) \quad (6.2.28)$$
$$\mathscr{A}\psi = \psi' \quad \mathscr{D}(\mathscr{A}) = \{\phi \in C_D: \phi' \in C_D\}.$$

The operator $\mathscr{A}$ has the following properties:
(a) $\overline{\mathscr{D}(\mathscr{A})} = C_D$,
(b) $\mathscr{R}(I - \lambda\mathscr{A}) = C_D$ for all sufficiently small $\lambda > 0$,
(c) $\exists \omega \in R: \mathscr{A} - \omega I$ is dissipative in Plant's norm $\|\cdot\|_\omega$.

The proof of these properties is rather difficult. It relies on some results of Dyson and Villella-Bresan (1976, 1979).

Due to Theorem 6.2.1 we establish that the operator $\mathscr{A}$ is an infinitesimal generator of a linear C_0-semigroup $\{S(t)\}_{t \geqslant 0}$ on C_D. If $x(\phi)(\cdot)$ is a solution of (6.2.26), then $S(t)\phi = x_t(\phi)$ for $t \geqslant 0$, $\phi \in C_D$. Turning now to the problem of the determination of the necessary and sufficient conditions for the EXS of this semigroup, we recall the result of Henry (1974) (Hale, 1977). Henry proved that if the matrix function μ in the operator L does not possess a singular part, i.e. L can be represented in the form

with
$$L(\phi) = \sum_{k=1}^{\infty} C_k \phi(-h_k) + \int_{-h}^{0} C(\theta) \phi(\theta) d\theta$$
$$0 < h_k \leqslant h \quad \sum_{k=1}^{\infty} |C_k| + \int_{-h}^{0} |C(\theta)| d\theta < \infty$$
(6.2.29)

then condition (6.2.13) is fulfilled. Thus, by Lemma 6.2.3 we have

$$\omega_0 = \sup\{\text{Re}\,\lambda: \lambda \in \sigma(\mathscr{A})\} = \sup\left\{\text{Re}\,\lambda: \det\left[I - \int_{-h}^{0} e^{\lambda\theta} d\mu(\theta)\right] = 0\right\}$$

where ω_0 is the type of the semigroup $\{S(t)\}_{t \geqslant 0}$. In this way Theorem 6.2.2 implies the following corollary.

Corollary 6.2.2 If condition (6.2.29) holds then the necessary and sufficient condition for the EXS of the semigroup $\{S(t)\}_{t \geqslant 0}$ is

$$\sup\left\{\text{Re}\,\lambda: \det\left[I - \int_{-h}^{0} e^{\lambda\theta} C(\theta) d\theta - \sum_{k=1}^{\infty} e^{-h_k \lambda} C_K\right] = 0\right\} < 0. \quad (6.2.30)$$

Lemma 6.2.7 is also true for the semigroup $\{S(t)\}_{t \geqslant 0}$ and the solution $x(\phi)(\cdot)$ of (6.2.26).

3. *Stability analysis of linear neutral functional-differential equations on the space* $C = C([-h, 0], R^n)$

Here we will also use Plant's weighted norm $\|\cdot\|_\omega$ equivalent to $\|\cdot\|$

Consider the class of linear functional differential equations described by the following initial-value problem:

$$\frac{d}{dt} Dx_t = Lx_t \quad t \geqslant 0 \quad (6.2.31)$$
$$x_0 = \phi \in C$$

6.2 Stability Theory of Abstract Linear Systems

where $x_t(\theta) = x(t+\theta)$ for a fixed $t \geq 0$, $\theta \in [-h, 0]$, $D\phi = \phi(0) - \int_{-h}^{0} d\mu(\theta)\phi(\theta)$,

$\underset{-h}{\overset{0}{\text{Var}}}\,\mu < \infty$, $\underset{-s}{\overset{0}{\text{Var}}}\,\mu \underset{s \to 0+}{\to} 0$, $L\phi = \int_{-h}^{0} d\eta(\theta)\phi(\theta)$, $\underset{-h}{\overset{0}{\text{Var}}}\,\eta < \infty$. Introducing the function $\psi(\theta, t) = x(t+\theta)$ we may formally rewrite (6.2.31) in the form

$$\frac{\partial \psi(\theta, t)}{\partial t} = \frac{\partial \psi(\theta, t)}{\partial \theta} \qquad t \geq 0 \quad \theta \in [-h, 0]$$

$$\psi(\theta, 0) = \phi(\theta) \qquad \theta \in [-h, 0] \text{ (initial condition)}$$

$$D\left[\frac{\partial}{\partial t}\psi(\cdot, t)\right] = L[\psi(\cdot, t)] \qquad t \geq 0.$$

This problem may in turn be rewritten as the abstract initial-value problem

$$\dot{\psi} = \mathscr{A}\psi$$
$$\psi(0) = \phi \in \mathscr{D}(\mathscr{A})$$
$$\mathscr{A}\psi = \psi' \quad \mathscr{D}(\mathscr{A}) = \{\psi \in C \colon \psi' \in C, \, D(\psi') = L(\psi)\} \qquad (6.2.32)$$

The operator $\mathscr{A}$ has the following properties:
(a) $\overline{\mathscr{D}(\mathscr{A})} = C$,
(b) $\mathscr{R}(I - \lambda\mathscr{A}) = C$ for every sufficiently small $\lambda > 0$,
(c) $\exists \omega \in R\colon \mathscr{A} - \omega I$ is dissipative in the norm $\|\cdot\|_\omega$.

The proof of these properties is a combination of the proofs of similar properties of the operators appearing in (6.2.22) and (6.2.26).

By Theorem 6.2.1 it follows that the operator $\mathscr{A}$ is an infinitesimal generator of a linear C_0-semigroup $\{S(t)\}_{t \geq 0}$ on C. The relationship between the semigroup $\{S(t)\}_{t \geq 0}$ and the solution $x(\phi)(\cdot)$ of problem (6.2.31) is $S(t)\phi = x_t(\varphi)$. Here the solution $x(\phi)(\cdot)$ is understood as a continuous function belonging to $C([-h, +\infty), R^n)$, such that the term $D[x_t(\phi)]$ is differentiable, coinciding on $[-h, 0]$ with ϕ and satisfying (6.2.31) for $t \geq 0$.

Henry (1974) has proved that if the operator $\phi(0) - D\phi$ is equal to the right-hand side of (6.2.29) and condition (6.2.30) holds, then (6.2.13) is fulfilled. Thus

$$\omega_0 = \sup\{\operatorname{Re}\lambda\colon \lambda \in \sigma(\mathscr{A})\}$$
$$= \sup\left\{\operatorname{Re}\lambda\colon \det\left[\lambda I - \lambda \int_{-h}^{0} e^{\lambda\theta} d\mu(\theta) - \int_{-h}^{0} e^{\lambda\theta} d\eta(\theta)\right] = 0\right\},$$

where ω_0 is the type of the semigroup $\{S(t)\}_{t \geq 0}$. By Theorem 6.2.2 this yields the following result.

Corollary 6.2.3 *If conditions (6.2.29) and (6.2.30) hold, then a necessary and sufficient condition for the EXS of the semigroup $\{S(t)\}_{t\geq 0}$ is*

$$\sup\left\{\operatorname{Re}\lambda:\ \det\left[\lambda I-\lambda\int_{-h}^{0}e^{\lambda\theta}C(\theta)\,d\theta-\lambda\sum_{k=1}^{\infty}e^{-\lambda h_k}C_k-\int_{-h}^{0}e^{\lambda\theta}d\eta(\theta)\right]=0\right\}<0.$$

(6.2.33)

It is not difficult to see that the result of Lemma 6.2.7 remains true for the semigroup $\{S(t)\}_{t\geq 0}$ and the solution $x(\phi)(\cdot)$ of (6.2.31).

4. Stability analysis of a class of linear functional-differential equations on the space H (Delfour's theory)

Let $L^2 = L^2(-h, 0; R^n)$ be a Hilbert space of Lebesgue square-integrable functions defined on $[-h, 0]$ taking values in R^n with an inner product $(\psi_1|\psi_2) = \int_{-h}^{0} \psi_1(\theta)^T \psi_2(\theta)\,d\theta$. Note that functions equal to each other almost everywhere are not distinguished. Define the space $H = R^n \times L^2$. This is a Hilbert space with the inner product

$$(u_1|u_2) = x_1^T x_2 + \int_{-h}^{0} \psi_1(\theta)^T \psi_2(\theta)\,d\theta \quad u_i = (x_i, \psi_i) \in H \quad (i = 1, 2).$$

Consider the following class of linear functional differential equations:

$$\dot{x}(t) = \sum_{i=0}^{N} A_i x(t-h_i) + \int_{-h}^{0} A(\theta) x(t+\theta)\,d\theta \quad t \geq 0$$

$$x(0) = x_0$$

$$x(\theta) = \phi(\theta) \quad \theta \in [-h, 0]$$

(6.2.34)

where

$$(x_0, \phi) \in H; A_i \in \mathscr{L}(R^n) \quad \forall i = 0, 1, 2, \ldots, N$$
$$-h = -h_N < -h_{N-1} < -h_{N-2} < \ldots < -h_1 < h_0 = 0 \quad h \in (0, +\infty),$$
$$A \in L^2(-h, 0; \mathscr{L}(R^n)).$$

This class of linear FDE on H was introduced by Delfour. Putting $\psi(\theta, t) = x(t+\theta)$ we may formally rewrite (6.2.34) in the form

$$\dot{x}(t) = \sum_{i=0}^{N} A_i \psi(-h_i, t) + \int_{-h}^{0} A(\theta)\psi(\theta, t)\,d\theta \quad t \geq 0$$

$$\frac{\partial \psi(\theta, t)}{\partial t} = \frac{\partial \psi(\theta, t)}{\partial \theta} \quad -h \leq \theta \leq 0 \quad t \geq 0$$

$$x(0) = x_0 \quad \psi(\theta, 0) = \phi(\theta) \quad -h \leq \theta \leq 0 \quad \text{(initial conditions)}$$

$$\psi(0, t) = x(t) \quad t \geq 0 \quad \text{(boundary condition).}$$

6.2 Stability Theory of Abstract Linear Systems

This can in turn be rewritten as an abstract initial-value problem

$$\dot{u} = \mathcal{A}u \quad t > 0$$
$$u(0) = u_0 \in \mathcal{D}(\mathcal{A})$$

where

$$u = (x, \psi) \in H \quad \mathcal{A}u = \left(\sum_{i=0}^{N} A_i \psi(-h_i) + \int_{-h}^{0} A(\theta)\psi(\theta)\,d\theta, \dot{\psi}\right)$$

$\dot{\psi}$ is a derivative of ψ,

$$\mathcal{D}(\mathcal{A}) = \{u \in H : \psi \in AC([-h, 0], R^n), \dot{\psi} \in L^2(-h, 0; R^n), \psi(0) = x\}$$
$$u_0 = (x_0, \phi) \tag{6.2.35}$$

The operator $\mathcal{A}$ has the following properties:
(i) $R(I - \lambda\mathcal{A}) = H$ for all sufficiently small $\lambda > 0$,
(ii) $\exists \omega \in R: \mathcal{A} - \omega I$ is dissipative in the norm

$$\|u\|_e^2 = x^T x + \int_{-h}^{0} \gamma(\theta)\psi(\theta)^T \psi(\theta)\,d\theta \quad u = (x, \psi) \in H$$

$$\gamma(\theta) = \theta + h + i \quad \theta \in [-h_{N-i+1}, -h_{N-i}] \quad (i = 0, 1, 2, \ldots, N).$$

Proof of (i) We want to show that for sufficiently small $\lambda > 0$ and for every $\hat{u} = (\hat{x}, \hat{\psi}) \in H$ the equation

$$u - \lambda\mathcal{A}u = \hat{u}$$

has a solution $u = (x, \psi) \in \mathcal{D}(\mathcal{A})$. We write this equation in the form of a system of equations

$$x - \lambda \sum_{i=0}^{N} A_i \psi(-h_i) - \lambda \int_{-h}^{0} A(\theta)\psi(\theta)\,d\theta = \hat{x}$$

$$\psi - \lambda\dot{\psi} = \hat{\psi}.$$

Since for $u \in \mathcal{D}(\mathcal{A})$ we have $\psi(0) = x$, the solution of the second equation is

$$\psi(\theta) = e^{\frac{\theta}{\lambda}} x + \frac{1}{\lambda} \int_{0}^{0} e^{\frac{\theta-\tau}{\lambda}} \hat{\psi}(\tau)\,d\tau.$$

Substituting this formula into the first equation we obtain

$$\left[I - \lambda \sum_{i=0}^{N} A_i e^{-\frac{h_i}{\lambda}} - \lambda \int_{-h}^{0} A(\theta) e^{\frac{\theta}{\lambda}}\,d\theta\right] x$$

$$= \hat{x} + \int_{-h}^{0} A(\theta) \int_{0}^{0} e^{\frac{\theta-\tau}{\lambda}} \hat{\psi}(\tau)\,d\tau\,d\theta + \sum_{i=0}^{N} A_i \int_{-h_i}^{0} e^{-\frac{h_i+\tau}{\lambda}} \hat{\psi}(\tau)\,d\tau.$$

As
$$\left|\sum_{i=0}^{N} A_i e^{-\frac{h_i}{\lambda}} + \int_{-h}^{0} A(\theta) e^{\frac{\theta}{\lambda}} d\theta\right| \leqslant \varkappa$$

where
$$\varkappa = \sum_{i=0}^{N} |A_i| + \int_{-h}^{0} |A(\theta)| d\theta$$

we establish that for every $\lambda \in (0, \varkappa^{-1})$ the matrix in the left-hand side is nonsingular and so x is determined in an unique way.

Proof of (ii) Directly from the definition of γ, $\gamma(-h) = 1 \leqslant \gamma(\theta) \leqslant h+N = \gamma(0)$, and therefore

$$x^T x + \int_{-h}^{0} \psi(\theta)^T \psi(\theta) d\theta \leqslant x^T x + \int_{-h}^{0} \gamma(\theta) \psi(\theta)^T \psi(\theta) d\theta$$

$$\leqslant (h+N) \left(x^T x + \int_{-h}^{0} \psi(\theta)^T \psi(\theta) d\theta \right).$$

We conclude that $\|\cdot\|_e$ and $\|\cdot\|$ are equivalent norms in H. The norm $\|\cdot\|_e$ satisfies the parallelogram law and this defines a new inner product in H

$$(u_1|u_2)_e = x_1^T x_2 + \int_{-h}^{0} \gamma(\theta) \psi_1(\theta)^T \psi_2(\theta) d\theta \quad u_i = (x_i, \psi_i) \in H.$$

According to (6.2.1′) an operator $\mathscr{A} - \omega I$ is dissipative for some $\omega \in R$ if $(u|\mathscr{A}u)_e \leqslant \omega \|u\|_e^2 \; \forall u \in \mathscr{D}(\mathscr{A})$. It remains to prove that such a number ω in fact exists.

$$u \in \mathscr{D}(\mathscr{A}) \Rightarrow (u|\mathscr{A}u)_e = \sum_{i=0}^{N} x^T A_i \psi(-h_i) + \int_{-h}^{0} x^T A(\theta) \psi(\theta) d\theta +$$

$$+ \int_{-h}^{0} \gamma(\theta) \psi(\theta)^T \dot{\psi}(\theta) d\theta$$

$$= x^T A_0 x + \tfrac{1}{2}(h+N) x^T x + \sum_{i=1}^{N} [x^T A_i - \tfrac{1}{2}\psi(-h_i)^T] \psi(-h_i) +$$

$$+ \int_{-h}^{0} [x^T A(\theta) - \psi(\theta)^T] \psi(\theta) d\theta \leqslant \omega |x|^2 \leqslant \omega \|u\|_e^2$$

with
$$\omega = |A_0| + \tfrac{1}{2}(h+N) + \tfrac{1}{2} \sum_{i=1}^{N} |A_i|^2 + \tfrac{1}{2} \int_{-h}^{0} |A(\theta)|^2 d\theta.$$

Thus the operator $\mathscr{A} - \omega I$ is dissipative in the norm $\|\cdot\|_e$. □

6.2 Stability Theory of Abstract Linear Systems

By Lemma 6.2.1 and Theorem 6.2.1 we now establish that the generator $\mathscr{A}$ is an infinitesimal generator of a linear C_0-semigroup $\{T(t)\}_{t \geq 0}$ on H, $T(t)u = (x(t), x_t)$, $\forall t \geq 0$, $\forall u \in H$. The R^n-component of this pair is the solution of (6.2.34) on $[0, +\infty)$ understood in the following sense. For every $T > 0$ the function $x \in L^2(-h, T; R^n)$ is absolutely continuous on $[0, T]$ and satisfies (6.2.34) almost everywhere on $[0, T]$, moreover $x(0) = x_0$, $x(\theta) = \phi(\theta)$ almost everywhere on $[-h, 0]$. In turn, the L^2-component is the segment x_t, $x_t(\theta) = x(t+\theta)$, $\theta \in [-h, 0]$ of this solution for $t \geq 0$.

The semigroup $\{T(t)\}_{t \geq 0}$ is compact for each $t \geq h$ (Delfour and Mitter, 1972). Hence after an appropriate time rescaling we conclude that $T(1) \in \mathscr{L}(H)$ is a compact operator, but this means that the continuous spectrum of $T(1)$ is empty or $\{0\}$, thus Condition (6.2.13) is fulfilled. Now, due to Lemma 6.2.3

$$\omega_0 = \sup\{\operatorname{Re}\lambda : \lambda \in \sigma(\mathscr{A})\}$$

$$= \sup\left\{\operatorname{Re}\lambda : \det\left[\lambda I - \sum_{i=0}^{N} e^{-\lambda h_i} A_i - \int_{-h}^{0} A(\theta)e^{\lambda\theta}d\theta\right] = 0\right\}.$$

Applying Theorem 6.2.2 and the considerations preceding Corollary 6.2.1 we obtain the following corollary.

Corollary 6.2.4 *A necessary and sufficient condition for the EXS of the semigroup* $\{T(t)\}_{t \geq 0}$ *is*

$$\left\{\lambda \in C : \det\left[\lambda I - \sum_{i=0}^{N} e^{-\lambda h_i} A_i - \int_{-h}^{0} A(\theta)e^{\lambda\theta}d\theta\right] = 0\right\} \subset \{\lambda \in C : \operatorname{Re}\lambda < 0\}. \tag{6.2.36}$$

Notice that this result is quite analogous to that of Corollary 6.2.1.

It turns out that a result similar to that of Lemma 6.2.7 holds, as can be seen in what follows.

Lemma 6.2.8 (Delfour, McCalla and Mitter, 1975) *Let* $\{T(t)\}_{t \geq 0}$ *be the semigroup described above and let* $x(u)(\cdot)$ *be a solution to* (6.2.34) *for* $u = (x_0, \phi) \in H$. *Then a necessary and sufficient condition for the (EXS) of the semigroup* $\{T(t)\}_{t \geq 0}$ *is* $\exists M \geq 1, \exists \Omega < 0$ *such that*

$$|x(u)(t)| \leq Me^{\Omega t}\|u\| \quad \forall u \in H \quad \forall t \geq 0. \tag{6.2.37}$$

Proof. Necessity $|x(u)(t)| \leq \|T(t)u\| \leq Me^{\Omega t}\|u\|$ $\forall u \in H$, $\forall t \geq 0$ where $M \geq 1$, $\Omega < 0$.

Sufficiency If $t \in [0, h]$, $u \in \mathscr{D}(\mathscr{A})$ then

$$\|T(t)u\|^2 = |x(u)(t)|^2 + \int_{-h}^{-t} |\phi(t+\theta)|^2 d\theta + \int_{-t}^{0} |x(u)(t+\theta)|^2 d\theta \leq M_1^2 \|u\|^2$$

where
$$M_1^2 = M^2 + 1 + M^2 \frac{e^{2\Omega h} - 1}{2\Omega}.$$

For $t \geq h$, $u \in \mathcal{D}(\mathcal{A})$ relation (6.2.37) gives
$$\|T(t)u\|^2 = |x(u)(t)|^2 + \int_{-h}^{0} |x(u)(t+\theta)|^2 d\theta \leq M_2^2 \|u\|^2 e^{2\Omega t}$$
where
$$M_2^2 = M^2 \frac{2\Omega + 1 - e^{-2\Omega h}}{2\Omega}.$$

Combining both estimates one obtains
$$\|T(t)u\|^2 \leq M_3^2 e^{2\Omega t} \|u\|^2 \quad \forall t \geq 0 \quad \forall u \in \mathcal{D}(\mathcal{A})$$
$$M_3^2 = \max\{M_1^2, M_2^2 e^{-2\Omega h}\} \quad M_3 \geq 1.$$

By the continuity of $T(t)$ for every fixed $t \geq 0$, continuity of norm $\|\cdot\|$ and due to the density of $\mathcal{D}(\mathcal{A})$ in H this yields the estimate
$$\|T(t)\| \leq M_3 e^{\Omega t} \quad \forall t \geq 0. \quad \square$$

System (6.2.35) admits an essential simplification of Theorem 6.2.3 because we have the following lemma.

Lemma 6.2.9 Let $\{T(t)\}_{t \geq 0}$ be the semigroup generated by the operator $\mathcal{A}$ as in (6.2.35), and Q be a real $(n \times n)$-matrix, $Q = Q^T > 0$. Then $\{T(t)\}_{t \geq 0}$ is EXS if and only if there exists an $\mathcal{H} \in \mathcal{L}(H)$, $\mathcal{H} = \mathcal{H}^*$, $(u|\mathcal{H}u) \geq 0$ $\forall u \in H$, such that
$$(\mathcal{A}u|\mathcal{H}u) + (u|\mathcal{H}\mathcal{A}u) = -x^T Q x \quad \forall u = (x, \psi) \in \mathcal{D}(\mathcal{A}) \subset H. \quad (6.2.38)$$

Proof. *Necessity* The thesis is a corollary of Lemma 6.2.5 after substituting $\mathcal{P}u = (Qx, 0)$ in it; it may additionally be observed that $(u|\mathcal{H}u) > 0$ for every $u \in H$ of the form $u = (x, 0)$, $x \neq 0$.

Sufficiency As in the proof of Theorem 6.2.3 we get the inequality
$$\int_0^\infty |x(u)(t)|^2 dt \leq c\|u\|^2 \quad \forall u \in \mathcal{D}(\mathcal{A}) \text{ (even on } H)$$

for some $c > 0$. Hence by the Fubini–Tonelli theorem
$$u \in \mathcal{D}(\mathcal{A}) \Rightarrow \int_0^\infty \|T(t)u\|^2 dt$$
$$= \int_0^\infty |x(u)(t)|^2 dt + \int_0^\infty \left(\int_{-h}^{0} |x(u)(t+\theta)|^2 d\theta \right) dt$$

6.2 Stability Theory of Abstract Linear Systems

$$\leqslant c||u||^2 + \int_0^\infty \int_\tau^{\tau+h} |x(u)(t)|^2 dt\, d\tau\, c||u||^2 + h \int_0^\infty |x(u)(t)|^2 dt$$

$$+ h \int_{-h}^0 |x(u)(t)|^2 dt \leqslant (c + hc + h)||u||^2 < \infty.$$

This estimate is true also on H. Finally by Lemma 6.2.6 it follows that $\{T(t)\}_{t \geqslant 0}$ is EXS. □

We claim now that the solution of problem (6.2.34) can be written in the form

$$x(t) = X(t)x_0 + \int_{-h}^0 \Psi(t, \theta)\phi(\theta)d\theta \qquad t \geqslant 0 \qquad (6.2.39)$$

where the $(n \times n)$-matrix $X(t)$ is a solution of the problem

$$\dot{X}(t) = \sum_{i=0}^N X(t - h_i) A_i + \int_{-h}^0 X(t + \theta) A(\theta) d\theta \qquad t > 0$$

$$X(0) = I \qquad (6.2.40)$$

$$X(t) = 0 \qquad t < 0$$

and

$$\Psi(t, \theta) = \sum_{i=1}^N \begin{cases} X(t - \theta - h_i) A_i & t \geqslant \theta + h_i \geqslant 0 \\ 0 & \text{otherwise} \end{cases} + \int_{\max\{-h, \theta - t\}}^\theta X(t - \theta + \xi) A(\xi) d\xi$$

(6.2.41)

We prove formulas (6.2.39)–(6.2.41) by the Laplace transform technique. After its application to (6.2.40) we easily obtain

$$\hat{X}(s) = \left[sI - \sum_{i=0}^N e^{-sh_i} A_i - \int_{-h}^0 e^{s\theta} A(\theta) d\theta \right]^{-1}$$

for the Laplace transform of $X(\cdot)$.

On the other hand, elementary calculations give the Laplace transform of the solution of (6.2.34)

$$\hat{x}(s) = \hat{X}(s) x_0 + \sum_{i=1}^N \int_{-h_i}^0 e^{-s(\theta + h_i)} \hat{X}(s) A_i \phi(\theta) d\theta$$

$$+ \int_{-h}^0 \hat{X}(s) A(\xi) \left(\int_\xi^0 \phi(\theta) e^{-s(\theta - \xi)} d\theta \right) d\xi.$$

Now we apply the Fubini–Tonelli theorem to the last term whence

$$x(t) = X(t)x_0 + \sum_{i=1}^{N} \int_{-h_i}^{0} X(t-\theta-h_i)A_i\phi(\theta)\,d\theta$$

$$+ \int_{-h}^{0} \left[\int_{-h}^{\theta} X(t-\theta+\xi)A(\xi)\,d\xi\right] \phi(\theta)\,d\theta = X(t)x_0 +$$

$$+ \sum_{i=1}^{N} \left\{ \begin{array}{ll} \int_{-h}^{0} X(t-\theta-h_i)A_i\phi(\theta)\,d\theta & t \geqslant \theta + h_i \geqslant 0 \\ 0 & \text{otherwise} \end{array} \right\} +$$

$$+ \int_{-h}^{0} \int_{\max\{-h,\theta-t\}}^{\theta} X(t-\theta+\xi)A(\xi)\phi(\theta)\,d\xi\,d\theta = X(t)x_0 + \int_{-h}^{0} \Psi(t,\theta)\phi(\theta)\,d\theta.$$

We have already accomplished a characterization of the EXS of the semigroup $\{T(t)\}_{t \geqslant 0}$ in terms of the spectrum $\sigma(\mathcal{A})$ of the generator $\mathcal{A}$—see Corollary 6.2.4. Now we shall characterize the (EXS) differently, employing Lemma 6.2.9 and formulas (6.2.39)–(6.2.41). Assume for the sake of simplicity $Q = I$ in Lemma 6.2.9 and let $\{T(t)\}_{t \geqslant 0}$ be EXS. According to the proofs of Lemmas 6.2.5 and 6.2.9 the unique selfadjoint solution to the Lyapunov operator equation (6.2.38) is an operator $\mathcal{H} \in \mathcal{L}(H)$, $\mathcal{H} = \mathcal{H}^*$ such that

$$(u|\mathcal{H}u) = \int_{0}^{\infty} x(t)^{\mathrm{T}} x(t)\,dt \quad u = (x_0, \phi) \in H.$$

Keeping in mind (6.2.39) and changing the order of integration according to the Fubini–Tonelli theorem we obtain, after some elementary transformations

$$(u|\mathcal{H}u) = \left((x_0, \phi) \Big| \left(H^{00}x_0 + \int_{-h}^{0} H^{10}(\theta)^{\mathrm{T}}\phi(\theta)\,d\theta, H^{10}(\cdot)x_0 \right.\right.$$

$$\left.\left. + \int_{-h}^{0} H^{11}(\cdot, \sigma)\phi(\sigma)\,d\sigma\right)\right)$$

where

$$H^{00} = \int_{0}^{\infty} X(t)^{\mathrm{T}} X(t)\,dt \qquad (6.2.42)$$

$$H^{10}(\theta) = \int_{0}^{\infty} \Psi(t,\theta)^{\mathrm{T}} X(t)\,dt \quad \theta \in [-h, 0] \qquad (6.2.43)$$

$$H^{11}(\theta, \sigma) = \int_{0}^{\infty} \Psi(t,\theta)^{\mathrm{T}} \Psi(t,\sigma)\,dt \quad \theta, \sigma \in [-h, 0]. \qquad (6.2.44)$$

6.2 Stability Theory of Abstract Linear Systems

Due to the fact that $\mathcal{H}$ is selfadjoint we get

$$\mathcal{H}u = \left(H^{00}x_0 + \int_{-h}^{0} H^{10}(\theta)^T\phi(\theta)\,d\theta,\; H^{10}(\cdot)x_0 + \int_{-h}^{0} H^{11}(\cdot,\sigma)\phi(\sigma)\,d\sigma\right)$$

$$u = (x_0, \phi) \in H. \tag{6.2.45}$$

Notice that the function Ψ in (6.2.41) may only have discontinuities of the form

$$\begin{aligned}\Psi(t, (-h_i)+) - \Psi(t, (-h_i)-) &= X(t)A_i \\ \Psi(t, (t-h_i)+) - \Psi(t, (t-h_i)-) &= -A_i\end{aligned} \quad (i = 1,\ldots,N) \quad t \geq 0 \tag{6.2.46}$$

In consequence, the function H^{10} is piecewise absolutely continuous with possible jumps only at $\theta = -h_i$, $i = 1, \ldots, N-1$

$$H^{10}((-h_i)+) - H^{10}((-h_i)-) = A_i^T H^{00}. \tag{6.2.47}$$

Since $\Psi(t, -h) = X(t)A_N$, we have from (6.2.43)

$$H^{10}(-h) = A_N^T H^{00}. \tag{6.2.48}$$

Let us now consider the function H^{11}. It immediately follows from (6.2.44) that

$$H^{11}(\theta, \sigma) = H^{11}(\sigma, \theta)^T \quad \theta, \sigma \in [-h, 0]. \tag{6.2.49}$$

The function $\theta \mapsto H^{11}(\theta, \sigma)$ may only have discontinuities at $\theta = -h_i$, $i = 1, \ldots, N-1$

$$H^{11}((-h_i)+, \sigma) - H^{11}((-h_i)-, \sigma) = A_i^T H^{10}(\sigma)^T \quad \sigma \in [-h, 0]. \tag{6.2.50}$$

The boundary value of this function is easily calculated as

$$H^{11}(-h, \sigma) = A_N^T H^{10}(\sigma)^T \quad \sigma \in [-h, 0]. \tag{6.2.51}$$

The jumps of $\sigma \mapsto H^{11}(\theta, \sigma)$ and the boundary value $H^{11}(\theta, -h)$ are given by the symmetry condition (6.2.49).

The operator $\mathcal{H} \in \mathcal{L}(H)$ defined by (6.2.45) satisfies the Lyapunov operator equation (6.2.38) which for the problem under consideration has the form

$$(\mathcal{A}u|\mathcal{H}u) + (u|\mathcal{H}\mathcal{A}u) = -x_0^T x_0 \quad \forall u = (x_0, \phi) \in \mathcal{D}(\mathcal{A}). \tag{6.2.52}$$

On the basis of (6.2.35) and (6.2.45) we obtain an explicit form of this equation:

$$x_0^T(A_0^T H^{00} + H^{00}A_0 + I)x_0 +$$

$$+ \sum_{i=1}^{N} [x_0^T H^{00} A_i \phi(-h_i) + \phi(-h_i)^T A_i^T H^{00} x_0] +$$

$$+ \int_{-h}^{0} [x_0^T H^{10}(\theta)^T \dot\phi(\theta) + \dot\phi(\theta)^T H^{10}(\theta) x_0]\,d\theta +$$

$$+ \int_{-h}^{0} x_0^T [A_0^T H^{10}(\theta)^T + H^{00} A(\theta)] \phi(\theta) d\theta +$$

$$+ \int_{-h}^{0} \phi(\theta)^T [A(\theta)^T H^{00} + H^{10}(\theta) A_0] x_0 d\theta +$$

$$+ \sum_{i=1}^{N} \int_{-h}^{0} [\phi(-h_i)^T A_i^T H^{10}(\theta)^T \phi(\theta) + \phi(\theta)^T H^{10}(\theta) A_i \phi(-h_i)] d\theta +$$

$$+ \int_{-h}^{0} \int_{-h}^{0} [\dot{\phi}(\theta)^T H^{11}(\theta, \sigma) \phi(\sigma) + \dot{\phi}(\sigma)^T H^{11}(\theta, \sigma)^T \phi(\theta)] d\theta d\sigma +$$

$$+ \int_{-h}^{0} \int_{-h}^{0} \phi(\theta)^T [A(\theta)^T H^{10}(\sigma)^T + H^{10}(\theta) A(\sigma)] \phi(\sigma) d\theta d\sigma = 0$$

$$\forall (x_0, \phi) \in \mathscr{D}(\mathscr{A}).$$

Integration by parts and the formulas (6.2.47)–(6.2.51) give after elementary calculations

$$x_0^T G^{00} x_0 + x_0^T \int_{-h}^{0} G^{10}(\theta)^T \phi(\theta) d\theta + \int_{-h}^{0} \phi(\theta)^T G^{10}(\theta) d\theta x_0 +$$

$$+ \int_{-h}^{0} \int_{-h}^{0} \phi(\theta)^T G^{11}(\theta, \sigma) \phi(\sigma) d\theta d\sigma = 0 \quad \forall (x_0, \phi) \in \mathscr{D}(\mathscr{A}) \quad (6.2.53)$$

where

$$G^{00} = A_0^T H^{00} + H^{00} A_0 + H^{10}(0) + H^{10}(0)^T + I$$

$$G^{10}(\theta) = -\frac{dH^{10}(\theta+)}{d\theta} + H^{10}(\theta) A_0 + A(\theta)^T H^{00} + H^{11}(\theta, 0) \quad (6.2.54)$$

$$G^{11}(\theta, \sigma) = \frac{\partial H^{11}(\theta+, \sigma)}{\partial \theta} + \frac{\partial H^{11}(\theta, \sigma+)}{\partial \sigma} + A(\theta)^T H^{10}(\sigma)^T + H^{10}(\theta) A(\sigma).$$

The derivatives are right-hand derivatives. Notice that these are well defined in the classical non-distributional sense for every $\theta, \sigma \in [-h, 0]$. The left hand side of (6.2.53) is a quadratic form defined on $\mathscr{D}(\mathscr{A})$. However, it is also well defined and continuous on the whole space H. Since $\mathscr{D}(\mathscr{A})$ is dense in H equation (6.2.53) is valid on all elements of H. Thus x_0 and ϕ may be chosen independently and, since $G^{00} = (G^{00})^T$ and $G^{11}(\theta, \sigma) = G^{11}(\sigma, \theta)^T$ we finally obtain

$$G^{00} = 0 \quad G^{10}(\theta) = 0 \quad G^{11}(\theta, \sigma) = 0 \quad \theta, \sigma \in [-h, 0]. \quad (6.2.55)$$

6.2 Stability Theory of Abstract Linear Systems

Summarizing what we have said so far, if a semigroup $\{T(t)\}_{t \geq 0}$ generated by $\mathcal{A}$ of the form (6.2.35) is EXS then for $Q = I \in \mathcal{L}(R^n)$ the unique solution $\mathcal{H} \in \mathcal{L}(H)$, $\mathcal{H} = \mathcal{H}^*$, $(u|\mathcal{H}u) \geq 0 \; \forall u \in H$ of the Lyapunov equation (6.2.38) has the form

$$\mathcal{H}u = \left(H^{00}x + \int_{-h}^{0} H^{10}(\theta)^T\psi(\theta)\,d\theta, H^{10}(\theta)x + \int_{-h}^{0} H^{11}(\theta,\sigma)\psi(\sigma)\,d\sigma\right),$$

$$u = (x,\psi) \in H$$

where H^{00}, $H^{10}(\theta)$, $H^{11}(\theta,\sigma)$ are the $(n \times n)$-matrices with the properties described above. Delfour was the first to obtain this result (Delfour, 1972; see also Delfour et al., 1975).

It is easy to see that this result admits an inversion in the following sense. If there exist matrices B^{00}, $B^{10}(\theta)$, $B^{11}(\theta,\sigma)$ with properties as those of matrices H^{00}, $H^{10}(\theta)$, $H^{11}(\theta,\sigma)$ respectively, and furthermore if for the operator defined by

$$\mathcal{H}u = \left(B^{00}x + \int_{-h}^{0} B^{10}(\theta)^T\psi(\theta)\,d\theta, B^{10}(\theta)x + \int_{-h}^{0} B^{11}(\theta,\sigma)\psi(\sigma)\,d\sigma\right),$$

$u = (x,\psi) \in H$ we have $(u|\mathcal{H}u) \geq 0 \; \forall u \in H$, then the semigroup $\{T(t)\}_{t \geq 0}$ is EXS. Indeed, the operator $\mathcal{H}$ is a solution of (6.2.38). This can be established similarly as for the operator $\mathcal{H}$ given by (6.2.45). Moreover $\mathcal{H}$ has the properties described in Lemma 6.2.9. Thus we have proved the following lemma.

Lemma 6.2.10 *A necessary and sufficient condition for the semigroup $\{T(t)\}_{t \geq 0}$ as above to be EXS, is the existence of H^{00}, H^{10}, H^{11} such that:*

(i) $H^{00} \in \mathcal{L}(R^n)$, $H^{00} = (H^{00})^T$; $H^{10}: [-h,0] \ni \theta \mapsto H^{10}(\theta) \in \mathcal{L}(R^n)$—*a piecewise absolutely continuous function, having jumps only at the points $\theta = -h_i$, $i = 1,2,\ldots,N-1$ of height $A_i^T H^{00}$, respectively; $H^{11}: [-h,0] \times [-h,0] \ni (\theta,\sigma) \mapsto H^{11}(\theta,\sigma) \in \mathcal{L}(R^n)$, $H^{11}(\theta,\sigma) = H^{11}(\sigma,\theta)^T \in \mathcal{L}(R^n)$. For a fixed $\sigma \in [-h,0]$, $H^{11}(\sigma,\cdot)$ is a piecewise absolutely continuous function having jumps only at the points $\theta = -h_i$, $i = 1,2,\ldots,N-1$, of heights $A_i^T H^{10}(\sigma)^T$, respectively. The elements H^{00}, H^{10}, H^{11} satisfy the following system of equations:*

$$A_0^T H^{00} + H^{00}A_0 + H^{10}(0) + H^{10}(0)^T + I = 0 \quad (6.2.56)$$

$$\frac{dH^{10}(\theta)}{d\theta} = A(\theta)^T H^{00} + H^{10}(\theta)A_0 + H^{11}(\theta,0) \quad \theta \in [-h,0]$$

$$\theta \neq -h_i \quad i = 1,\ldots,N-1 \quad H^{10}(-h) = A_N^T H^{00} \quad (6.2.57)$$

$$\frac{\partial H^{11}(\theta,\sigma)}{\partial \theta} + \frac{\partial H^{11}(\theta,\sigma)}{\partial \sigma} + A(\theta)^T H^{10}(\sigma)^T + H^{10}(\theta)A(\sigma) = 0$$

$$(\theta, \sigma) \in [-h, 0] \times [-h, 0] \quad \theta \neq -h_i \quad i = 1, \ldots, N-1 \quad \sigma \neq -h_j$$
$$j = 1, \ldots, N-1,$$
$$H^{11}(-h, \sigma) = A_N^T H^{10}(\sigma)^T \quad H^{11}(\theta, -h) = H^{10}(\theta) A_N. \quad (6.2.58)$$

(ii) *For the operator* $\mathcal{H}$ *defined as follows*:

$$\mathcal{H}u = \left(H^{00}x + \int_{-h}^{0} H^{10}(\theta)^T \psi(\theta) d\theta, \; H^{10}(\cdot)x + \int_{-h}^{0} H^{11}(\cdot, \sigma) \psi(\sigma) d\sigma \right)$$

$$u = (x, \phi) \in H \quad (6.2.59)$$

we have

$$(u|\mathcal{H}u) \geq 0 \quad \forall u = (x, \psi) \in H. \quad (6.2.60)$$

Notice that the boundary-value problem (6.2.58) can be solved (for proof see Delfour *et al.* (1975). The solution has the form

$$H^{11}(\theta, \sigma) = \begin{cases} H^{10}(\theta - \sigma - h) A_N & \theta \geq \sigma \\ A_N^T H^{10}(\sigma - \theta - h)^T & \theta \leq \sigma \end{cases} +$$

$$+ \sum_{i=1}^{N-1} \begin{cases} A_i^T H^{10}(\theta - \sigma - h_i)^T & -h \leq \sigma - \theta - h_i \quad -h_i < \theta \\ 0 & \text{otherwise} \end{cases}$$

$$+ \sum_{j=1}^{N-1} \begin{cases} H^{10}(\theta - \sigma - h_j) A_j & -h \leq \theta - \sigma - h_j \quad -h_j < \sigma \\ 0 & \text{otherwise} \end{cases}$$

$$+ \int_{-h}^{\sigma} \begin{cases} H^{10}(\theta - \sigma + \xi) A(\xi) & -h \leq \theta - \sigma + \xi \\ 0 & \text{otherwise} \end{cases} d\xi +$$

$$+ \int_{-h}^{0} \begin{cases} A(\xi)^T H^{10}(\sigma - \theta + \xi)^T & -h \leq \sigma - \theta + \xi \\ 0 & \text{otherwise} \end{cases} d\xi.$$

The first term corresponds to the solution of a homeogeneous boundary problem (6.2.58) with given boundary conditions, and the remaining terms are a solution of a nonhomeogeneous boundary problem (6.2.58) with null boundary conditions.

Some remarks on the equations of Delfour's theory in the case of a system with one lumped delay

If system (6.2.34) is of a particular form with $N = 1$, $A(\theta) \equiv 0$, i.e.

$$\dot{x}(t) = A_0 x(t) + A_1 x(t-h) \quad (6.2.61)$$

6.2 Stability Theory of Abstract Linear Systems

then the system of equations (6.2.56)–(6.2.58) reduces to the form

$$A_0^T H^{00} + H^{00} A_0 + H^{10}(0) + H^{10}(0)^T + I = 0 \qquad (6.2.62a)$$

$$\frac{dH^{10}(\theta)}{d\theta} = H^{10}(\theta) A_0 + H^{11}(\theta, 0) \qquad \theta \in [-h, 0] \qquad (6.2.62b)$$

$$H^{10}(-h) = A_1^T H^{00}$$

$$\frac{\partial H^{11}(\theta, \sigma)}{\partial \theta} + \frac{\partial H^{11}(\theta, \sigma)}{\partial \sigma} = 0 \qquad (\theta, \sigma) \in [-h, 0] \times [-h, 0] \qquad (6.2.62c)$$

$$H^{11}(-h, \sigma) = A_1^T H^{10}(\sigma)^T \qquad H^{11}(\theta, -h) = H^{10}(\theta) A_1.$$

A general solution of (6.2.62c) without the boundary conditions has the form $H^{11}(\theta, \sigma) = \varphi(\theta - \sigma)$, where φ is an absolutely continuous function. Taking into account the boundary conditions one obtains

$$H^{11}(\theta, \sigma) = \begin{cases} A_1^T H^{10}(\sigma - \theta - h)^T & \theta - \sigma \leq 0 \\ H^{10}(\theta - \sigma - h) A_1 & \theta - \sigma \geq 0. \end{cases} \qquad (6.2.63)$$

In order to determine H^{00} and H^{10} the following boundary problem has to be solved:

$$\frac{dH^{10}(\theta)}{d\theta} = H^{10}(\theta) A_0 + A_1^T H^{10}(-h-\theta)^T \qquad \theta \in [-h, 0] \qquad (6.2.64)$$

$$H^{10}(-h) = A_1^T H^{00} \qquad H^{10}(0) + H^{10}(0)^T = -I - A_0^T H^{00} - H^{00} A_0.$$

This problem can be solved by means of a method that makes use of the results by Castelan and Infante (1977). We substitute

$$\beta(\theta) = H^{10}(-h-\theta)^T \qquad \theta \in [-h, 0] \qquad (6.2.65)$$

into (6.2.64) and get the boundary-value problem

$$\frac{dH^{10}(\theta)}{d\theta} = H^{10}(\theta) A_0 + A_1^T \beta(\theta) \qquad (6.2.66a)$$

$$\frac{d\beta(\theta)}{d\theta} = -H^{10}(\theta) A_1 - A_0^T \beta(\theta) \qquad (6.2.66b)$$

$$H^{10}(0) + H^{10}(0)^T = -I - A_0^T H^{00} - H^{00} A_0 \qquad (6.2.66c)$$

$$\beta(0) = H^{00} A_1. \qquad (6.2.66d)$$

When we have obtained the solution $H^{10} = H^{10}(\theta, H^{00})$ of this system we determine the unknown matrix H^{00} from the equation

$$H^{10}(-h, H^{00}) = H^{00} A_1 = \beta(0). \qquad (6.2.67)$$

The solution of system (6.2.66) can be found in an explicit form by the application of the tensor (Kronecker) product of matrices, defined as follows:

$$C = [c_{kl}]_{\substack{k=1,2,\ldots,p \\ l=1,2,\ldots,q}} \Rightarrow C \otimes D \stackrel{df}{=} [c_{kl} D]_{\substack{k=1,2,\ldots,p \\ l=1,2,\ldots,q}}.$$

Together with the tensor product of matrices we define the operation

$$\operatorname{col} C \stackrel{\mathrm{df}}{=} [c_{11} c_{12} \ldots c_{1q} c_{21} c_{22} \ldots c_{2q} \ldots c_{p1} c_{p2} \ldots c_{pq}]^{\mathrm{T}}.$$

The basic relationship connecting the above operations is (Lancaster, 1969)

$$\operatorname{col}[CYD] = (C \otimes D^{\mathrm{T}}) \operatorname{col} Y. \qquad (6.2.68)$$

Bearing these notions in mind we can rewrite (6.2.66) and (6.2.67) in the form

$$\left. \begin{array}{c} \begin{bmatrix} \dfrac{\mathrm{d}\operatorname{col} H^{10}(\theta)}{\mathrm{d}\theta} \\ \dfrac{\mathrm{d}\operatorname{col} \beta(\theta)}{\mathrm{d}\theta} \end{bmatrix} = \begin{bmatrix} I \otimes A_0^{\mathrm{T}} & A_1^{\mathrm{T}} \otimes I \\ -(I \otimes A_1^{\mathrm{T}}) & -(A_0^{\mathrm{T}} \otimes I) \end{bmatrix} \begin{bmatrix} \operatorname{col} H^{10}(\theta) \\ \operatorname{col} \beta(\theta) \end{bmatrix} \\ \operatorname{col}[H^{10}(0) + H^{10}(0)^{\mathrm{T}}] = -\operatorname{col} I - [(A_0^{\mathrm{T}} \otimes I) + (I \otimes A_0^{\mathrm{T}})] \operatorname{col} H^{00} \\ \operatorname{col} \beta(0) = (I \otimes A_1^{\mathrm{T}}) \operatorname{col} H^{00} \end{array} \right\} \quad (6.2.69)$$

$$\operatorname{col} H^{10}(-h, H^{00}) = (I \otimes A_1^{\mathrm{T}}) \operatorname{col} H^{00} (= \operatorname{col} \beta(0)). \qquad (6.2.70)$$

Various generalizations of the theory above for systems with one lumped delay have been discussed (Infante and Castelan, 1978; Datko, 1980; Castelan and Infante, 1979; Castelan, 1980). The last work discusses the system of equations of Delfour's theory for a system with multiple but commensurate lumped delays. Another way of obtaining the system of equations of Delfour's theory is also presented through the theory of stability of discrete systems. Infante and Slemrod (1972) were the first to use this approach.

5. *Stability analysis for a class of neutral systems on space H*

The definition of space H is given in the previous section. Consider now the following class of neutral systems:

$$\begin{array}{ll} \dot{v}(t) = Ax(t) + (AC + B)x(t-h) & t > 0 \\ v(t) = x(t) - Cx(t-h) & t > 0 \\ v(0) = v_0 \quad x(\theta) = \phi(\theta) \text{ a. e. on } [-h, 0] \\ (v_0, \phi) \in H \quad A, B, C \in \mathscr{L}(R^n) \quad 0 < h < +\infty. \end{array} \qquad (6.2.71)$$

Substituting $\psi(\theta, t) = x(t+\theta)$ we may formally rewrite (6.2.71) in the form

$$\begin{array}{l} \dot{v}(t) = Av(t) + (AC + B)\psi(-h, t) \quad t \geqslant 0 \\ \dfrac{\partial \psi(\theta, t)}{\partial t} = \dfrac{\partial \psi(\theta, t)}{\partial \theta} \quad t \geqslant 0 \quad -h \leqslant \theta \leqslant 0. \\ v(0) = v_0 \quad \psi(\theta, 0) = \phi(\theta) \quad -h \leqslant \theta \leqslant 0 \text{ (initial conditions)} \\ \psi(0, t) = v(t) - C\psi(-h, t) \quad t \geqslant 0 \text{ (boundary condition).} \end{array} \qquad (6.2.72)$$

6.2 Stability Theory of Abstract Linear Systems

This problem can in turn be rewritten as an abstract initial-value problem

$$\begin{cases} \dot{u} = \mathscr{A}u & t > 0 \\ u(0) = u_0 \in \mathscr{D}(\mathscr{A}) \end{cases} \quad (6.2.73)$$

where $u = (v, \psi) \in H$, $\mathscr{A}u = (Av + (AC+B)\psi(-h), \dot{\psi})$, $\mathscr{D}(\mathscr{A}) = \{u = (v, \psi) \in H: \psi$ is absolutely continuous in $[-h, 0]$, $\dot{\psi} \in L^2(-h, 0; R^n)$, $v = \psi(0) - C\psi(-h)\}$, $u(0) = u_0 = (v_0, \phi)$.

The operator $\mathscr{A}$ in (6.2.73) has the following properties (Grabowski, 1983):
(i) $\mathscr{R}(I - \lambda \mathscr{A}) = H$ for all sufficiently small $\lambda > 0$,
(ii) $\exists \omega \in R: \mathscr{A} - \omega I$ is dissipative in the norm

$$\|u\|_e^2 = v^T v + \int_{-h}^{0} \psi(\theta)^T \left(I - \frac{\theta}{h} C^T C\right) \psi(\theta) \, d\theta, \quad u = (v, \psi) \in H.$$

Notice that this norm is equivalent to $\|\cdot\|$.

Proof of (i) We want to prove that for all sufficiently small $\lambda > 0$ the equation

$$u - \lambda \mathscr{A} u = \hat{u}, \quad \hat{u} = (\hat{v}, \hat{\psi}) \in H$$

has a solution $u \in \mathscr{D}(\mathscr{A})$. The full form of this equation is

$$v - \lambda A v - \lambda (AC+B)\psi(-h) = \hat{v}$$
$$\psi - \lambda \dot{\psi} = \hat{\psi}.$$

The solution of the second equation in the space $W^{1,2}(-h, 0; R^n)$ is

$$\psi(\theta) = e^{\frac{\theta}{\lambda}} \delta + \frac{1}{\lambda} \int_0^{\theta} e^{\frac{\theta - \tau}{\lambda}} \hat{\psi}(\tau) \, d\tau$$

where δ is a vector in R^n. As we seek a solution $u \in \mathscr{D}(\mathscr{A})$, δ is chosen in such a manner that

$$v = \psi(0) - C\psi(-h) = \left(I - e^{-\frac{h}{\lambda}} C\right)\delta + \frac{1}{\lambda} \int_{-h}^{0} e^{\frac{-h-\tau}{\lambda}} \hat{\psi}(\tau) \, d\tau,$$

$$(I - \lambda A)v - \lambda(AC+B)\psi(-h) = \hat{v} = \left(I - \lambda A - e^{-\frac{h}{\lambda}} C - \lambda e^{-\frac{h}{\lambda}} B\right)\delta +$$
$$+ \frac{1}{\lambda}(\lambda B + C) \int_{-h}^{0} e^{\frac{-h-\tau}{\lambda}} \hat{\psi}(\tau) \, d\tau.$$

For all sufficiently small numbers $\lambda > 0$ the matrix $\lambda A + e^{-\frac{h}{\lambda}} C + \lambda e^{-\frac{h}{\lambda}} B$ has norm less than 1 and therefore the equation

$$\left(I - \lambda A - e^{-\frac{h}{\lambda}} C - \lambda e^{-\frac{h}{\lambda}} B\right)\delta = \frac{1}{\lambda}(\lambda B + C) \int_{-h}^{0} e^{\frac{-h-\tau}{\lambda}} \hat{\psi}(\tau) \, d\tau + \hat{v} \quad (6.2.74)$$

has an unique solution δ^*. This in turn means that $u = (v, \psi)$, where

$$v = (I - e^{-\frac{h}{\lambda}} C)\delta^* - \frac{1}{\lambda} C \int_{-h}^{0} e^{\frac{-h-\tau}{\lambda}} \hat{\psi}(\tau) d\tau$$

$$\psi(\theta) = e^{\frac{\theta}{\lambda}} \delta^* + \frac{1}{\lambda} \int_{-h}^{0} e^{\frac{\theta-\tau}{\lambda}} \hat{\psi}(\tau) d\tau \quad (6.2.75)$$

is for all sufficiently small $\lambda > 0$ a solution to the equation $u - \lambda \mathscr{A} u = \hat{u}$, belonging to $\mathscr{D}(\mathscr{A})$.

Proof of (ii) The norm $\|\cdot\|_e$ satisfies the parallelogram law and thus defines a new inner product in H

$$(u_1|u_2)_e = v_1^T v_2 + \int_{-h}^{0} \psi_1(\theta)^T \left(I - \frac{\theta}{h} C^T C\right) \psi_2(\theta) d\theta,$$

$u_i = (v_i, \psi_i) \in H$, $i = 1, 2$. According to (6.1.1') $\mathscr{A} - \omega I$ is dissipative if there exists an $\omega \in R$ such that $(u|\mathscr{A}u)_e \leqslant \omega \|u\|_e^2 \; \forall u \in \mathscr{D}(\mathscr{A})$. This will be now checked

$$u \in \mathscr{D}(\mathscr{A}) \Rightarrow (u|\mathscr{A}u)_e = v^T A v + v^T (AC + B)\psi(-h) +$$

$$+ \int_{-h}^{0} \psi(\theta)^T \left(I - \frac{\theta}{h} C^T C\right) \dot{\psi}(\theta) d\theta$$

$$= v^T (A + \tfrac{1}{2} I) v + v^T (AC + B + C)\psi(-h) - \tfrac{1}{2} \psi(-h)^T \psi(-h) +$$

$$+ \frac{1}{2h} \int_{-h}^{0} \psi(\theta)^T C C^T \psi(\theta) d\theta \leqslant \omega \|u\|_e^2$$

where $\omega = \max \{|A| + \tfrac{1}{2} + \tfrac{1}{2}|AC + B + C|^2, \tfrac{1}{2} h^{-1} |C|^2\}$. $\square$

By Lemma 6.2.1 $\overline{\mathscr{D}(\mathscr{A})} = H$ and all hypotheses of Theorem 6.2.1 are satisfied, therefore $\mathscr{A}$ generates a linear semigroup $\{T(t)\}_{t \geqslant 0}$, $T(t)u_0 = (v(t), x_t)$ for $u_0 \in H$, where $x_t(\theta) = x(t+\theta)$ for a fixed $t \geqslant 0$, and $\theta \in [-h, 0]$. An element $T(t)u_0$ may be obtained by applying the method of steps to (6.2.72).

Let us also notice that relationships (6.2.75) give the unique solution of the equation $(\lambda^{-1} I - \mathscr{A})u = \lambda^{-1} \hat{u}$ for every $\hat{u} \in H$ and those $\lambda \in C$, $\lambda \neq 0$ for which (6.2.74) has a unique solution δ. Hence, replacing λ^{-1} by s and multiplying the right-hand sides of (6.2.75) by s^{-1} we get explicit formulas for the resolvent $(sI - \mathscr{A})^{-1}$ of the operator $\mathscr{A}$

$$(sI - \mathscr{A})^{-1}(v_0, \phi)$$
$$= ((I - e^{-hs} C)\hat{X}(s)(v_0 + e^{-sh}(Cs + B)\mathscr{L}_F(\phi)) - e^{-hs} C \mathscr{L}_F(\phi), \phi_0)$$
$$= ((I - e^{-hs} C)\hat{X}(s) v_0 + e^{-hs}[(I - e^{-hs} C)\hat{X}(s) B$$
$$+ (A + e^{-hs} B)\hat{X}(s) C] \mathscr{L}_F(\phi), \phi_0) \quad (6.2.76)$$

6.2 Stability Theory of Abstract Linear Systems

where
$$\hat{X}(s) = (sI - se^{-hs}C - A - e^{-hs}B)^{-1}$$

$$\mathscr{L}_F(\phi) = \int_{-h}^{0} e^{-s\theta}\phi(\theta)d\theta \quad \text{(the finite Laplace transform of } \phi\text{)}$$

$$\phi_0(\theta) = e^{s\theta}\hat{X}(s)v_0 + e^{-s(\theta-h)}\hat{X}(s)(sC+B)\mathscr{L}_F(\phi) + \int_{\theta}^{0} e^{s(\theta-\tau)}\phi(\tau)d\tau.$$

Evidently, the spectrum of $\mathscr{A}$ is a purely point one and has the form
$$\sigma(\mathscr{A}) = \{s \in C: \det(sI - se^{-hs}C - A - e^{-hs}B) = 0\}.$$

Moreover, since for fixed $s \in C\setminus\sigma(\mathscr{A})$ the operators $\mathscr{L}_F: L^2(-h, 0; C^n) \to C^n$, $C^n \ni v_0 \mapsto (\theta \mapsto e^{s\theta}\hat{X}(s)v_0) \in L^2(-h, 0; C^n)$ and $\phi \mapsto e^{s(\cdot)} \star \phi \in \mathscr{L}(L^2(-h, 0; C^n))$ are compact then for $s \notin \sigma(\mathscr{A})$, $(sI - \mathscr{A})^{-1}$ (the resolvent of $\mathscr{A}$) is a compact operator. Formula (6.2.76) determines the Laplace transform of $T(\cdot)u = (v(u)(\cdot), x.(u))$, $u = (v_0, \phi)$. This enables us to find the following representation for $v(u)(\cdot)$:

$$v(u)(t) = [X(t) - CX(t-h)]v_0 + \int_{-h}^{0} \{[X(t-\theta-h) - CX(t-2h-\theta)]B +$$
$$+ [AX(t-h-\theta) + BX(t-\theta-2h)]C\}\phi(\theta)d\theta \quad t \geq 0 \quad (6.2.77)$$

where the matrix-valued fundamental solution of (6.2.71) satisfies
$$\dot{X}(t) - \dot{X}(t-h)C = \dot{X}(t)A + X(t-h)B \text{ for all } t \geq 0 \quad t \neq kh$$
$$k = 0, 1, 2, \ldots \quad X(0) = I \quad X(t) = 0 \text{ for } t < 0 \quad (6.2.78)$$

$X(t) - X(t-h)C$ is continuous for $t \geq 0$.

By $\hat{X}$ we denote the Laplace transform of X.

Lemma 6.2.11 *In the above notation, the following conditions are equivalent*
(i) $|\lambda(C)| < 1$ \hfill (6.2.79a)
(this means that all eigenvalues of the matrix C have moduli strictly less than 1*)*
and
$$\text{Re}\,\sigma(\mathscr{A}) < 0 \quad (6.2.79b)$$
(this means that $\sigma(\mathscr{A}) \subset \{s \in C: \text{Re } s < 0\}$),
  (ii) $\{T(t)\}_{t \geq 0}$ *is an EXS semigroup,*
  (iii) *There exist $M \geq 1$, $\Omega < 0$ such that*
$$|v(u)(t)| \leq Me^{\Omega t}\|u\| \quad \forall t \geq 0 \quad \forall u \in H \quad (6.2.80)$$
and $|\lambda(C)| < 1$,

(iv) *There exist* $M_1 \geq 1$, $\Omega_1 < 0$ *such that*
$$|x(u)(t)| \leq M_1 e^{\Omega_1 t}[|v_0| + \sup_{\theta \in [-h, 0]} |\phi(\theta)|]$$
$$\forall t \geq 0 \quad \forall u = (v_0, \phi) \in H \quad (6.2.81)$$
where ϕ is a bounded and measurable function,
(v) *There exist* $k \geq 1$, $\beta < 0$ *such that*
$$|X(t)| \leq k e^{\beta t} \quad \forall t \geq 0. \quad (6.2.82)$$

Proof (i) $\Rightarrow$ (ii) Since $\operatorname{Re}\sigma(\mathscr{A}) < 0$ then by Lemma 6.2.4 it remains to prove that $(sI - \mathscr{A})^{-1}$, the resolvent of $\mathscr{A}$, is uniformly bounded in the right closed, complex half-plane. It follows from (6.2.79a) that
$$|\lambda(C)| < 1 \leq |e^{sh}|$$
for every complex s, $\operatorname{Re} s \geq 0$. Consequently, there exists $\varepsilon > 0$ such that $\|(I - e^{-sh}C)x\| \geq \varepsilon \|x\|$ for every $x \in C^n$, and all complex s, $\operatorname{Re} s \geq 0$. Now the estimate
$$\|s(I - e^{-sh}C)x - Ax - e^{-sh}Bx\| \geq |s| \|(I - e^{-sh}C)x\| - \|Ax\| - \|Bx\|$$
$$\geq [|s|\varepsilon - \|A\| - \|B\|]\|x\|$$
for every $x \in C^n$ and all complex s, $\operatorname{Re} s \geq 0$, $|s| > \dfrac{1}{\varepsilon}[\|A\| + \|B\|]$, jointly with (6.2.79b) imply the uniform boundedness of $\hat{X}(s)$ in the right closed complex half-plane. Thus, our claim easily follows from (6.2.76).

(ii) $\Rightarrow$ (iii) We have only to show that $|\lambda(C)| < 1$, as (6.2.80) readily follows from EXS. Suppose that $\sigma(\mathscr{A}) = \phi$. Then $\hat{X}(s)$ is an entire function and consequently there exists $T > 0$ such that $X(t) = 0$ for all $t \geq T$. Applying the method of steps to (6.2.78) one can establish that the jumps of $X(t)$ at $t = kh$ are equal to C^k, $k = 0, 1, 2, \ldots$ Hence for sufficiently large k, $C^k = 0$. Thus $\lambda(C)$, the spectrum of C is $\{0\}$ and $|\lambda(C)| < 1$ holds.

Suppose now that $\sigma(\mathscr{A}) \neq \phi$. Then 0 is not the Picard's exceptional value for the entire function $s \mapsto \det(sI - se^{-sh}C - A - e^{-sh}B)$ (i.e. for the characteristic quasipolynomial). The quasipolynomial can be represented in a form of polynomial with respect to $z = e^{-sh}$, with free term equal to $\det(sI - A)$. A necessary and sufficient condition for the quasipolynomial to have an n-th order pole at infinity is therefore $\det[I - zC - s^{-1}(zB + A)] = \det(I - s^{-1}A)$ for every $s, z \in C$, $s \neq 0$. By taking the limit for $|s| \to \infty$ this necessarily implies that $\det(wI - C) = w^n (w = z^{-1}) \; \forall w \in C$. Hence, in the case of n-th order pole at $\{\infty\}$ $\lambda(C) = \{0\}$ and $|\lambda(C)| < 1$. If ∞ is an essential singularity then by Picard's Second Theorem (Sidorov, Fedoryuk and Shabunin, 1985, Theorem 7, p. 136) the quasipolynomial has infinitely many zeros, which can be arranged in a sequence $\{s_j\}, |s_j| \to \infty$. If $a \in C$, $a \neq 0$ is an eigenvalue of C, then for the sequence $\{z_k\}$,

6.2 Stability Theory of Abstract Linear Systems

$$z_k = \frac{\ln|a|}{h} + i\left(\frac{\arg a + 2k\pi}{h}\right) \quad k = 0, 1, 2, \ldots$$

we have $|z_k| \to \infty$ and

$$\det\left[e^{z_k h}\left(I - \frac{1}{z_k}A\right) - \left(C + \frac{1}{z_k}B\right)\right] = \det\left[(aI - C) - \frac{a}{z_k}A + \frac{1}{z_k}B\right] \to$$

$$\to \det(aI - C) = 0 \text{ as } k \to \infty.$$

Hence $\{z_k\}$ is asymptotically convergent to some subsequence of $\{s_j\}$. Since by EXS, Lemma 6.2.3 and Theorem 6.2.2 $\sup_j \operatorname{Re} s_j < 0$ then $\operatorname{Re} z_k < 0$ and hence $|\lambda(C)| < 1$.

(iii) $\Rightarrow$ (iv) If $|\lambda(C)| < 1$ then there exist $m \geq 1$, $r \in (0, 1)$ such that

$$\left|C^{\left[\frac{t}{h}\right]}\right| \leq mr^{\left[\frac{t}{h}\right]} = m^{\left[\frac{t}{h}\right]\ln r} < me^{\ln r\left(\frac{t}{h}-1\right)} = \frac{m}{r}e^{\frac{t\ln r}{h}} \quad \forall t \geq 0$$

where $\left[\frac{t}{h}\right]$ denotes the greatest integer less than $\frac{t}{h}$.

For any real Ω_1, $\max\left\{\frac{\ln r}{h}, \Omega\right\} < \Omega_1 < 0$ we have

$$|v(u)(t)| \leq M\|u\|e^{\Omega t} \leq M\|u\|e^{\Omega_1 t} \quad \forall t \geq 0, \quad \forall u \in H$$

and $0 < re^{-h\Omega_1} < 1$.

Assume now that the initial condition $u = (v_0, \phi)$ is such that ϕ is a bounded measurable function. Since $x(u)(t)$ is the solution of the following initial-value problem:

$$x(t) = Cx(t-h) + v(t) \quad t \geq 0$$
$$x(\theta) = \phi(\theta) \quad -h \leq \theta \leq 0$$

then by the method of steps

$$x(u)(t) = C^{\left[\frac{t}{h}\right]+1}\phi\left(t - h - h\left[\frac{t}{h}\right]\right) + \sum_{j=0}^{\left[\frac{t}{h}\right]} C^j v(t-jh) \quad t \geq 0.$$

Hence,

$$|x(u)(t)| \leq mrr^{\left[\frac{t}{h}\right]}\sup_{\theta \in [-h,0]}|\phi(\theta)| + mM\|u\|e^{\Omega_1 t}\sum_{j=0}^{\infty}(re^{-\Omega_1 h})^j$$

$$\leq me^{\Omega_1 t}\left\{\sup_{\theta \in [-h,0]}|\phi(\theta)| + \frac{M}{1-re^{-\Omega_1 h}}[|v_0| + \|\phi\|_{L^2(-h,0;R^n)}]\right\}$$

$$\leq M_l e^{\Omega_1 t}[|v_0| + \sup_{\theta \in [-h,0]}|\phi(\theta)|] \quad \forall t \geq 0$$

where

$$M_l = m\left(1 + \frac{M\max\{\sqrt{h}, 1\}}{1 - re^{-\Omega_1 h}}\right) \geq 1$$

and therefore (6.2.81) holds.

(iv) ⇒ (i) Taking the initial condition $u = (v_0, 0)$, $v_0 \in R^n$ and using the method of steps, observe that the function $x(u)(\cdot)$ has jumps of the height $C^k v_0$ at points kh, $k = 0, 1, 2, \ldots$ Hence, by (6.2.81) $C^k v_0 \to 0$ as $k \to \infty$ for every $v_0 \in R^n$ and therefore (6.2.79a) holds.

We now claim that (6.2.79b) also holds. The proof goes by contradiction. Suppose that there is a $\lambda \in C$: $\operatorname{Re}\lambda \geqslant 0$, $\lambda \in \sigma(\mathscr{A})$. Then there is a $c \in C^n$ such that $x(u)(t) = e^{\lambda t}c + e^{\bar\lambda t}\bar c$ is a solution of (6.2.72) with initial condition $u = (2\operatorname{Re}[(I - e^{-\bar\lambda h}C)c], 2\operatorname{Re}(e^{\bar\lambda \theta}c))$. The R^n-norm of this solution is $2e^{\operatorname{Re}\lambda t}|c|$ and therefore it does not decay to zero as t tends to infinity. However, this is contrary to (6.2.81). Thus our claim is true.

(iv) ⇒ (v) This implication is obvious.

(v) ⇒ (iii) As we have seen the jumps of $X(t)$ at $t = kh$, $k = 0, 1, 2, \ldots$ are equal to C^k and therefore (6.2.82) implies $|\lambda(C)| < 1$.

By (6.2.77) and (6.2.82),

$$|v(u)(t)| \leqslant [|X(t)| + |X(t-h)||C|]|v_0| + \int_{-h}^{0} [|X(t-h-\theta)|(|B| + |A||C|) +$$
$$+ 2|C||B||X(t-2h-\theta)|]|\phi(\theta)|\,d\theta \leqslant [|X(t)| + |C||X(t-h)|]|v_0| +$$
$$+ \{[|B| + |A||C|] \max_{\theta \in [-h, 0]} |X(t-h-\theta)| + 2||B|C| \max_{\theta \in [-h, 0]} |X(t-2h-\theta)|\}$$

$$\int_{-h}^{0} |\phi(\theta)|\,d\theta \leqslant |v_0|[ke^{-\beta t} + |C|ke^{-\beta(t-h)}] + \{(|B| + |A||C|)ke^{-\beta(t-h)} +$$
$$+ 2|B||C|ke^{-\beta(t-2h)}\} \varkappa \sqrt{\int_{-h}^{0} |\phi(\theta)|^2\,d\theta} \leqslant ke^{-\beta t}\{|v_0| \cdot (1 + |C|e^{\beta h}) +$$
$$+ [(|B| + |A||C|)e^{\beta h} + 2|B||C|e^{2h\beta}]\varkappa\|\phi\|_{L^2}\} \leqslant \tilde k e^{\beta t}\|u\|_H$$

where

$$\tilde k = k\sqrt{(1 + |C|e^{\beta h})^2 + \varkappa^2[(|B| + |A||C|)e^{\beta h} + 2|B||C|e^{2\beta h}]^2}.$$

and (6.2.80) holds. □

6.3 Stability of Linear Systems with Nonlinear Perturbations (Lyapunov's First Method)

In this section, a generalization of Lyapunov's first method for systems with delay will be presented. This generalization is achieved using two approaches, which are also applied in the analysis of finite-dimensional systems. The first is founded on the use of the variation-of-constants formula. In Section 6.3.1 we derive the variation-of-constants formula for linear neutral systems. Using this formula we give important estimates for the fundamental solution.

6.3 Stability of Linear Systems with Nonlinear Perturbations

In Section 6.3.2 we apply these results to derive a version of Lyapunov's first method for a wide class of systems with delay. The second approach is based on the use of a Lyapunov functional for unperturbed linear systems to the stability analysis of perturbed nonlinear systems.

6.3.1 Variation-of-constants form a for neutral linear functional-differential equations

We shall discuss homogeneous linear neutral functional-differential equations of the form

$$\frac{d}{dt} Dx_t = Lx_t \quad t \geq 0$$
$$x_0 = \phi \in C. \tag{6.3.1}$$

The notation and assumptions are as in Section 6.2.3.

It has been shown that $\{T(t)\}_{t \geq 0}$ is a linear C_0-semigroup on C where $T(t)\phi = x_t(\phi)$, $x_t(\phi)(\theta) = x(\phi)(t+\theta)$, $x(\phi)(\cdot)$ is the solution of (6.3.1). Assume now that $G \in C([0, +\infty), R^n)$, $F \in L^1([0, +\infty), R^n)$ and consider the nonhomogeneous system

$$\frac{d}{dt}[Dx_t - G(t)] = Lx_t + F(t) \quad t \geq 0$$
$$x_0 = \phi \in C. \tag{6.3.2}$$

Lemma 6.3.1 *There exist $a, b \in R$ such that for each $t \geq 0$*

$$\|x_t\| \leq be^{at}\left\{\|\phi\| + \sup_{0 \leq u \leq t} |G(u) - G(0)| + \int_0^t |F(s)| ds\right\}. \tag{6.3.3}$$

Proof Consider the integral equation satisfied by the solution to (6.3.2)

$$x(t) = \phi(0) - \int_{-h}^{0} d\mu(\theta)\phi(\theta) + \int_{-h}^{0} d\mu(\theta)\begin{cases} \phi(t+\theta) & t+\theta \leq 0 \\ x(t+\theta) & t+\theta \geq 0 \end{cases} +$$

$$+ G(t) - G(0) + \int_0^t Lx_s \, ds + \int_0^t F(s) \, ds. \tag{6.3.4}$$

Let $t \in [0, s]$ and s be sufficiently small, such that $\underset{-s}{\overset{0}{\text{Var}}} \mu < 1$ and $s \leq \frac{h}{2}$ (due to $\underset{-s}{\overset{0}{\text{Var}}} \mu \to 0$ such an s exists). By (6.3.4) we have

$$|x(t)| \leq ||\phi|| + ||\phi|| \underset{-h}{\overset{0}{\operatorname{Var}}} \mu + \sup_{0 \leq u \leq t} |G(u) - G(0)| + \int_0^t ||L|| \, ||x_s|| \, ds +$$

$$+ \int_0^t |F(s)| \, ds + \left| \int_{-s}^0 d\mu(\theta) \begin{cases} \phi(t+\theta) & t+\theta \leq 0 \\ x(t+\theta) & t+\theta \geq 0 \end{cases} \right| + \left| \int_{-h}^s \right|.$$

$-h \leq \theta \leq -s \Rightarrow t-h \leq t+\theta \leq t-s \leq 0$ and the last term is estimated by $\underset{-h}{\overset{0}{\operatorname{Var}}} \mu ||\phi||$. Consequently, for $\xi \in [-h, 0]$ we have

$$|x(t+\xi)| \leq \begin{cases} |\phi(t+\xi)| & t+\xi \leq 0 \\ [1+2\underset{-h}{\overset{0}{\operatorname{Var}}}\mu]||\phi|| + \sup_{0 \leq u \leq t} |G(u)-G(0)| + \int_0^t |F(s)| \, ds + \\ + ||L|| \int_0^t ||x_s|| \, ds + \left| \int_{-s}^0 d\mu(\theta) \begin{cases} x(t+\xi+\theta) & t+\xi+\theta \geq 0 \\ \phi(t+\xi+\theta) & t+\xi+\theta \leq 0 \end{cases} \right| & t+\xi \geq 0. \end{cases}$$

In the last term $0 \leq t \leq s$, $\xi \leq 0$, $t+\xi \geq 0$ hence $\xi \geq -s$. But $-s \leq \theta \leq 0$, thus $\xi + \theta \geq -2s \geq -h$ and due to this the term under the sign of absolute value is not greater than $||x_t||$. From the above inequality we thus obtain

$$||x_t|| \leq [1+2\underset{-h}{\overset{0}{\operatorname{Var}}}\mu]||\phi|| + \sup_{0 \leq u \leq t} |G(u)-G(0)| + \int_0^t |F(s)| \, ds +$$

$$+ \underset{-s}{\overset{0}{\operatorname{Var}}}\mu ||x_t|| + ||L|| \int_0^t ||x_s|| \, ds \qquad t \in [0, s]$$

or

$$||x_t|| \leq \delta \left[\alpha ||\phi|| + \sup_{0 \leq u \leq t} |G(u)-G(0)| + \int_0^t |F(s)| \, ds \right] + \beta \int_0^t ||x_s|| \, ds$$

$$\delta = \frac{1}{1 - \underset{-s}{\overset{0}{\operatorname{Var}}}\mu} \qquad \alpha = 1 + 2\underset{-h}{\overset{0}{\operatorname{Var}}}\mu \qquad \beta = \delta ||L||.$$

Applying Gronwall's inequality one obtains

$$||x_t|| \leq \delta \left[\alpha ||\phi|| + \sup_{0 \leq u \leq t} |G(u)-G(0)| + \int_0^t |F(s)| \, ds \right] e^{\beta t} \qquad t \in [0, s]. \quad (6.3.5)$$

6.3 Stability of Linear Systems with Nonlinear Perturbations

Similar arguments lead to the following inequality:

$$||x_t|| \leq \delta e^{\beta(t-\tau)}\left[\alpha||x_\tau|| + \sup_{\tau \leq u \leq t} |G(u) - G(\tau)| + \int_\tau^t |F(s)|\,ds\right]$$

$$t \in [\tau, \tau+s] \quad \tau \geq 0. \tag{6.3.6}$$

Let γ be such that $\gamma > \dfrac{\ln \alpha\delta}{s} + \beta \geq \beta$ (as $\alpha \geq 1$, $\delta \geq 1$). We claim that

$$\forall k \in N: ||x_t|| \leq \delta e^{\gamma t}[\alpha||\phi|| + e^{\frac{t\ln 3}{s}} \sup_{0 \leq u \leq t} |G(u) - G(0)| +$$

$$+ \int_0^t |F(s)|\,ds] \quad \forall t \in [0, ks]. \tag{6.3.7}$$

The proof of estimate (6.3.7) goes by induction with respect to $k \in N$. It follows from (6.3.5) that for $k = 1$ (6.3.7) holds. Assume now that (6.3.7) is valid for some $k \in N$. We shall prove that (6.3.7) holds also for $k+1$. Notice that since (6.3.7) is valid for $k = 1$ it remains to show that (6.3.7) holds for $t \in [s, (k+1)s]$. Let us take $\tau = t-s \geq 0$ for the proof. By virtue of (6.3.6)

$$||x_t|| \leq \delta e^{\beta s}\left[\alpha||x_{t-s}|| + \sup_{t-s \leq u \leq t} |G(u) - G(t-s)| + \int_{t-s}^t |F(s)|\,ds\right] \tag{6.3.8}$$

$t \in [s, (k+1)s] \Rightarrow t-s \in [0, ks]$, hence by the assumption that (6.3.7) holds for $k \in N$ we have

$$||x_{t-s}|| \leq \delta e^{-\gamma(t-s)}\left[\alpha||\phi|| + e^{\frac{(t-s)\ln 3}{s}} \sup_{0 \leq u \leq t-s} |G(u) - G(0)| + \int_0^{t-s} |F(s)|\,ds\right]$$

$$\leq \frac{1}{\alpha} e^{\gamma t} e^{-\beta s}\left[\alpha||\phi|| + e^{\frac{(t-s)\ln 3}{s}} \sup_{0 \leq u \leq t-s} |G(u) - G(0)| + \int_0^{t-s} |F(s)|\,ds\right]. \tag{6.3.9}$$

Substituting (6.3.9) into (6.3.8) we obtain

$$||x_t|| \leq \delta e^{\beta s}\Big\{e^{\gamma t} e^{-\beta s}\Big[\alpha||\phi|| + e^{\frac{(t-s)\ln 3}{s}} \sup_{0 \leq u \leq t-s} |G(u) - G(0)| +$$

$$+ \int_0^{t-s} |F(s)|\,ds\Big] + \sup_{t-s \leq u \leq t} |G(u) - G(t-s)| + \int_{t-s}^t |F(s)|\,ds\Big\}$$

$$= \delta e^{\gamma t}\Big\{\alpha||\phi|| + e^{\frac{(t-s)\ln 3}{s}} \sup_{0 \leq u \leq t-s} |G(u) - G(0)| +$$

$$+\mathrm{e}^{-t(\gamma-\beta)}\sup_{t-s\leqslant u\leqslant t}|G(u)-G(t-s)|+\int_0^{t-s}|F(s)|ds+$$

$$+\mathrm{e}^{-t(\gamma-\beta)}\int_{t-s}^t|F(s)|ds\Big\} \leqslant \delta\mathrm{e}^{\gamma t}\Big\{\alpha||\phi||+\mathrm{e}^{\frac{t\ln 3}{s}}\Big[\tfrac{1}{3}\sup_{0\leqslant u\leqslant t-s}|G(u)-G(0)|+$$

$$+\mathrm{e}^{\frac{-t\ln 3}{s}}\sup_{t-s\leqslant u\leqslant t}|G(u)-G(t-s)|\Big]+\int_0^t|F(s)|ds\Big\}.$$

As

$$\sup_{0\leqslant u\leqslant t-s}|G(u)-G(0)| \leqslant \sup_{0\leqslant u\leqslant t}|G(u)-G(0)|$$

$$\sup_{t-s\leqslant u\leqslant t}|G(u)-G(t-s)| \leqslant \sup_{t-s\leqslant u\leqslant t}|G(u)-G(0)|+|G(t-s)-G(0)|$$

$$\leqslant 2\sup_{0\leqslant u\leqslant t}|G(u)-G(0)|$$

and $t \geqslant s$ one obtains

$$||x_t|| \leqslant \delta\mathrm{e}^{\gamma t}\Big\{\alpha||\phi||+\mathrm{e}^{\frac{t\ln 3}{s}}\sup_{0\leqslant u\leqslant t}|G(u)-G(0)|+\int_0^t|F(s)|ds\Big\}.$$

Thus we have proved that (6.3.7) holds. The estimate (6.3.3) follows directly from (6.3.7) and $a = \gamma + \dfrac{\ln 3}{s}$, $b = \alpha\delta$. $\square$

One of the consequences of the growth rate estimate (6.3.3) is that since $G \in C([0, +\infty), R^n)$ and $F \in L^1([0, +\infty), R^n)$, the solution of problem (6.3.2) has a Laplace transform. This enables one to derive the variation-of-constants formula by the Laplace transform technique. Let $\mathscr{L}[G] = \hat{G}$, $\mathscr{L}[F] = \hat{F}$, $\mathscr{L}[x] = \hat{x}$ be the Laplace transforms of the functions G, F and the solution x. Hence

$$s\Big[\hat{x}(s)-\hat{G}(s)-\int_{-h}^0 d\mu(\theta)\int_0^\infty x(t+\theta)\mathrm{e}^{-st}\Big]+\int_{-h}^0 d\mu(\theta)\phi(\theta)-\phi(0)+G(0)$$

$$= \hat{F}(s)+\int_{-h}^0 d\eta(\theta)\int_0^\infty x(t+\theta)\mathrm{e}^{-st}dt.$$

Taking into account the fact that $x(\theta) = \phi(\theta)$ for $\theta \in [-h, 0]$, this yields

$$\Big[sI-s\int_{-h}^0 d\mu(\theta)\mathrm{e}^{s\theta}-\int_{-h}^0 d\eta(\theta)\mathrm{e}^{s\theta}\Big]\hat{x}(s) = s\hat{G}(s)-G(0)+\hat{F}(s)+$$

6.3 Stability of Linear Systems with Nonlinear Perturbations

$$+\phi(0)+s\int_{-h}^{0}d\mu(\theta)e^{s\theta}\int_{\theta}^{0}\phi(\tau)e^{-s\tau}d\tau+$$

$$+\int_{-h}^{0}d\eta(\theta)e^{s\theta}\int_{\theta}^{0}\phi(\tau)e^{-s\tau}d\tau-\int_{-h}^{0}d\mu(\theta)\phi(\theta). \tag{6.3.10}$$

Notice that

$$\hat{X}(s) = \left[sI - s\int_{-h}^{0}d\mu(\theta)e^{s\theta} - \int_{-h}^{0}d\eta(\theta)e^{s\theta}\right]^{-1} \tag{6.3.11}$$

is the Laplace transform of $X(t)$, the solution of the following initial-value problem:

$$DX_t = I + \int_0^t LX_s\,ds \quad t \geq 0$$

$$X_0(\theta) = \begin{cases} I & \theta = 0 \\ 0 & \theta \in [-h, 0) \end{cases} \tag{6.3.12}$$

where $X_t(\theta) = X(t+\theta)$ for $t \geq 0$ and $\theta \in [-h, 0]$. The above observation enables us to rewrite (6.3.10) in the form

$$\hat{x}(s) = \hat{X}(s)[s\hat{G}(s) - G(0)] + \hat{X}(s)\hat{F}(s) + \hat{X}(s)\left[\phi(0) - \int_{-h}^{0}d\mu(\theta)\phi(\theta) + \right.$$

$$\left. + s\int_{-h}^{0}d\mu(\theta)e^{s\theta}\int_{\theta}^{0}\phi(\tau)e^{-s\tau}d\tau + \int_{-h}^{0}d\eta(\theta)e^{s\theta}\int_{\theta}^{0}\phi(\tau)e^{-s\tau}d\tau\right]. \tag{6.3.13}$$

It is known from the theory of linear C_0-semigroups (Balakrishnan, 1976) that the value of a resolvent of the infinitesimal generator of the semigroup $\{T(t)\}_{t \geq 0}$, related to the solution of (6.3.1), on the element ϕ is the Laplace transform of the function $t \mapsto T(t)\phi$. It follows from the relationship (6.2.32) that the infinitesimal generator of the semigroup $\{T(t)\}_{t \geq 0}$ has the form

$$\mathscr{A}: (\mathscr{D}(\mathscr{A}) \subset C) \to C, \quad \mathscr{A}\psi = \dot{\psi}, \quad \mathscr{D}(\mathscr{A}) = \{\psi \in C: \dot{\psi} \in C, D\dot{\psi} = L\psi\} \tag{6.3.14}$$

The resolvent $(sI - \mathscr{A})^{-1}$ of the operator $\mathscr{A}$ is determined in the following manner: $(sI - \mathscr{A})^{-1}\phi = \psi$, $\phi \in C \Leftrightarrow \psi$ is the unique solution of the equation $s\psi - \mathscr{A}\psi = \phi$.

In virtue of (6.3.14) the full form of this equation is as follows

$$s\psi(\theta) - \dot{\psi}(\theta) = \phi(\theta)$$

$$\psi(0) \text{ is such that } D[s\psi - \phi] = L\psi$$

hence
$$\psi(\theta) = e^{s\theta}\psi(0) + \int_{\theta}^{0} e^{s(\theta-\tau)}\phi(\tau)\,d\tau \qquad (6.3.15a)$$

$$\psi(0) = \hat{X}(s)\left[\phi(0) + s\int_{-h}^{0} d\mu(\theta)\int_{\theta}^{0} e^{s(\theta-\tau)}\phi(\tau)\,d\tau + \right.$$
$$\left. + \int_{-h}^{0} d\eta(\theta)\int_{\theta}^{0} e^{s(\theta-\tau)}\phi(\tau)\,d\tau - \int_{-h}^{0} d\mu(\theta)\phi(\theta)\right]. \qquad (6.3.15b)$$

In accordance with the interpretation of formulas (6.3.15) given above we know that $\mathscr{L}[(T(\cdot)\phi)(0)] = \psi(0)$, where $\psi(0)$ is determined by (6.3.15b). We conclude from a comparison of (6.3.15b) with (6.3.13) that the last term in (6.3.13) is the Laplace transform of the function $t \mapsto (T(t)\phi)(0)$, that is it corresponds to the solution $x(\phi)(\cdot)$ of the homogeneous problem (6.3.1) restricted to $[0, +\infty)$.

Interpreting the two first terms in (6.3.13) as the transforms of appropriate convolutions and employing the inverse Laplace transform we obtain

$$x(t) = X(0)G(t) - X(t)G(0) - \int_{0}^{t} [d_{\xi}X(t-\xi)]G(\xi) +$$
$$+ \int_{0}^{t} X(t-\xi)F(\xi)\,d\xi + (T(t)\phi)(0) \qquad t \geq 0. \qquad (6.3.16)$$

The relationship (6.3.16) is called the variation-of-constants formula. The variation-of-constants formula can be carried over to the state space C. That is to say

$$t+\theta \geq 0 \Rightarrow x_t(\theta) = x(t+\theta) = X(0)G(t+\theta) - X(t+\theta)G(0) -$$
$$- \int_{0}^{t+\theta} [d_{\xi}X(t+\theta-\xi)]G(\xi) + \int_{0}^{t+\theta} X(t+\theta-\xi)F(\xi)\,d\xi + (T(t+\theta)\phi)(0)$$

$\xi \in [t+\theta, t] \Rightarrow t+\theta-\xi \leq 0$, hence the upper limits of integration can be replaced by t. Moreover, $X(0)G(t+\theta) = X_0(\theta)G(t)$ and
$$(T(t+\theta)\phi)(0) = x_{t+\theta}(\phi)(0) = x(\phi)(t+\theta) = x_t(\phi)(\theta) = (T(t)\phi)(\theta).$$
Taking these facts into account one obtains

$$x_t(\theta) = X_0(\theta)G(t) - X_t(\theta)G(0) - \int_{0}^{t} [d_{\xi}X(t+\theta-\xi)]G(\xi) +$$
$$+ \int_{0}^{t} X(t+\theta-\xi)F(\xi)\,d\xi + (T(t)\phi)(\theta).$$

6.3 Stability of Linear Systems with Nonlinear Perturbations

in both integrals $\xi \geq 0$, hence for $t+\theta < 0$ it holds $t+\theta-\xi < 0$ but it is then obvious that $X(t+\theta-\xi) = 0$, thus the above formula is valid also when $t+\theta \leq 0$. Finally,

$$x_t = X_0 G(t) - X_t G(0) - \int_0^t [d_\xi X_{t-\xi}] G(\xi) + \int_0^t X_{t-\xi} F(\xi) d\xi + T(t)\phi \quad t \geq 0. \tag{6.3.17}$$

For further considerations we need the lemma formulated below, whose proof includes an interesting application of the variation-of-constants formula (6.3.16).

Lemma 6.3.2 (Henry, Cruz and Hale) *If the semigroup $\{T(t)\}_{t \geq 0}$ related to the solutions of the homogeneous problem (6.3.1) is EXS, i.e. if there exist $K \geq 1$, $\alpha > 0$ such that*

$$\|T(t)\phi\| = \|x_t(\phi)\| = \sup_{\theta \in [-h,0]} |x(\phi)(t+\theta)| \leq K e^{-\alpha t} \|\phi\| \quad \forall \phi \in C, \quad \forall t \geq 0,$$

$x(\phi)(\cdot)$ *is the solution to (6.3.1), then the following estimates for the fundamental solution $X(t)$ hold:*

$$\exists K_1 \geq 1: \operatorname*{Var}_{t-1}^{t} X = \operatorname*{Var}_{0}^{1} X(t-\cdot) \leq K_1 e^{-\alpha t} \quad \forall t \geq 0 \tag{6.3.18}$$

$$\operatorname*{Var}_{0}^{s} X(t-\cdot) \leq K_1 \frac{e^\alpha}{e^\alpha - 1} e^{-\alpha(t-s)} \quad \forall s \in [0,t] \quad \forall t \geq 0 \tag{6.3.19}$$

$$\exists K_2 \geq 1: |X(t)| \leq K_2 e^{-\alpha t} \quad \forall t \geq 0. \tag{6.3.20}$$

Proof Consider the initial-value problem (6.3.2) in which $F = 0$, $G \in \mathcal{H}$ $\stackrel{\mathrm{df}}{=} \{H \in C([0,+\infty), R^n): |H(t)| \leq 1 \ \forall t \geq 0, H(0) = 0, H$ is constant on $[1, +\infty)\}$ with zero initial condition. In accordance with (6.3.16) the solution of this problem has the form

$$x(0, G)(t) = X(0)G(t) - X(t)G(0) - \int_0^t [d_s X(t-s)] G(s)$$

$$= X(0)G(t) - \int_0^t [d_s X(t-s)] G(s) = -\int_0^{t+} [d_s X(t-s)] G(s). \tag{6.3.21}$$

Here $+$ means that the jump of $X(t-\cdot)$ at the point $s = t$ is included in the domain of integration. In virtue of (6.3.20) we have

$$x_1(\theta) \stackrel{\mathrm{df}}{=} x(0, G)(1+\theta) = \begin{cases} 0 & 1+\theta \leq 0 \\ -\int_0^{(1+\theta)+} [d_s X(t+\theta-s)] G(s) & 1+\theta \geq 0 \end{cases}$$

$$= -\int_0^{1+} [d_s X(1-s+\theta)] G(s) \quad \theta \in [-h, 0].$$

Furthermore,

$$\|x_1\| = k \leqslant \max_{s \in [0,1]} |G(s)| \overset{+1}{\underset{0}{\mathrm{Var}}} X(1-\cdot) < \infty$$

because the fundamental solution is of bounded variation on the bounded interval contained in $[-h, +\infty)$. As the function $G \in \mathscr{H}$ is constant on $[1, +\infty)$, $x(0, G)(\cdot)$ satisfies the equation $\dfrac{d}{dt} Dx_t = Lx_t$ on $[1, +\infty)$ and hence we have

$$x(0, \overset{\downarrow}{\underset{\uparrow\text{—initial conditions—}\uparrow}{G}})(t) = x(x_1, \overset{\downarrow}{0})(t-1).$$
$$ \text{forcing terms}$$

From EXS it follows that

$$\|x_\tau(x_1)\| \leqslant e^{-\alpha\tau} kK \quad \forall \tau \geqslant 0.$$

In particular for $\tau = t-1 \geqslant 0$ this yields

$$\|x_t(0, G)\| \leqslant kKe^\alpha e^{-\alpha t} \quad \forall t \geqslant 1. \tag{6.3.22}$$

Interpreting the integral $\int_0^1 [d_s X(t-s)] G(s)$ as a functional on the space $C([0, 1], R^n)$ and taking into account the fact that the dual space to $C([0, 1], R^n)$ is $BV([0, 1], R^n)$ — the space of functions of bounded variation defined on $[0, 1]$ taking values in R^n with variation as a norm, one obtains

$$\underset{t-1}{\overset{t}{\mathrm{Var}}} X = \underset{0}{\overset{1}{\mathrm{Var}}} X(t - \cdot) \leqslant \sup_G \left| \int_0^{1+} [d_s X(t-s)] G(s) \right| \tag{6.3.23}$$

where the supremum is taken on the unit ball in the space of continuous functions defined on $[0, 1]$ with values in R^n vanishing at zero. Notice that for $t \geqslant 1$ and in accordance with (6.3.16) we have

$$x(0, G)(t) = X(0)G(t) - \int_0^t [d_s X(t-s)] G(s) = X(0)G(t) -$$

$$- \int_0^1 - \int_1^t [d_s X(t-s)] G(s).$$

The integration-by-parts formula can be applied to the last term, thus

$$x(0, G)(t) = X(t-1)G(1) - \int_0^1 [d_s X(t-s)] G(s) = - \int_0^1 [d_s X(t-s)] G(s).$$

5.3 Stability of Linear Systems with Nonlinear Perturbations 163

Due to this and the definition of $\mathcal{H}$, (6.3.23) can be rewritten as

$$\underset{t-1}{\overset{t}{\mathrm{Var}}}X = \underset{0}{\overset{1}{\mathrm{Var}}}X(t-\cdot) \leqslant \sup_{G\in\mathcal{H}}\left|\int_0^{1+}|[d_sX(t-s)]G(s)|\right| =$$

$$= \sup_{G\in\mathcal{H}}|x(0,G)(t)| \quad \text{for } t \geqslant 1. \tag{6.3.24}$$

Considering the appropriate maximizing sequence we obtain (6.3.18) from (6.3.24) and (6.3.22) with $K_1 = kK$.

The estimate (6.3.19) follows from (6.3.18). Namely,

$$\underset{1}{\overset{2}{\mathrm{Var}}}X(t-\cdot) = \underset{0}{\overset{1}{\mathrm{Var}}}X(t-1-\cdot) \leqslant K_1 e^{-\alpha(t-1)}$$

$$\underset{2}{\overset{3}{\mathrm{Var}}}X(t-\cdot) = \underset{0}{\overset{1}{\mathrm{Var}}}X(t-2-\cdot) \leqslant K_1 e^{-\alpha(t-2)}$$

$$\cdots\cdots\cdots\cdots\cdots\cdots\cdots\cdots\cdots\cdots\cdots\cdots\cdots\cdots$$

$$\underset{N-1}{\overset{N}{\mathrm{Var}}}X(t-\cdot) = \underset{0}{\overset{1}{\mathrm{Var}}}X(t-N+1-\cdot) \leqslant K_1 e^{-\alpha(t-N+1)}$$

$$\underset{N}{\overset{s}{\mathrm{Var}}}X(t-\cdot) \leqslant \underset{s-1}{\overset{s}{\mathrm{Var}}}X(t-\cdot) \leqslant \underset{0}{\overset{1}{\mathrm{Var}}}X(t-s+1-\cdot) \leqslant K_1 e^{-\alpha(t-s+1)}.$$

Hence,

$$\underset{0}{\overset{s}{\mathrm{Var}}}X(t-\cdot) \leqslant K_1\left[\sum_{i=1}^{N}\underset{i-1}{\overset{i}{\mathrm{Var}}}X(t-\cdot)+\underset{s-1}{\overset{s}{\mathrm{Var}}}X(t-\cdot)\right]$$

$$\leqslant K_1\left[\sum_{i=1}^{N}e^{-\alpha(t-i+1)}+e^{-\alpha(t-s+1)}\right] = K_1 e^{-\alpha(t+1)}\left[\sum_{i=1}^{N}e^{\alpha i}+e^{\alpha s}\right]$$

$$\leqslant K_1 e^{-\alpha(t+1)}\left[\frac{e^{\alpha(N+1)}-e^{\alpha}}{e^{\alpha}-1}+e^{\alpha s}\right] \leqslant K_1 e^{-\alpha(t+1)}\left[\frac{e^{\alpha(s+1)}-e}{e^{\alpha}-1}+e^{\alpha s}\right]$$

$$= K_1 e^{-\alpha t}\left[\frac{e^{\alpha s}-1}{e^{\alpha}-1}+e^{\alpha(s-1)}\right] \leqslant K_1 e^{-\alpha(t-s)}\left[\frac{1}{e^{\alpha}-1}+e^{-\alpha}\right]$$

$$\leqslant K_1 \frac{e^{\alpha}}{e^{\alpha}-1}e^{-\alpha(t-s)} \quad \forall s \in [0,t] \quad \forall t \geqslant 0$$

and (6.3.19) holds.

Now we give the proof of inequality (6.3.20). To this end let us consider the initial-value problem (6.3.2) with $G = 0$,

$$F \in h \overset{\mathrm{df}}{=} \left\{H \in L^1([0,+\infty), R^n) \colon \int_0^1 |H(s)|ds = 1, H(t) = 0 \text{ for } t > 1\right\}$$

with zero initial condition. Denote by $x(0, F)(\cdot)$ the solution to this problem. In accordance with the variation-of-constants formula (6.3.16) we have $x(0, F)(t) = \int_0^t X(t-s)F(s)\,ds$. Notice that $x(0, F)(\cdot)$ is also the solution of the equation $\dfrac{d}{dt}\left[Dx_t - \int_0^t F(s)\,ds\right] = Lx_t$, $t \geq 0$ with zero initial condition and the function $t \mapsto G(t) = \int_0^t F(s)\,ds$ belongs to $\mathscr{H}$.

Indeed, since $F \in L^1([0, +\infty), R^n)$, G is absolutely continuous on $[0, +\infty)$,

$$|G(t)| = \left|\int_0^t F(s)\,ds\right| \leq \int_0^t |F(s)|\,ds = \int_0^1 |F(s)|\,ds = 1 \qquad G(0) = 0.$$

Repeating the arguments which led to the estimate (6.3.22) we now obtain

$$|x(0, F)(t)| \leq K_1 e^{\alpha} e^{-\alpha t} \qquad \forall t \geq 1. \tag{6.3.25}$$

Interpreting the integral $\int_0^1 X(t-s)F(s)\,ds$ as a functional on the space $L^1([0, 1], R^n)$ and taking into consideration the fact that the space $L^\infty([0, 1], R^n)$ is its dual space we have

$$\operatorname*{ess\,sup}_{[0,1]} |X(t-\cdot)| = \operatorname*{ess\,sup}_{[t-1,t]} |X| = \sup_{s \in [t-1, t]} |X(s)|$$

$$= \sup_F \left|\int_0^1 X(t-s)F(s)\,ds\right| \tag{6.3.26}$$

where the supremum is taken on the unit ball in the space $L^1([0, 1], R^n)$. By definition of h, for $t > 1$

$$x(0, F)(t) = \int_0^1 X(t-s)F(s)\,ds + \int_1^t = \int_0^1 X(t-s)F(s)\,ds$$

thus the supremum in the right-hand side of (6.3.26) can be replaced by

$$\sup_{F \in h} \left|\int_0^1 X(t-s)F(s)\,ds\right| = \sup_{F \in h} |x(0, F)(t)|.$$

Hence

$$|X(t)| \leq \sup_{s \in [t-1, t]} |X(s)| = \sup_{F \in h} |x(0, F)(t)| \leq \text{(for } t \geq 1) K_1 e^{\alpha} e^{-\alpha t}$$

but this is equivalent to (6.3.20). $\square$

6.3 Stability of Linear Systems with Nonlinear Perturbations

6.3.2 Lyapunov's first method for hereditary systems

Our objective is to study the relationship between the exponential stability of the linear neutral differential equation

$$\frac{d}{dt} Dx_t = Lx_t \quad t \geq 0 \qquad (6.3.27)$$

and the perturbed equation

$$\frac{d}{dt}[Dx_t - G(x_t)] = Lx_t + F(x_t) \quad t \geq 0 \qquad (6.3.28)$$

where

$$D\phi = \phi(0) - \int_{-h}^{0} d\mu(\theta)\phi(\theta) \quad \overset{0}{\underset{-h}{\text{Var}}}\mu < \infty \quad \overset{0}{\underset{-s}{\text{Var}}}\mu \to 0, \quad s \to 0$$

$$L\phi = \int_{-h}^{0} d\eta(\theta)\phi(\theta) \quad \overset{0}{\underset{-h}{\text{Var}}}\eta < \infty.$$

Throughout this section, we assume that $G: C \to R^n$ and $F: C \to R^n$ satisfy suitable smoothness conditions to ensure that a solution of (6.3.28) exists through each point $\phi \in C = C([-h, 0], R^n)$, that it is unique, depends continuously upon ϕ, and can be continued to the right as long as the trajectory remains in a bounded set in C.

We can now prove the following theorem.

Theorem 6.3.1 (Hale and Izé, 1971; Hale and Cruz, 1969) *Suppose that G and F satisfy the following condition*:

$$\forall \varepsilon > 0 \; \exists \delta = \delta(\varepsilon) > 0: \; |F(\phi)| \leq \varepsilon \|\phi\| \quad |G(\phi)| \leq \varepsilon \|\phi\| \quad \forall \phi \in C$$
$$\|\phi\| < \delta. \qquad (6.3.29)$$

If system (6.3.27) *is EXS then the equilibrium point* $0 \in C$ *of* (6.3.28) *is uniformly asymptotically stable*, i.e.

$$\forall \lambda > 0 \; \exists \varkappa = \varkappa(\lambda) > 0: \; \|\phi\| < \varkappa \Rightarrow \|x_t\| < \lambda \quad \forall t \geq 0 \qquad (6.3.30)$$

(uniform stability) and

$$\exists \eta > 0: \; \forall \mu > 0 \; \exists t_0 = t_0(\mu) \geq 0: \; \|\phi\| < \eta \Rightarrow \|x_t\| < \mu \qquad (6.3.31)$$

$\forall t \geq t_0$ *(uniform attractivity).*

Proof (Part 1) Let us derive the estimate for the growth rate of solutions of (6.3.28) in a neighbourhood of an equilibrium point. Setting the functions

$t \mapsto G(x_t)$, $t \mapsto F(x_t)$ in the place of functions G, F in the variation-of-constants formula (6.3.17) we obtain

$$x_t = X_0 G(x_t) - X_t G(\phi) - \int_0^t [d_s X_{t-s}] G(x_s) + \int_0^t X_{t-s} F(x_s) \, ds + T(t)\phi$$

$$X_0 = \begin{cases} 0 & \theta \in [-h, 0) \\ I & \theta = 0. \end{cases} \tag{6.3.32}$$

It follows from the EXS the system (6.3.27) and Lemma 6.3.2 that there exist numbers $M \geq 1$ and $\Omega > 0$ such that

$$\|T(t)\phi\| \leq M\|\phi\|e^{-\Omega t} \quad \forall t \geq 0 \quad \forall \phi \in C$$

$$\|X_t\| \leq Me^{-\Omega t} \quad \forall t \geq 0 \tag{6.3.33}$$

$$\operatorname*{Var}_0^s X_{t-\cdot} \leq Me^{-\Omega(t-s)} \quad \forall s \in [0, t] \quad \forall t \geq 0.$$

In virtue of (6.3.32), (6.3.33), (6.3.29) and by continuity of the solution of (6.3.28) if follows that

$$\forall \varepsilon > 0: \|x_t\| \leq \varepsilon\|x_t\| + M\varepsilon\|\phi\|e^{-\Omega t} + \varepsilon \operatorname*{Var}_0^t X_{t-\cdot} \cdot \sup_{0 \leq s \leq t} \|x_s\| +$$

$$+ \varepsilon \sup_{0 \leq s \leq t} \|x_s\| \int_0^t \|X_{t-s}\| ds + M\|\phi\|e^{-\Omega t} \leq M\|\phi\|(1+\varepsilon)e^{-\Omega t} +$$

$$+ \frac{\varepsilon M}{\Omega}(1+2\Omega) \sup_{0 \leq s \leq t} \|x_s\|, \tag{6.3.34}$$

$\forall t \in [0, t_1]$ where $t_1 = t_1(\phi) > 0$ is such that $\|x_t(\phi)\| < \delta$ $\forall t \in [0, t_1]$ and $\delta = \delta(\varepsilon) > 0$ is the number which corresponds to ε by (6.3.29).

(Part 2) *Uniform stability of the equilibrium point*

Fix arbitrary $\lambda > 0$. We find the number $\delta = \delta(\varepsilon) > 0$ related to

$$\varepsilon = \frac{\Omega}{2M(1+2\Omega)} > 0. \tag{6.3.35}$$

We claim that for

$$\varkappa = \frac{\min\{\delta, \lambda\}(1+2\Omega)}{2[\Omega + 2M(1+2\Omega)]} \quad (< \min\{\delta, \lambda\}) \tag{6.3.36}$$

(6.3.30) holds. Suppose for proof by contradiction that this is not the case. Then, since $\varkappa < \lambda$, for an arbitrary fixed $\phi \in C$, $\|\phi\| < \varkappa$ a moment $t_2 > 0$ exists such that for $t \in [0, t_2)$ we have $\|x_t(\phi)\| < \lambda$ and $\|x_{t_2}(\phi)\| = \lambda$. Note that $\varkappa < \min\{\delta, \lambda\} \leq \lambda$, see Figure 6.3.1, hence the moment $t_3 > 0$ exists,

6.3 Stability of Linear Systems with Nonlinear Perturbations

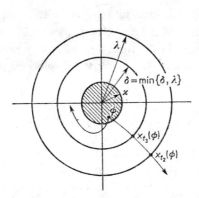

Figure 6.3.1 An auxiliary diagram for the proof of Theorem 6.3.1

$t_3 \leq t_2$ at which $\|x_{t_3}(\phi)\| = \min\{\delta, \lambda\}$ and $\|x_t(\phi)\| < \min\{\delta, \lambda\}$ for $t \in [0, t_3)$. However, $\min\{\delta, \lambda\} \leq \delta$, therefore for $x_t(\phi)$ the estimate (6.3.34) is true with an arbitrary $t \in [0, t_3)$. In particular, for $t = t_3$ we obtain

$$\min\{\delta, \lambda\} = \|x_{t_3}\| \leq M\|\phi\|e^{-\Omega t_3}(1+\varepsilon) + \frac{\varepsilon M}{\Omega}(1+2\Omega)\min\{\delta, \lambda\}$$

$$< M \frac{\min\{\delta, \lambda\}(1+2\Omega)}{2[\Omega+2M(1+2\Omega)]} \cdot \frac{\Omega+2M(1+2\Omega)}{2M(1+2\Omega)} + \frac{(1+2\Omega)\Omega\min\{\delta, \lambda\}}{2\Omega(1+2\Omega)}$$

$$= \min\{\delta, \lambda\},$$

which contradics the previously listed properties of the solution $x_t(\phi)$. In fact we have proved that the solution $x_t(\phi)$ starting from the ball with radius $\min\{\delta, \lambda\}$ (hence also from the ball with radius λ), never reaches its boundary and so (6.3.30) obviously holds. For ε as in (6.3.35) the estimate (6.3.34) can be considerably improved:

$$\forall \lambda > 0 \quad \exists \varkappa = \varkappa(\lambda) \text{ (defined by (6.3.36))}: \|\phi\| < \varkappa$$

$$\Rightarrow \|x_t\| \leq M\|\phi\|e^{-\Omega t} \frac{\Omega+2M(1+2\Omega)}{2M(1+2\Omega)} + \frac{\lambda}{2} \quad \forall t \geq 0. \quad (6.3.37)$$

(Part 3) *Uniform attractivity of the equilibrium point*

In particular, by (6.3.37), for $\lambda = 1$ there exists $\varkappa = \varkappa(1) > 0$ such that

$$\|x_t\| \leq \frac{\Omega+2M(1+2\Omega)}{2(1+2\Omega)} \|\phi\|e^{-\Omega t} + \frac{1}{2} \quad \forall t \geq 0.$$

Hence
$\exists \Delta < 1 \quad \exists T > 0$ (independent of ϕ): $\|x_t\| \leq \|\phi\|$

$$\forall t \geq T \quad \forall \phi \in C \quad \|\phi\| < \varkappa \quad (6.3.38)$$

and $\varkappa$ is as in (6.3.36) with substitution $\lambda = 1$. The proof goes on by similar arguments as at the end of the proof of Lemma 6.2.5. An arbitrary number $t \geqslant T$ can be represented in the form $t = mT(1+\varrho)$, where $m \in N$, $\varrho \in [0, 1]$. Now in virtue of the semigroup property of solutions, by the Schwarz inequality and (6.3.28) we obtain

$$\|x_t\| = \|x_{mT+m\varrho T}\| = \|(x_{T+\varrho T})^m\| \leqslant \Delta^m \|\phi\| = \|\phi\| e^{m \ln \Delta}$$
$$= \|\phi\| e^{-t\left(\frac{-\ln \Delta}{T+\varrho T}\right)} \leqslant \varkappa e^{-t\left(\frac{-\ln \Delta}{T+\varrho T}\right)} \quad \forall t \geqslant T.$$

Taking $\eta = \varkappa(1) > 0$ and $t_0 = \max\left\{(T+\varrho T)\dfrac{\ln \varkappa(1) - \ln \mu}{-\ln \Delta},\ 0\right\} \geqslant 0$ we easily find that (6.3.31) holds. □

Remark 6.3.1 It follows from the proof that the solutions decay to zero at an exponential rate. Since for the linear system (6.3.27) EXS is equivalent to the uniform asymptotic stability (UAS), Theorem 6.3.1 can be interpreted as a sufficient condition for the preservation of local EXS or local UAS by the equilibrium point, independently of nonlinear perturbations of a rather general type.

For various modifications and generalizations of Theorem 6.3.1 see Hale (1977), Hale and Ize (1971), Hale and Martinez-Amores (1977) and Chukwu (1981).

6.3.3 Stability of a class of abstract linear systems with a nonlinear perturbation

The purpose of this section is to present a linearization theorem for a class of abstract linear systems with nonlinear perturbation. In the following theorem the satisfaction of rather strong hypotheses is required, especially with regard to nonlinear perturbation. However, the thesis guarantees not only the preservation of EXS of the equilibrium point in the perturbed system, but also the preservation of strong solutions of an abstract differential equation.

We need some additional notions and results for the presentation of the results.

Definition 6.3.1 Let $\mathscr{X}$ be a Banach space with norm $\|\cdot\|$ and let C be an arbitrary subset. A family $\{S(t)\}_{t \geqslant 0}$ of continuous operators $S(t)\colon C \to C$ is called a C_0-semigroup on C if
  (i) the mapping $S(\cdot)u\colon [0, +\infty) \to C$ is continuous for each fixed $u \in C$,
  (ii) $S(0)u = u \quad \forall u \in C$,
  (iii) $S(t+\tau) = S(t)S(\tau) \quad \forall t, \tau \geqslant 0$.
If, furthermore

6.3 Stability of Linear Systems with Nonlinear Perturbations

(iv) $\|S(t)u_1 - S(t)u_2\| \leq e^{\omega t}\|u_1 - u_2\|$ $\forall u_1, u_2 \in C$, $\forall t \geq 0$, then the C_0-semigroup $\{S(t)\}_{t \geq 0}$ is called the *semigroup of quasi-contractions* (contractions if $\omega = 0$).

The C_0-semigroups of quasi-contractions are a natural generalization of the linear C_0-semigroups (compare Definition 6.3.1 with Theorem 6.2.1). The basic results regarding the nonlinear C_0-semigroups of quasi-contraction were obtained by Crandall (1971), and Crandall and Liggett (1971). Since their paper several contributions to this problem have been made, in particular, by Miyadera (1971) and Reich (1981).

Theorem 6.3.2 (Crandall, Liggett and Miyadera) *Let* $\mathcal{A}: (\mathcal{D}(\mathcal{A}) \subset \mathcal{X}) \to \mathcal{X}$, $\mathcal{X}$ *a Banach space with norm* $\|\cdot\|$, *be a nonlinear operator having the following properties*:

$$\exists \omega \in R: (\mathcal{A} - \omega I) \text{ is dissipative (Definition 6.2.2)} \tag{6.3.39}$$

$$\exists \lambda_0 > 0: \forall \lambda \in (0, \lambda_0): \overline{\mathcal{D}(\mathcal{A})} \subset \mathcal{R}(I - \lambda \mathcal{A}). \tag{6.3.40}$$

Then $\forall u \in \overline{\mathcal{D}(\mathcal{A})}$, $\forall t \geq 0$ $\exists \lim_{n \to \infty} \left(I - \frac{t}{n}\mathcal{A}\right)^{-n} u$ *and the formula*

$$S(t)u = \lim_{n \to \infty}\left(I - \frac{t}{n}\mathcal{A}\right)^{-n} u \quad u \in \overline{\mathcal{D}(\mathcal{A})} \quad t \geq 0 \tag{6.3.41}$$

called Hille's exponential formula, determines a C_0-*semigroup of quasi-contractions on* $\overline{\mathcal{D}(\mathcal{A})}$ *with Lipschitz constant* $e^{\omega t}$.

Moreover, if $\mathcal{A}$ *is closed and the state space is reflexive* (Balakrishnan, 1976), *then* $S(\cdot)u_0$ *is the strong solution* (Definition 6.2.3) *of the following abstract initial-value problem*:

$$\begin{aligned} \dot{u}(t) &= \mathcal{A}u \quad t > 0 \\ u(0) &= u_0 \in \mathcal{D}(\mathcal{A}) \end{aligned} \tag{6.3.42}$$

on $[0, +\infty)$.

Various applications of Theorem 6.3.2 to FDE are presented in Flaschka and Leitman (1975), Plant (1977), Webb (1974, 1976), Walker (1980), Crandall (1971). Here we give a new result involving Theorem 6.3.2. A very similar result, derived in a different way, is given in Gorbunov and Shikhov (1975).

Theorem 6.3.3 *Let* $\mathcal{X}$ *be a Hilbert space with inner product* $\langle \cdot, \cdot \rangle$. *Consider the initial-value problem*

$$\begin{aligned} \dot{u}(t) &= \mathcal{A}u + \mathcal{N}(u) \quad t > 0 \\ u(0) &= u_0 \in \mathcal{D}(\mathcal{A}) \end{aligned} \tag{6.3.43}$$

where $\mathscr{A}: (\mathscr{D}(\mathscr{A}) \subset \mathscr{X}) \to \mathscr{X}$ is a linear operator with the following properties:

$$\mathscr{R}(I-\lambda\mathscr{A}) = \mathscr{X} \text{ for all sufficiently small } \lambda > 0 \qquad (6.3.44)$$

$$\exists \omega \in R: \langle u, \mathscr{A}u \rangle \leqslant \omega \langle u, u \rangle \quad \forall u \in \mathscr{D}(\mathscr{A}) \qquad (6.3.45)$$

the linear C_0-semigroup $\{T(t)\}_{t \geqslant 0}$ generated by $\mathscr{A}$ is EXS (6.3.46)

and the nonlinear operator $\mathscr{N}: \mathscr{X} \ni u \mapsto \mathscr{N}(u) \in \mathscr{X}$ has the following properties:

$$\frac{\|\mathscr{N}(u)\|}{\|u\|} \xrightarrow[u \to 0]{} 0, \qquad (6.3.47)$$

$$\forall R > 0 \quad \exists L = L(R) > 0: \|\mathscr{N}(u_1) - \mathscr{N}(u_2)\| \leqslant L\|u_1 - u_2\|$$

$\forall u_1, u_2 \in \overline{K(0, R)}$ where $\overline{K(0, R)}$ denotes a closed ball centred at $0 \in \mathscr{X}$ with the radius R. (6.3.48)

Thesis: In a sufficiently small neighbourhood of the point $0 \in \mathscr{X}$ the right-hand side of (6.3.43) generates a C_0-semigroup of nonlinear operators, being the family of strong solutions of (6.3.43). This semigroup is also EXS.

Proof In virtue of Theorem 6.2.1, Lemmas 6.2.1, 6.2.2 and hypotheses (6.3.44), (6.3.45) we establish that $\mathscr{A}$ is the generator of a linear C_0-semigroup $\{T(t)\}_{t \geqslant 0}$ on $\overline{\mathscr{D}(\mathscr{A})} = \mathscr{X}$. Now, by (6.3.46) and Theorem 6.2.3 there exists

$$\mathscr{H} = \mathscr{H}^* \in \mathscr{L}(\mathscr{X}) \quad \langle u, \mathscr{H}u \rangle \geqslant 0 \quad \forall u \in \mathscr{X}$$
$$\langle \mathscr{A}u, \mathscr{H}u \rangle + \langle u, \mathscr{H}\mathscr{A}u \rangle = -\langle u, u \rangle \quad \forall u \in \mathscr{D}(\mathscr{A}). \qquad (6.3.49)$$

Consider the functional

$$V(u) = \langle u, u \rangle + (1+2|\omega|)\langle u, \mathscr{H}u \rangle. \qquad (6.3.50)$$

It follows directly from (6.3.50) that

$$V(u) \geqslant \langle u, u \rangle = \|u\|^2 \qquad (6.3.51)$$
$$V(u) \leqslant \|u\|^2 [1+\|\mathscr{H}\|(1+2|\omega|)]. \qquad (6.3.52)$$

Next, by (6.3.49) and (6.3.45) we obtain

$$\lambda > 0 \Rightarrow V(u - \lambda\mathscr{A}u - \lambda\mathscr{N}(u)) - V(u) = \lambda^2 \langle \mathscr{A}u + \mathscr{N}, \mathscr{A}u + \mathscr{N} \rangle -$$
$$- \lambda \langle u, \mathscr{A}u \rangle - \lambda \langle \mathscr{A}u, u \rangle - \lambda \langle \mathscr{N}, u \rangle - \lambda \langle u, \mathscr{N} \rangle +$$
$$+ (1+2|\omega|)\lambda^2 \langle \mathscr{A}u + \mathscr{N}, \mathscr{H}(\mathscr{A}u + \mathscr{N}) \rangle - \lambda(1+2|\omega|)[\langle u, \mathscr{H}\mathscr{A}u \rangle +$$
$$+ \langle u, \mathscr{H}\mathscr{N} \rangle + \langle \mathscr{A}u, \mathscr{H}u \rangle + \langle \mathscr{N}, \mathscr{H}u \rangle] \geqslant -2\lambda \langle u, \mathscr{N} \rangle +$$
$$+ \lambda \langle u, u \rangle - 2\lambda(1+2|\omega|)\langle u, \mathscr{H}\mathscr{N} \rangle \geqslant \lambda \|u\|^2 \left\{ 1 - 2[1+\|\mathscr{H}\|(1+ \right.$$
$$\left. + 2|\omega|)]\frac{\|\mathscr{N}(u)\|}{\|u\|} \right\} \quad \forall u \in \mathscr{D}(\mathscr{A}) \setminus \{0\}. \qquad (6.3.53)$$

6.3 Stability of Linear Systems with Nonlinear Perturbations

It follows from (6.3.47) that $\forall \varepsilon > 0 \ \exists \delta = \delta(\varepsilon) > 0$: $\|u\| < \delta \Rightarrow \dfrac{\|\mathcal{N}(u)\|}{\|u\|} \leq \varepsilon$.

In particular, for some $\delta > 0$ we have

$$\frac{\|\mathcal{N}(u)\|}{\|u\|} \leq \frac{1}{4[1+\|\mathcal{H}\|(1+2|\omega|)]} \quad \forall u, \ \|u\| < \delta.$$

Taking this into account in (6.3.53) we obtain

$$\lambda > 0 \Rightarrow V(u - \lambda \mathcal{A} u - \lambda \mathcal{N}(u)) - V(u) \geq \tfrac{1}{2}\lambda \|u\|^2 \geq 0$$
$$\forall u \in \overline{K(0, \delta)} \cap \mathcal{D}(\mathcal{A}). \tag{6.3.54}$$

(6.3.50) implies that the level-set $\{u \in \mathcal{X} : V(u) \leq l\}$ is contained in the ball $K(0, \sqrt{l})$. In particular, if $l < \delta^2$, then it lies in $K(0, \delta)$. Furthermore it is readily seen that due to (6.3.51) an open ball centred at $0 \in \mathcal{X}$ can be inscribed into the level-set of V for all $l > 0$.

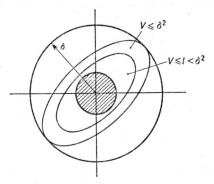

Figure 6.3.2 An auxiliary diagram for the proof of Theorem 6.3.1

Let $\hat{\mathcal{A}}$ be the maximal restriction of the operator $u \mapsto \mathcal{A}u + \mathcal{N}(u)$ to the set $\mathcal{D}(\hat{\mathcal{A}}) = \{u \in \mathcal{X} : V(u) \leq l < \delta^2\} \cap \mathcal{D}(\mathcal{A})$. We shall prove that $\hat{\mathcal{A}}$ defined in this manner has the following properties:

$$\overline{\mathcal{D}(\hat{\mathcal{A}})} \subset \mathcal{R}(I - \lambda \hat{\mathcal{A}}) \text{ for all sufficiently small } \lambda > 0 \tag{6.3.55}$$
$$\exists \Omega \in R: \ \hat{\mathcal{A}} - \Omega I \text{ is dissipative.} \tag{6.3.56}$$

For the proof of (6.3.55) let us take arbitrary $v \in \overline{\mathcal{D}(\hat{\mathcal{A}})} = \{u \in \mathcal{X} : V(u) \leq l < \delta^2\}$. Obviously

$$V(v) \leq l. \tag{6.3.57}$$

We shall prove that for all sufficiently small $\lambda > 0$ we have $v \in \mathcal{R}(I - \lambda \hat{\mathcal{A}})$ or, equivalently, that for all sufficiently small $\lambda > 0$ the equation

$$v = u - \lambda \mathcal{A} u - \lambda \mathcal{N}(u) \tag{6.3.58}$$

has a solution $u \in \mathscr{D}(\hat{\mathscr{A}})$, that is $u \in \mathscr{D}(\mathscr{A})$ and $V(u) \leq l$. Making use of (6.3.44), (6.3.45) and Remark 6.2.2 it is easy to show that the resolvent of $\mathscr{A}$, i.e. the mapping $(I - \lambda \mathscr{A})^{-1}: \mathscr{X} \to \mathscr{D}(\mathscr{A})$ is well-defined and satisfies the Lipschitz condition with Lipschitz constant $\dfrac{1}{1 - \lambda \omega}$ $\forall \lambda > 0$, $\lambda \omega < 1$. Thus, for such λ, (6.3.58) can be rewritten in the form

$$u = (I - \lambda \mathscr{A})^{-1} v + \lambda (I - \lambda \mathscr{A})^{-1} \mathscr{N}(u). \qquad (6.3.59)$$

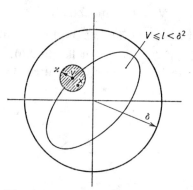

Figure 6.3.3 An auxiliary diagram for the proof of Theorem 6.3.1

Consider the ball $\overline{K(v, \varkappa)}$ where $\varkappa > 0$ is small enough to ensure that $\overline{K(v, \varkappa)} \subset \overline{K(0, \delta)}$. Obviously such a number exists, as the level-set $\{u \in \mathscr{X}: V(u) \leq l < \delta^2\}$ is strongly immersed in the ball $K(0, \delta)$. Consider further the mapping

$$\Psi: \overline{K(v, \varkappa)} \ni u \mapsto \Psi(u) \stackrel{\text{df}}{=} (I - \lambda \mathscr{A})^{-1} v + \lambda (I - \lambda \mathscr{A})^{-1} \mathscr{N}(u) \in \mathscr{D}(\mathscr{A}).$$

On the one hand, by assumption (6.3.48)

$$\|v - \Psi(u)\| = \|[I - (I - \lambda \mathscr{A})^{-1}] v - \lambda (I - \lambda \mathscr{A})^{-1} \mathscr{N}(u)\|$$

$$= \lambda \|(I - \lambda \mathscr{A})^{-1} (v + \mathscr{N}(u))\| \leq \frac{\lambda}{1 - \lambda \omega} \|v + \mathscr{N}(u)\|$$

$$\leq \frac{\lambda}{1 - \lambda \omega} [\|v\| + \|\mathscr{N}(u)\|] \leq \frac{\lambda}{1 - \lambda \omega} [\delta + \delta L(\delta)].$$

We see that λ can always be diminished enough to ensure $\dfrac{\lambda \delta (1 + L(\delta))}{1 - \lambda \omega} \leq \varkappa$ and with such $\lambda > 0$ we have $\Psi: \overline{K(v, \varkappa)} \to \overline{K(v, \varkappa)} \cap \mathscr{D}(\mathscr{A})$. On the other, by assumption (6.3.48)

6.3 Stability of Linear Systems with Nonlinear Perturbations

$$\|\Psi(u_1)-\Psi(u_2)\| = \lambda\|(I-\lambda\mathscr{A})^{-1}[\mathscr{N}(u_1)-\mathscr{N}(u_2)]\| \leq \frac{\lambda}{1-\lambda\omega}\|\mathscr{N}(u_1)-\mathscr{N}(u_2)\| \leq \frac{\lambda L(\delta)}{1-\lambda\omega}\|u_1-u_2\| \quad \forall u_1, u_2 \in \overline{K(v,\varkappa)}.$$

By further diminishing λ we can obtain $\dfrac{\lambda L(\delta)}{1-\lambda\omega} < 1$ and then Ψ is a strict contraction. In accordance with the Banach fixed point theorem, Ψ has a unique fixed point in $\overline{K(v,\varkappa)} \cap \mathscr{D}(\mathscr{A})$. But every fixed point of Ψ is the solution of (6.3.59). Now to finish the proof, it remains to show that this fixed point also satisfies the inequality $V(u) \leq l$. As $u \in \overline{K(v,\varkappa)} \cap \mathscr{D}(\mathscr{A}) \subset \overline{K(0,\delta)} \cap \mathscr{D}(\mathscr{A})$, we have by (6.3.54)

$$V(u-\lambda\mathscr{A}u-\lambda\mathscr{N}(u))-V(u) = V(v)-V(u) \geq \tfrac{1}{2}\lambda\|u\|^2 \geq 0$$

which together with (6.3.57) yields $V(u) \leq l$. Let us take for the proof of (6.3.56) arbitrary $u_1, u_2 \in \mathscr{D}(\hat{\mathscr{A}}) = \{u \in \mathscr{X}: V(u) \leq l\} \cap \mathscr{D}(\mathscr{A})$. We then have the following estimate:

$$\|(u_1-\lambda\hat{\mathscr{A}}u_1)-(u_2-\lambda\hat{\mathscr{A}}u_2)\| = \|u_1-\lambda\mathscr{A}u_1-\lambda\mathscr{N}(u_1)-u_2+\lambda\mathscr{A}u_2+\lambda\mathscr{N}(u_2)\|$$
$$\geq \|(I-\lambda\mathscr{A})(u_1-u_2)\|-\lambda\|\mathscr{N}(u_1)-\mathscr{N}(u_2)\|.$$

Now, in virtue of the assumptions (6.3.45), (6.3.48) and Remark 6.2.2

$$\|(u_1-\lambda\hat{\mathscr{A}}u_1)-(u_2-\lambda\hat{\mathscr{A}}u_2)\| \geq (1-\lambda\omega)\|u_1-u_2\|-\lambda L(\delta)\|u_1-u_2\|,$$

thus operator $\hat{\mathscr{A}}-(\omega+L(\delta))I$ is dissipative. Of course $\Omega = \omega+L(\delta)$ in (6.3.56). The operator $\mathscr{A}$ is closed as the infinitesimal generator of the linear semigroup $\{T(t)\}_{t \geq 0}$ on $\mathscr{X}$ (Balakrisknan, 1976). Hence, $\hat{\mathscr{A}}$ also is closed as the sum of closed and continuous operators. It now follows from Theorem 6.3.2 that through (6.3.41) $\hat{\mathscr{A}}$ generates a semigroup of quasi-contractions $\{\mathscr{S}(t)\}_{t \geq 0}$ on $\mathscr{D}(\hat{\mathscr{A}})$, that is, on $\{u \in \mathscr{X}: V(u) \leq l < \delta^2\}$. This semigroup simultaneously provides the solutions to (6.3.43). It remains to show that $\{\mathscr{S}(t)\}_{t \geq 0}$ is EXS. Note that for $u \in \mathscr{D}(\hat{\mathscr{A}})$ the function $[0,+\infty) \ni t \mapsto v(t) = V[\mathscr{S}(t)u]$ is right-hand side differentiable and we have

$$\frac{d^+v(t)}{dt} = \bigl(\operatorname{grad}V(\mathscr{S}(t)u)\bigr)\bigl(\mathscr{A}\mathscr{S}(t)u+\mathscr{N}(\mathscr{S}(t)u)\bigr). \qquad (6.3.60)$$

Repeating once more the estimates which lead to (6.3.54) we obtain

$$\bigl(\operatorname{grad}V(u)\bigr)\bigl(\mathscr{A}u+\mathscr{N}(u)\bigr) \leq -\tfrac{1}{2}\|u\|^2 \quad \forall u \in \mathscr{D}(\hat{\mathscr{A}}). \qquad (6.3.61)$$

We now derive a differential inequality from (6.3.60), (6.3.61), (6.3.52)

$$\frac{d^+v(t)}{dt} \leq -\tfrac{1}{2}\|\mathscr{S}(t)u\|^2 \leq \frac{-1}{2[1+\|\mathscr{H}\|(1+2|\omega|)]}v(t) = -2\varepsilon v(t)$$

$$v(0) = V(u).$$

Applying the theorem on differential inequalities (Hartman, 1964) we have the estimate

$$V[\mathscr{S}(t)u] = v(t) \leq e^{-2\varepsilon t}V(u) \quad \forall t \geq 0 \quad \forall u \in \mathscr{D}(\hat{\mathscr{A}}).$$

Taking the inequalities (6.3.51), (6.3.52) into account we obtain

$$\|\mathscr{S}(t)u\|^2 \leq e^{-2\varepsilon t}\|u\|^2 \frac{1}{4\varepsilon} \quad \forall t \geq 0 \quad \forall u \in \mathscr{D}(\hat{\mathscr{A}}).$$

As the estimate

$$\|\mathscr{S}(t)u\| \leq e^{-\varepsilon t}\|u\|\frac{1}{2\sqrt{\varepsilon}} \quad \forall t \geq 0 \quad \forall u \in \overline{\mathscr{D}(\hat{\mathscr{A}})}$$

follows readily from the one above, the proof of Theorem 6.3.3 is completed. □

6.3.4 Examples

Example 1

Consider again system (6.1.7), rewriting it now in the form of a system of equations

$$\dot{x}_1(t) = x_2(t) \qquad (6.3.62)$$

$$\dot{x}_2(t) = -\frac{a}{h}x_2(t) - \frac{b}{h}\sin[x_1(t-h)].$$

Interpreting (6.3.62) as FDE on the space $C = C([-h, 0], R^2)$ we find that it has the form (6.3.28) with

$$\mu(\theta) = \begin{bmatrix} 0 & 0 \\ 0 & 0 \end{bmatrix} \quad \forall \theta \in [-h, 0]$$

$$\eta(\theta) = \begin{bmatrix} 0 & -h \leq \theta \leq 0 & \begin{cases} 0 & -h \leq \theta < 0 \\ 1 & \theta = 0 \end{cases} \\ \begin{cases} 0 & \theta = -h \\ -\dfrac{b}{h} & -h < \theta \leq 0 \end{cases} & \begin{cases} 0 & -h \leq \theta < 0 \\ -\dfrac{a}{h} & \theta = 0 \end{cases} \end{bmatrix}.$$

$$\phi = \begin{bmatrix} \phi_1 \\ \phi_2 \end{bmatrix} \in C \quad G(\phi) = \begin{bmatrix} 0 \\ 0 \end{bmatrix} \quad F(\phi) = \begin{bmatrix} 0 \\ -\dfrac{b}{h}\{\sin[\phi_1(-h)] - \phi_1(-h)\} \end{bmatrix}.$$

6.3 Stability of Linear Systems with Nonlinear Perturbations

We now prove that condition (6.3.29) is fulfilled. To do this we choose an arbitrary $\varepsilon > 0$. Due to the obvious properties of the since function there exists $\delta(\varepsilon) > 0$

$$|\sin[\phi_1(-h)] - \phi_1(-h)| < \varepsilon|\phi_1(-h)| \quad \text{if } |\phi_1(-h)| < \delta. \quad (6.3.63)$$

We claim that such δ may be substituted into (6.3.29). Indeed, $\phi \in C$, $\|\phi\| < \delta \Rightarrow |\phi_1(-h)| \leqslant \max_{\theta \in [-h, 0]} |\phi(\theta)| = \|\phi\| < \delta$ and by (6.3.62) we have

$$|F(\phi)| = \frac{b}{h}|\sin\phi_1(-h) - \phi_1(-h)| < \frac{\varepsilon b}{h}|\phi_1(-h)| \leqslant \frac{\varepsilon b}{h}\|\phi\|,$$

consequently (6.3.29) holds.

It now follows from Theorem 6.3.1 that the equilibrium point $0 \in C$ in system (6.3.62) is uniformly asymptotically stable (UAS) if the companion linearized system (of the form (6.3.27)) is exponentially stable (EXS). By virtue of Corollary 6.2.1, however, it is necessary and sufficient for the EXS of the linearized system that

$$\left\{\lambda \in C: \det\left[\lambda I - \int_{-h}^{0} e^{\lambda\theta} d\eta(\theta)\right] = \lambda^2 + \frac{a}{h}\lambda + \frac{b}{h}e^{-\lambda h} = 0\right\} \subset \{\lambda \in C: \operatorname{Re}\lambda < 0\}. \quad (6.3.64)$$

Theorem 6.3.3 cannot formally be applied to the analysis of system (6.3.62), as an abstract system (6.3.43) on a Hilbert space $\mathscr{X} = R^2 \times L^2([-h, 0], R^2)$ with inner product

$$\langle u_1, u_2 \rangle = x_1^T x_2 + \int_{-h}^{0} \psi_1(\theta)^T \psi_2(\theta) d\theta \quad u_i = (x_i, \psi_i) \in \mathscr{X} \quad (6.3.65)$$

as the perturbation $\mathscr{N}(u)$ includes an operator $L^2 \ni \phi \mapsto \phi_1(-r) \in R$ that is not well-defined, and thus the rather strict requirements on $\mathscr{N}(u)$ imposed in Theorem 6.3.3 are not satisfied.

Example 2

Consider again the mathematical model of the nuclear reactor dynamics in the form (Goryachenko, 1971)

$$\begin{aligned}\dot{x}_1(t) &= -ax_2(t-h) - b[e^{x_1(t)} - 1] \\ \dot{x}_2(t) &= -x_2(t) + [e^{x_1(t)} - 1].\end{aligned} \quad (6.3.66)$$

Some results related to the stability of system (6.3.66) have been derived in Example 3, Section 6.1.

As in Example 1, (6.3.66) can be interpreted as a FDE (6.3.28) on the space $C = C([-h, 0], R^2)$ with

$$\mu(\theta) = \begin{bmatrix} 0 & 0 \\ 0 & 0 \end{bmatrix} \quad \forall \theta \in [-h, 0],$$

$$\eta(\theta) = \begin{bmatrix} \begin{cases} 0 & -h \leqslant \theta < 0 \\ -b & \theta = 0 \end{cases} & \begin{cases} a & \theta = -h \\ 0 & -h < \theta \leqslant 0 \end{cases} \\ \begin{cases} 0 & -h \leqslant \theta < 0 \\ 1 & \theta = 0 \end{cases} & \begin{cases} 0 & -h \leqslant \theta < 0 \\ -1 & \theta = 0 \end{cases} \end{bmatrix},$$

$$\phi = \begin{bmatrix} \phi_1 \\ \phi_2 \end{bmatrix} \in C \quad G(\phi) = \begin{bmatrix} 0 \\ 0 \end{bmatrix} \quad F(\phi) = \begin{bmatrix} -b\{e^{\phi_1(0)} - 1 - \phi_1(0)\} \\ e^{\phi_1(0)} - 1 - \phi_1(0) \end{bmatrix}.$$

Utilizing properties of the function $y \mapsto e^y - 1 - y$ it can be proved, as in Example 1, that condition (6.3.29) holds and hence the EXS of the linearized system follows the UAS of the equilibrium point of the nonlinear system. By Corollary 6.2.1 EXS takes place if and only if

$$\left\{ \lambda \in C : \det\left[\lambda I - \int_{-h}^{0} e^{\lambda \theta} d\eta(\theta) \right] = \lambda^2 + \lambda(b+1) + b + ae^{-\lambda h} = 0 \right\}$$

$$\subset \{\lambda \in C : \operatorname{Re} \lambda < 0\} \tag{6.3.67}$$

Methods which allow us to check whether (6.3.64) or (6.3.67) holds, will be presented in the next section.

System (6.3.66) can be also interpreted as an abstract system (6.3.43) on the Hilbert space $\mathscr{X} = R^2 \times L^2([-h, 0], R^2)$ with inner product (6.3.65), and

$$u = \begin{bmatrix} x_1 \\ x_2 \\ \psi_1 \\ \psi_2 \end{bmatrix} \begin{matrix} \} x \in R^2 \\ \\ \} \psi \in L^2 \end{matrix} \quad \mathscr{A}u = \begin{bmatrix} -bx_1 - a\psi_2(-h) \\ x_1 - x_2 \\ \dot{\psi}_1 \\ \dot{\psi}_2 \end{bmatrix}, \quad \mathscr{N}(u) = \begin{bmatrix} -bF(x_1) \\ F_d x_1) \\ 0 \\ 0 \end{bmatrix},$$

$$F(y) = e^y - 1 - y$$

$$\mathscr{D}(\mathscr{A}) = \{u = (x, \psi) \in \mathscr{X} : \psi \in AC([-h, 0], R^2), \quad \dot{\psi} \in L^2([-h, 0], R^2),$$

$$\psi(0) = x\}.$$

It follows from the results given in Section 6.2 that all hypotheses of Theorem 6.3.3 related to the operator $\mathscr{A}$ are fulfilled. Moreover, the necessary and sufficient condition for the EXS of the semigroup generated by $\mathscr{A}$ has again the form (6.3.67). It is easy to establish that the nonlinear perturbation $\mathscr{N}(u)$ also satisfies appropriate hypotheses of Theorem 6.3.3. Consequently under condition (6.3.67) the local EXS of the equilibrium point $0 \in \mathscr{X}$ takes place in the nonlinear system.

Remark 6.3.2 System (6.3.66) is a particular case of a so-called Lur'e control system on a Hilbert space $(\mathscr{X}, \langle \cdot, \cdot \rangle)$

$$\dot{u}(t) = \mathscr{A}u + F(\langle c, u \rangle)b$$
$$u(0) = u_0 \in \mathscr{D}(\mathscr{A})$$
(6.3.68)

in which $\mathscr{A}$ satisfies the hypotheses of Theorem 6.3.3; $b, c \in \mathscr{X}$; $F: R \to R$ is a locally-Lipschitzian function such that $\left|\dfrac{F(y)}{y}\right|_{y \to 0} \to 0$.

It is not difficult to check that under the above assumptions the nonlinear perturbation $\mathscr{N}(u) = F(\langle c, u \rangle)b$ satisfies the hypotheses of Theorem 6.3.3 too.

6.4 Criteria for the Location of Characteristic Function Zeros in Left Complex Half-plane

It follows from the results presented in Section 6.2 that in the stability analysis of linear time-delay systems, the conditions under which the system eigenvalues (zeros of the characteristic function) have negative real parts, are of great importance. The most effective criteria of this kind will be discussed in this section.

The main tool is the principle of argument, presented in Section 6.4.1. In Section 6.4.2 we obtain a generalization of the Mikhaylov–Leonhard criterion for retarded functional-differential equations (RFDE). As all characteristic functions appearing in Section 6.2 are entire functions it seems quite natural to apply the existing theory of entire function zero distribution, for the derivation of conditions for the location of the spectrum in the left complex half-plane. A survey of results of this kind is given in Levin's excellent monograph (1964). Unfortunately, the majority of the results presented there have little practical applicability because of the difficulties involved in the verification of hypotheses of the respective theorems. Thus in practice Pontryagin's criterion (Pontryagin, 1942), which we derive in Section 6.4.3, or some similar condition have to be used to yield effective results. It deals with the characteristic functions which are reducible to quasipolynomials and permits one to obtain the necessary and sufficient condition for the zeros of quasipolynomial to be in the left half-plane in the case of lumped and commensurable delays. In Chapter 7 Pontryagin's criterion is applied to a stability analysis of the simplest classical automatic control systems.

6.4.1 The principle of argument

The following theorem, which is a fundamental result in the theory of complex-variable functions, is called the principle of argument. For sketch of a proof see for example Sidorov, Fedoryuk and Shabunin (1985, Section 30).

Theorem 6.4.1 (i) Let $\Omega \subset C$ be a domain (i.e. Ω is an open set and its two arbitrary points can be connected by a broken line contained in Ω), whose boundary $\partial\Omega$ is a closed, rectifiable Jordan curve (i.e. it is of finite length and has no multiple points).
(ii) Let $f: C \ni z \mapsto f(z) \in C$ be a meromorphic function (i.e. a single-valued analytic function which has no singular points except poles) and let there be no zeros or poles on $\partial\Omega$.
Then

$$\frac{1}{2\pi i} \int_{\partial\Omega^+} \frac{f'(z)}{f(z)} \, dz = N - P = \frac{1}{2\pi} \sum_{m \geq 0} \{\arg f[z(t_m-)] - \arg f[z(t_m+)]\}$$

(6.4.1)

where the symbol $+$ denotes that the curve $\partial\Omega$ is oriented counter-clockwise; f' is the derivative of f; N is the total number of zeros of the function f in Ω counting their multiplicities; P is the total number of poles of the function f in Ω, counting their multiplicities; $\arg z \in [0, 2\pi)$ is the principal value of $\text{Arg } z$, the argument of $z \in C$; $z: [a, b] \ni t \mapsto z(t) \in \partial\Omega$ is an arbitrary continuous parametrization of the curve such that as t increases from a to b then $z(t)$ moves once around $\partial\Omega$ in the counter-clockwise direction; $z(a-) \stackrel{df}{=} z(a)$, $z(b+) \stackrel{df}{=} z(b)$; t_m is a point of discontinuity of the function $[a, b] \ni t \mapsto \arg f[z(t)] \in [0, 2\pi)$.

The summation in (6.4.1) is over all discontinuity points t_m.

Notice that discontinuities of the function $[a, b] \ni t \mapsto \arg f[z(t)] \in [0, 2\pi)$ occur for those $t \in [a, b]$ for which $\text{Im} f[z(t)] = 0$ and $\text{Re} f[z(t)] > 0$ simultaneously. In the stability analysis of linear finite-dimensional systems the principle of argument has been used to obtain Mikhaylov–Leonhard and Nyquist criteria (Parks and Hahn 1981). For simple systems with retardation similar results have been obtained by Krall (1964).

In Section 6.4.2 we present a generalization of the Mikhaylov–Leonhard criterion for a wide class of time-delay systems.

4.2 Application of the principle of argument to the location of characteristic function zeros

The systems investigated in Section 6.2, Examples 1 and 4, have the following characteristic function:

$$C \ni s \mapsto f(s) = \det\left[sI - \int_{-h}^{0} e^{s\theta} d\eta(\theta)\right] \in C,$$

$$\eta: [-h, 0] \to \mathscr{L}(R)^n, \quad \operatorname*{Var}_{-h}^{0} \eta < \infty. \tag{6.4.2}$$

On the basis of Theorem 6.4.1, we derive a theorem which permits the determination of the number of zeros of function (6.4.2) in the right complex plane. Consider the set Ω_R as in Figure 6.4.1.

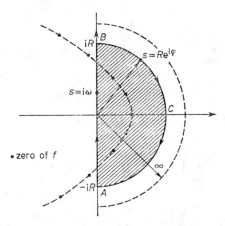

Figure 6.4.1 The set Ω_R used to obtain the generalized Mikhaylov criterion

The investigated function is an entire one and thus it has no singular points in C (it does have in $\bar{C} = C \cup \{\infty\}$). Assume that f has no zeros on the imaginary axis. It follows from Section 6.2, Example 3, that if R is sufficiently large, then no zeros of f lie on the arc BCA either. Therefore all hypotheses of Theorem 6.4.1 are fulfilled if $\Omega = \Omega_R$, and so we have

$$N_R = -\frac{1}{2\pi} \sum_k \{\arg f[s_R(t_k-)] - \arg f[s_R(t_k+)]\} \tag{6.4.3}$$

where N_R is the total number of zeros of the function f in Ω_R (counting their multiplicities); $s_R: [a, b] \ni t \mapsto s_R(t) \in \partial\Omega_R$ is an arbitrary continuous parametrization of $\partial\Omega_R$ such that as t increases from a to b then $s_R(t)$ moves

round once $\partial\Omega_R$ in the clockwise direction; $s_R(a-) \stackrel{df}{=} s_R(a)$, $s_R(b+) \stackrel{df}{=} s_R(b$
t_k is a point of discountinuity of the function $t \mapsto \arg f[s_R(t)]$.

The summation in (6.4.3) is over all discontinuity points. Notice that

$$f(Re^{i\varphi}) = \det\left[Re^{i\varphi}I - \int_{-h}^{0} e^{R\theta e^{i\varphi}} d\eta(\theta)\right]$$

$$= R^n e^{in\varphi} \det\left[I - \int_{-h}^{0} \frac{e^{\theta R(\cos\varphi + i\sin\varphi) - i\varphi}}{R} d\eta(\theta)\right]. \quad (6.4.4)$$

As

$$\left|\int_{-h}^{0} \frac{e^{\theta R(\cos\varphi + i\sin\varphi) - i\varphi}}{R} d\eta(\theta)\right|$$

$$\leq \mathop{\mathrm{Var}}_{-h}^{0} \eta \cdot \max_{\theta \in [-h, 0]} \frac{|e^{\theta R\cos\varphi} e^{i(\theta R\sin\varphi - \varphi)}|}{R} \leq \mathop{\mathrm{Var}}_{-h}^{0} \eta \cdot \frac{1}{R} \cdot \max_{\theta \in [-h, 0]} e^{R\theta \cos\varphi}$$

$$\varphi \in \left[-\frac{\pi}{2}, \frac{\pi}{2}\right] \Rightarrow R\cos\varphi \geq 0 \Rightarrow \theta R\cos\varphi \leq 0 \Rightarrow e^{R\theta \cos\varphi} \leq 1$$

hence

$$\left|I - \int_{-h}^{0} \frac{e^{\theta R(\cos\varphi + i\sin\varphi) - i\varphi}}{R} d\eta(\theta)\right| \xrightarrow[R \to \infty]{} 1 \quad (6.4.5)$$

uniformly with respect to $\varphi \in \left[-\frac{\pi}{2}, \frac{\pi}{2}\right]$.

On the one hand, using (6.4.5) in (6.4.4) we conclude that in the right close s-half-plane the function f behaves like the function $s \mapsto s^n$ for sufficient large $|s|$. On the other, $f(-i\omega) = \overline{f(i\omega)}$, $\forall \omega \in R$.

Taking into account these both facts, we obtain by (6.4.3)

$$N = \lim_{R \to \infty} N_R = -\frac{1}{\pi} \sum_m \{\arg f(i\omega_m -) - \arg f(i\omega_m +)\} + \frac{n}{2}$$

where N is the total number of zeros of the function f (counting their multiplicities) with positive real parts, ω_m is a point of discontinuity of the functio $[0, +\infty) \ni \omega \mapsto f(i\omega) \in \bar{C}$; $\arg f(i\infty +) \stackrel{df}{=} \arg e^{in\pi/2}$; $\arg f(i0-) \stackrel{df}{=} \arg f(0$ Note that the function f has no zeros on the imaginary axis if and only $f(i\omega) \neq 0$ $\forall \omega \in R$. Thus we have obtained the following theorem.

6.4 Criteria for the Location of Characteristic Function Zeros

Theorem 6.4.2 *The function* (6.4.2) *has exactly N zeros (counting their multiplicities) in the open right complex half-plane and has no zeros on the imaginary axis if and only if*

(i) $f(i\omega) \neq 0 \ \forall \omega \in R$,

(ii) $N = \dfrac{n}{2} - \dfrac{1}{\pi} \sum_m \{\arg f(i\omega_m -) - \arg f(i\omega_m +)\}$ *where ω_m is a point of discontinuity of the function* $[0, +\infty) \ni \omega \mapsto \arg f(i\omega) \in [0, 2\pi)$; $\arg f(i0-)$ $= \arg f(0)$, $\arg f(i\infty +) = \arg e^{in\pi/2}$; $\arg z$ *is the principal value of* $\text{Arg} z$, *the argument of* z. *The summation in* (ii) *is over all discontinuity points.*

Definition 6.4.1 *The curve* $\{f(i\omega) \in \overline{C}, \omega \in [0, +\infty]\}$, *oriented by the increasing values of the parameter ω, is called the Mikhaylov plot of function f.*

Examples of Mikhaylov plots for $n = 1, 2, 3, 4$ are depicted in Figure 6.4.2. To see their asymptotic behaviour for large values of ω better, the plots are

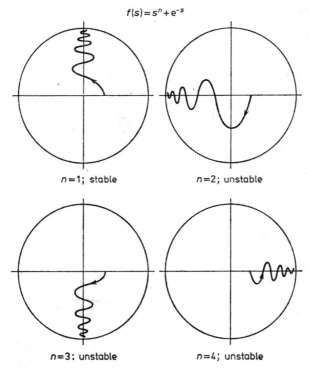

$f(s) = s^n + e^{-s}$

$n=1$; stable $n=2$; unstable

$n=3$; unstable $n=4$; unstable

Figure 6.4.2 Mikhaylov plots for $n = 1, 2, 3, 4$ in the w-complex domain

drawn in the plane of the complex variable $w = \dfrac{Ls}{\sqrt{1+|s|^2}}, L > 0$. This is the so-called Poincaré transformation, which maps the s-complex plane onto the circle $|w| \leq L$ in the w-complex plane. The imaginary and real axes in the s-plane are also mapped onto the appropriate segments of the same axes in the w-plane. The verification of these properties of the Poincaré transformation is left to the reader.

Example

Using Theorem 6.4.2 we establish the range of parameters $a, b, h > 0$ for which the implication (6.3.64) holds. Substituting $z = sh$, $\beta = hb > 0$ in (6.3.43) we reduce the problem to the question of whether the zeros of the function $C \ni z \mapsto f(z) = z^2 + az + \beta e^{-z} \in C$ have negative real parts. To obtain the Mikhaylov plot we set $z = i\omega$ and obtain

$$f(i\omega) = \beta\left[\left(\cos\omega - \frac{\omega^2}{\beta}\right) + i\left(\frac{a}{\beta} - \frac{\sin\omega}{\omega}\right)\right] \quad f(i0) = \beta + i0.$$

The analysis of the Mikhaylov plots for different parameters a, β will be carried out through a graphical discussion of solvability of the equations $\cos\omega = \dfrac{\omega^2}{\beta}$, $\dfrac{a}{\beta} = \dfrac{\sin\omega}{\omega}$, cf. Figure 6.4.3.

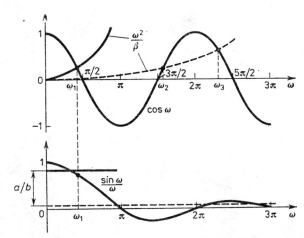

Figure 6.4.3 Graphical discussion of the equations: $\cos\omega = \dfrac{\omega^2}{\beta}$, $\dfrac{a}{\beta} = \dfrac{\sin\omega}{\omega}$

6.4 Criteria for the Location of Characteristic Function Zeros

ω_1 is the unique solution of the equation $\cos\omega = \dfrac{\omega^2}{\beta}$ in the interval $(0, \pi/2)$. Obviously such ω_1 exists.

If $\dfrac{\sin\omega_1}{\omega_1} < \dfrac{a}{\beta}$, then because $n = 2$ the Mikhaylov plot appears as in Figure 6.4.4a. The shape of the plot in the shaded regions may be somewhat different, depending on whether the equation $\cos\omega = \dfrac{\omega^2}{\beta}$ has a positive solution different from ω_1. By Theorem 6.4.2 we have $N = 0$ and all zeros of f have negative real parts. If $\dfrac{\sin\omega_1}{\omega_1} = \dfrac{a}{\beta}$ then the plot has the form shown in Figure 6.4.4b. Since it passes through the point $0 + i0 \in C$, the zeros of f occur on the imaginary axis. They are, of course, $\pm i\omega_1$.

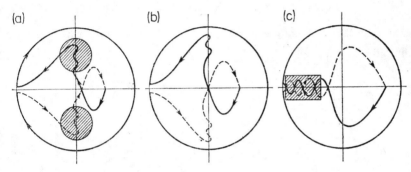

Figure 6.4.4 Different modes of Mikhaylov plots in the example under discussion

If $\dfrac{\sin\omega_1}{\omega_1} > \dfrac{a}{\beta}$ then the analysis of the Mikhaylov plots becomes more complicated and should be divided into subcases in which the equation $\cos\omega = \dfrac{\omega^2}{\beta}$ has respectively 1, 3, 5, 7, etc. solutions on the half-line $(0, +\infty)$ (the limit cases with 2, 4, 6, 8, etc. solutions are of little importance).

1. ω_1 is a unique solution of the equation $\dfrac{\omega^2}{\beta} = \cos\omega$ in $(0, +\infty)$.

The appropriate plot is depicted in Figure 6.4.4c. In the shaded region the plot may be again somewhat different from the one presented depending on the number of solutions of the equation $\dfrac{\sin\omega}{\omega} = \dfrac{a}{\beta}$. By Theorem 6.4.2 we establish that $N = 2$, that is the transition from the inequality $\dfrac{\sin\omega_1}{\omega_1} > \dfrac{a}{\beta}$

to the opposite one corresponds to the transmission of the conjugate pair of zeros of f from the left to the right complex halfplane.

2. The equation $\cos\omega = \dfrac{\omega^2}{\beta}$ has exactly three solutions in $(0, +\infty)$.

Using the graphical discussion of the solutions $\cos\omega = \dfrac{\omega^2}{\beta}$, $\dfrac{\sin\omega}{\omega} = \dfrac{a}{\beta}$ presented in Figure 6.4.3, we establish that three kinds of plots are possible. These are shown in Figure 6.4.5. Figure 6.4.5a corresponds to the inequality $\dfrac{\sin\omega_3}{\omega_3} < \dfrac{a}{\beta}$, where ω_3 is the solution to $\cos\omega = \dfrac{\omega^2}{\beta}$ in the interval $(\tfrac{3}{2}\pi, 2\pi)$, Figure 6.4.5b corresponds to the equality $\dfrac{\sin\omega_3}{\omega_3} = \dfrac{a}{\beta}$ and Figure 6.4.5c to $\dfrac{\sin\omega_3}{\omega_3} > \dfrac{a}{\beta}$. In the last case we obtain $N = 4$ by Theorem 6.4.2 which means that the next conjugate complex pair of zeros of f enters the right half-plane.

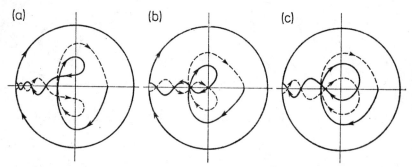

Figure 6.4.5 Other modes of Mikhaylov plots in the example under discussion

3. It may be observed that where there are 5, 7, 9, etc. solutions of the equation $\cos\omega = \dfrac{\omega^2}{\beta}$, possible plots are analogous but the transition from the inequality $\dfrac{\sin\omega_{2k+1}}{\omega_{2k+1}} < \dfrac{a}{\beta}$, $k = 2, 3, \ldots$ to the opposite one always corresponds to an increment of N by 2. As a result of each change of the inequality sign, the conjugate pairs of zeros move to the right half-plane one after the other.

Summarizing our discussion, we conclude that the inclusion (6.3.63) holds if and only if

$$\frac{\sin\omega_1}{\omega_1} < \frac{a}{bh} \tag{6.4.6}$$

6.4 Criteria for the Location of Characteristic Function Zeros

where ω_1 is the solution of the equation $\cos\omega = \dfrac{\omega^2}{bh}$ in an interval $(0, \pi/2)$.

In Example 1, Section 6.1, we obtained $1 < \dfrac{a}{hb}$ as a sufficient condition for (6.3.61) by the Lyapunov functional technique. This inequality of course implies (6.4.6).

6.4.3 Pontryagin's theory of quasipolynomial zero location

Since the proofs of the results presented in this section can be found only in Pontryagin's paper (1942) and are not easily available, we give their complete versions. We begin the presentation of Pontryagin's results on quasipolynomial zero location with the following basic definition.

Definition 6.4.2 Let $(z, t) \mapsto h(z, t)$ be a polynomial in two variables with complex coefficients

$$h(z, t) = \sum_{m,n \geq 0} a_{mn} z^m t^n \qquad a_{mn} \in C. \qquad (6.4.7)$$

The term $a_{rs} z^r t^s$ is called the principal term of the polynomial (6.4.7) if $a_{rs} \neq 0$ and for every other term $a_{mn} z^m t^n$, $a_{mn} \neq 0$ we have $r \geq m, s \geq n$ and at least one of these inequalities is strict. The function $C \ni z \mapsto h(z, e^z) \in C$ is called a quasipolynomial.

The characteristic function of a system which has a finite number of lumped and commensurable delays can be transformed into a quasipolynomial of the form (6.4.7) by multiplication by e^{ks} where k is the largest power of e^{-s} appearing in the characteristic function. Since the term e^{ks} vanishes nowhere in C, the original characteristic function and the quasipolynomial obtained from it have the same set of zeros.

Theorem 6.4.3 *If a polynomial of the form (6.4.7) does not possess a principal term, then the quasipolynomial $z \mapsto h(z, e^z)$ has infinitely many zeros with arbitrarily large positive real parts.*

Proof Let us consider an integer lattice (Figure 6.4.6). The dimensions of this lattice are determined by the maximal and minimal values of the indices m, n in (6.4.7). The appropriate coefficient a_{mn} of the polynomial h may be associated with each node of the lattice. The lack of a principal term means that zero coefficient corresponds to the upper right node of the lattice. Thus there always exists a straight line $\beta = \alpha n + m$ (α, β rational and positive) which passes through at least two nodes corresponding to the nonzero coefficients of the polynomial h, such that zero coefficients correspond to all nodes

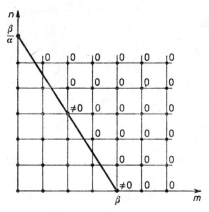

Figure 6.4.6 An auxiliary diagram for the proof of Theorem 6.4.3

lying in the half-plane $\beta > \alpha n + m$. Consider now a polynomial $\sum_{n \geqslant 0} a_{\beta-n\alpha, n} i^{\beta - n\alpha} z^n$.
By the above properties of the straight line $\beta = \alpha n + m$, this polynomial has at least one nonzero root z_0. We claim that there exists a sequence $\{\xi_k\}_{k \in N} \subset C$, $\xi_k \underset{k \to \infty}{\to} 0$, such that for almost all k the numbers

$$z_k = \alpha \ln 2k\pi + 2k\pi i + \ln z_0 + \xi_k \qquad (6.4.8)$$

are zeros of the quasipolynomial $z \mapsto h(z, e^z)$. For the proof we substitute a new variable $\xi \in C$ in the place of z,

$$z = \alpha \ln 2k\pi + 2k\pi i + \ln z_0 + \xi. \qquad (6.4.9)$$

This formula may also be written in the form

$$z = 2k\pi i \left(\frac{\ln z_0 + \xi + \alpha \ln 2k\pi}{2k\pi i} + 1 \right) = 2k\pi i (1 + \delta_1(\xi, k)). \qquad (6.4.10)$$

The sequence of analytic functions $\{\delta_1(\,\cdot\,, k)\}_{k \in N}$ is of course uniformly convergent to the zero function as $k \to \infty$ on an arbitrary compact subset of C. From (6.4.9) we also obtain

$$e^z = (2k\pi)^\alpha z_0 e^\xi. \qquad (6.4.11)$$

It now follows from the relationships (6.4.7), (6.4.10), (6.4.11) that

$$h(z, e^z) = \sum_{m,n} a_{mn} (2k\pi)^{\alpha n + m} i^m (1 + \delta_1)^m z_0^n e^{n\xi}$$

$$= \sum_n a_{\beta - \alpha n, n} (2k\pi)^\beta i^{\beta - n\alpha} z_0^n e^{n\xi} + (2k\pi)^\beta \delta_2(\xi, k)$$

6.4 Criteria for the Location of Characteristic Function Zeros

where

$$\delta_2(\xi, k) = \sum_{\substack{n \geq 0, m \neq 0 \\ \alpha n + m = \beta}} a_{mn} i^m z_0^n e^{n\xi} \left[\sum_{l=1}^{m} \binom{m}{l} \delta_1^l \right] +$$

$$+ \frac{1}{(2k\pi)^\beta} \sum_{\substack{m,n \\ \alpha n + m \neq \beta}} a_{mn} i^m (2k\pi)^{m+\alpha n} (1 + \delta_1)^m z_0^n e^{n\xi}.$$

It is obvious from the properties of the sequence $\{\delta_1(\cdot, k)\}$ and the fact that in the third sum $\beta > \alpha n + m$, that the sequence of analytic functions $\{\delta_2(\cdot, k)\}$ is also uniformly convergent to the zero function as $k \to \infty$ on an arbitrary compact subset of C. A number z of the form (6.4.9) is a zero of the quasipolynomial $z \mapsto h(z, e^z)$ if and only if

$$\sum_n a_{\beta-\alpha n, n} i^{\beta-n\alpha} z_0^n e^{n\xi} + \delta_2(\xi, k) = 0.$$

From the uniform convergence of the sequence $\{\delta_2(\cdot, k)\}$ to zero, and as 0 is the root of a function $\xi \mapsto \sum_n a_{\beta-\alpha n, n} i^{\beta-n\alpha} z_0^n e^{n\xi}$, it follows that there exists a sequence $\{\xi_k\}_{k \in N}$, $\xi_k \xrightarrow[k \to \infty]{} 0$ such that for almost all k we have

$$\sum_n a_{\beta-\alpha n, n} i^{\beta-n\alpha} z_0^n e^{n\xi_k} + \delta_2(\xi_k, k) = 0.$$

After substituting this sequence into (6.4.8) we obtain the sequence of roots of the quasipolynomial $z \mapsto h(z, e^z)$. In particular, (6.4.8) clearly implies that Theorem 6.4.3 holds. □

If the investigated system has a finite number of lumped and commensurable delays and may be described by any of the models discussed in Examples 1–5, Section 6.2, then the quasipolynomial derived from its characteristic function has a principal term. Indeed, the infinitesimal generator of an appropriate semigroup otherwise has infinitely many eigenvalues with arbitrarily large, positive real parts, which contradicts the fact that the spectrum of such an operator is located to the left of some straight line parallel to the imaginary axis.

We introduce a similar concept of principal term for polynomials in three variables of the form

$$f(z, u, v) = \sum_{m, n \geq 0} z^m \varphi_m^{(n)}(u, v) \qquad (6.4.12)$$

where $\varphi_m^{(n)}(u,v)$ is a homogeneous polynomial of degree n; $\varphi_m^{(n)}(1,\pm i) \neq 0$ $\forall m \geqslant 0, \forall n \geqslant 1$.

Definition 6.4.3 *An expression $z^r \varphi_r^{(s)}(u,v)$ is called the* principal term *of polynomial* (6.4.12), *provided that for every other term $z^m \varphi_m^{(n)}(u,v)$ we have $r \geqslant m, s \geqslant n$ and at least one of these inequalities is strict.*

The following result is analogous to that of Theorem 6.4.3.

Theorem 6.4.4 *If a polynomial f of the form* (6.4.12) *has no principal term, then the function $C \ni z \mapsto f(z, \cos z, \sin z) \in C$ has infinitely many zeros with arbitrarily large imaginary parts.*

Proof Substituting $z = i\xi, \xi \in C$ into the equation $f(z, \cos z, \sin z) = 0$ we obtain due to (6.4.12)

$$e^{-(\text{maximal index } n)\xi} \cdot h(\xi, e^\xi) = 0$$

where $h(\xi, e^\xi)$ is a quasipolynomial, corresponding to a polynomial $h(u,t)$ of the form (6.4.7), without a principal term. This equation has solutions equal to the roots of $h(\xi, e^\xi)$.

Using Theorem 6.4.3 we conclude that $h(\xi, e^\xi)$ has infinitely many roots with arbitrarily large positive real parts. As the transformation $z = i\xi$ maps the lower z-half-plane onto the right ξ-half-plane and conversely, we conclude that the function $z \mapsto f(z, \cos z, \sin z)$ has infinitely many zeros with arbitrarily large negative imaginary parts. The transformation $z = -i\xi$ which in turn maps the upper z-half-plane onto the right ξ-half-plane leads to the same conclusion for zeros with arbitrarily large positive imaginary parts. Thus the proof is completed. □

The polynomial (6.4.12) which has a principal term $z^r \varphi_r^{(s)}(u,v)$ can be represented in the form

$$f(z, u, v) = z^r \varphi_*^{(s)}(u,v) + \sum_{\substack{m < r \\ n \leqslant s}} z^m \varphi_m^{(n)}(u,v),$$

$$\varphi_*^{(s)}(u,v) = \sum_{n \leqslant s} \varphi_r^{(n)}(u,v). \qquad (6.4.13)$$

Lemma 6.4.1 *The function $C \ni z \mapsto \Phi_*^{(s)}(z) \in C$, defined by*

$$\Phi_*^{(s)}(z) = \varphi_*^{(s)}(\cos z, \sin z) \qquad (6.4.14)$$

has the following properties:

(i) *It is (2π)-periodic*,

(ii) *$\forall a \in R$ in the strip $a \leqslant \operatorname{Re} z < 2\pi + a$ it has exactly $2s$ zeros, counting their multiplicities*,

6.4 Criteria for the Location of Characteristic Function Zeros

(iii) *There exists an uncountable set of real numbers ε such that $\Phi_*^{(s)}(\varepsilon + iy) \neq 0$ $\forall y \in R$.*

Proof Property (i) is evident, as $\Phi_*^{(s)}(z)$ is a trigonometric polynomial of degree s with respect to both $\sin z$ and $\cos z$.

$$\Phi_*^{(s)}(z) = \varphi_*^{(s)}\left(\frac{e^{iz}+e^{-iz}}{2}, \frac{e^{iz}-e^{-iz}}{2i}\right)$$

$$= \sum_{n \leq s} \varphi_r^{(n)}\left(\frac{e^{iz}+e^{-iz}}{2}, \frac{e^{iz}-e^{-iz}}{2i}\right) = e^{izs}\varphi_r^{(s)}\left(\frac{1}{2}, \frac{-i}{2}\right) +$$

$$+ \ldots + e^{-izs}\varphi_r^{(s)}\left(\frac{1}{2}, \frac{i}{2}\right), \text{ with } \varphi_r^{(s)}\left(\frac{1}{2}, \frac{\pm i}{2}\right) \neq 0$$

by (6.4.12). Thus the equation $\Phi_*^{(s)}(z) = 0$ is equivalent to the system of equations

$$t^{2s}\varphi_r^{(s)}\left(\frac{1}{2}, \frac{-i}{2}\right) + \ldots + \varphi_r^{(s)}\left(\frac{1}{2}, \frac{i}{2}\right) = 0$$

$$e^{iz} = t.$$

The first of these equations has exactly $2s$ nonzero solutions: $t_1, t_2, \ldots, t_{2s}$. In turn, for each fixed t_j, $j = 1, 2, \ldots, 2s$ the solution of the equation $e^{iz} = t_j$ is $z_j = -i \ln|t_j| + \arg t_j + 2k\pi$. Hence, $\forall a \in R$ in the strip $a \leq \text{Re}\, z < a + 2\pi$ there are exactly $2s$ zeros of the function $z \mapsto \Phi_*^{(s)}(z)$, counting their multiplicities, and property (ii) is proved. Property (iii) follows from (i) and (ii). □

The following theorem plays a fundamental role in Pontryagin's theory.

Theorem 6.4.5 *If (i) Let f be a polynomial of the form (6.4.12), and $z^r \varphi_r^{(s)}(u, v)$ its principal term,*

ii) Let $\varepsilon \in R$ be a number such that $\Phi_*^{(s)}(\varepsilon + iy) \neq 0$ $\forall y \in R$, where $\Phi_*^{(s)}(\cdot)$ is the function defined by (6.4.14) (by Lemma 6.4.1 such a number always exists). Then in the strip $-2k\pi + \varepsilon \leq \text{Re}\, z \leq 2k\pi + \varepsilon$ the function $z \mapsto F(z) = f(z, \cos z, \sin z)$ has exactly $4sk + r$ zeros, starting from a sufficiently large $k \in N$. Thus the function F has only real zeros if and only if it has exactly $4ks + r$ real zeros in the interval $[-2k\pi + \varepsilon, 2k\pi + \varepsilon] \subset R$, starting from a sufficiently large $k \in N$.

Proof

$$F(z) = f(z, \cos z, \sin z) = z^r \Phi_*^{(s)}(z) + \sum_{\substack{m < r \\ n \leq s}} z^m \Phi_m^{(n)}(z)$$

$$= z^{(r)} \Phi_*^{(s)}(z) \left[1 + \sum_{\substack{m < r \\ n \leq s}} z^{m-r} \frac{\Phi_m^{(n)}(z)}{\Phi_*^{(s)}(z)}\right] \qquad (6.4.15)$$

where $\Phi_m^{(n)}(z) = \varphi_m^{(n)}(\cos z, \sin z)$. Now we prove that taking a sufficiently large $k \in N$, $b > 0$ we have

$$\left| \sum_{\substack{m < r \\ n \leq s}} z^{m-r} \frac{\Phi_m^{(n)}(z)}{\Phi_*^{(s)}(z)} \right| < 1 \quad \forall z \in \partial P_{kb} \qquad (6.4.16)$$

in the rectangular region

$$P_{kb} = \{z \in C: \ -2k\pi + \varepsilon \leq \operatorname{Re} z \leq 2k\pi + \varepsilon, \ |\operatorname{Im} z| \leq b\}.$$

Notice that is suffices to show that the expression $\left|\Phi_m^{(n)}(z) \dfrac{1}{\Phi_*^{(s)}(z)}\right|$ is bounded on ∂P_{kb}, as the term under modulus is multiplied by negative powers of z. We have

$$\Phi_m^{(n)}(z) = \varphi_m^{(n)}(\cos z, \sin z) = \varphi_m^{(n)}\left(\frac{e^{iz} + e^{-iz}}{2}, \frac{e^{iz} - e^{-iz}}{2i}\right)$$

$$= e^{-inz}\left[\varphi_m^{(n)}\left(\frac{1}{2}, \frac{i}{2}\right) + e^{iz}\alpha_1 + \ldots + e^{2izn}\alpha_{2n}\right]$$

$$= e^{-inz}\left[\varphi_m^{(n)}\left(\frac{1}{2}, \frac{i}{2}\right) + \delta_1(z)\right] \qquad (6.4.17a)$$

and $\delta_1(z) \underset{\operatorname{Im} z \to \infty}{\to} 0$ uniformly with respect to $\operatorname{Re} z$ taken from a bounded interval on the real axis: $\varphi_m^{(n)}(1, \pm i) \neq 0 \ \forall m, \ \forall n \geq 1$.

Similarly,

$$\Phi_m^{(n)}(z) = \varphi_m^{(n)}(\cos z, \sin z) = \varphi_m^{(n)}\left(\frac{e^{iz} + e^{-iz}}{2}, \frac{e^{iz} - e^{-iz}}{2i}\right)$$

$$= e^{inz}\left[\varphi_m^{(n)}\left(\frac{1}{2}, \frac{-i}{2}\right) + e^{iz}\beta_1 + \ldots + e^{2inz}\beta_{2n}\right]$$

$$= e^{inz}\left[\varphi_m^{(n)}\left(\frac{1}{2}, \frac{-i}{2}\right) + \delta_2(z)\right] \qquad (6.4.17b)$$

and $\delta_2(z) \underset{\operatorname{Im} z \to -\infty}{\to} 0$ uniformly in $\operatorname{Re} z$ taken from a bounded interval on the real axis. By the same arguments

$$\Phi_*^{(s)}(z) = e^{-isz}\left[\varphi_r^{(s)}\left(\frac{1}{2}, \frac{i}{2}\right) + \delta_3(z)\right] \qquad (6.4.18)$$

$\delta_3(z) \underset{\operatorname{Im} z \to \infty}{\to} 0$ uniformly in $\operatorname{Re} z$ taken from a bounded interval on the real axis, and

6.4 Criteria for the Location of Characteristic Function Zeros

$$\Phi_*^{(s)}(z) = e^{isz}\left[\varphi_r^{(s)}\left(\frac{1}{2}, \frac{-i}{2}\right) + \delta_4(z)\right] \quad (6.4.18b)$$

$\delta_4(z) \underset{\operatorname{Im} z \to -\infty}{\to} 0$ uniformly in $\operatorname{Re} z$ taken from a bounded interval on the real axis.

By virtue of Lemma 6.4.1, there is only a finite number of zeros of $\Phi_*^{(s)}(\cdot)$ in any bounded-width strip, parallel to the imaginary axis, then there always exists a sufficiently large $b > 0$, such that

$$\Phi_*^{(s)}(z) \neq 0 \quad \text{if} \quad |\operatorname{Im} z| \geq b. \quad (6.4.19a)$$

Moreover, by Lemma 6.4.1

$$\Phi_*^{(s)}(\pm 2k\pi + \varepsilon + iy) = \Phi_*^{(s)}(\varepsilon + iy) \neq 0 \quad \forall y \in R. \quad (6.4.19b)$$

Evidently, if k and b are sufficiently large then by (6.4.19) the expression $|\Phi_m^{(n)}(z)/\Phi_*^{(s)}(z)|$ is bounded on ∂P_{kb} by some constant, dependent in general upon k and b. Taking sufficiently large k we may release ourselves from the dependence upon b, since by (6.4.17)–(6.4.19) we have

$$z = \pm 2k\pi + \varepsilon + iy \Rightarrow \left|\frac{\Phi_m^{(n)}(z)}{\Phi_*^{(s)}(z)}\right| = e^{-(s-n)y} \left|\frac{\varphi_m^{(n)}\left(\frac{1}{2}, \frac{i}{2}\right) + \delta_1(z)}{\varphi_r^{(s)}\left(\frac{1}{2}, \frac{i}{2}\right) + \delta_2(z)}\right| \underset{y \to \infty}{\to} 0$$

$$\left.\begin{array}{l} z = x + ib \\ -2k\pi + \varepsilon \leq x \leq 2k\pi + \varepsilon \end{array}\right\} \Rightarrow \left|\frac{\Phi_m^{(n)}(z)}{\Phi_*^{(s)}(z)}\right| = e^{-(s-n)b} \left|\frac{\varphi_m^{(n)}\left(\frac{1}{2}, \frac{i}{2}\right) + \delta_1(z)}{\varphi_r^{(s)}\left(\frac{1}{2}, \frac{i}{2}\right) + \delta_3(z)}\right| \underset{b \to \infty}{\to} 0$$

$$\left.\begin{array}{l} z = x - ib \\ -2k\pi + \varepsilon \leq x \leq 2k\pi + \varepsilon \end{array}\right\} \Rightarrow \left|\frac{\Phi_m^{(n)}(z)}{\Phi_*^{(s)}(z)}\right| = e^{-(s-n)b} \left|\frac{\varphi_m^{(n)}\left(\frac{1}{2}, \frac{-i}{2}\right) + \delta_2(z)}{\varphi_r^{(s)}\left(\frac{1}{2}, \frac{-i}{2}\right) + \delta_3(z)}\right| \underset{b \to \infty}{\to} 0.$$

All these arguments imply that (6.4.16) holds, starting from some sufficiently large b.

However, by virtue of (6.4.16) we have

$$|F(z)| \geq \left|z^r \Phi_*^{(s)}(z) + \sum_{\substack{m < r \\ n \leq s}} z^m \Phi_m^{(n)}(z)\right| \geq |z^r \Phi_*^{(s)}(z)| -$$

$$- \left|\sum_{\substack{m < r \\ n \leq s}} z^m \Phi_m^{(n)}(z)\right| > 0 \quad \forall z \in \partial P_{kb}$$

and therefore F has no zeros on ∂P_{kb}.

Taking $\Omega = \text{Int} P_{kb}$ in Theorem 6.4.1, we obtain by (6.4.15)

$$N_b = \frac{1}{2\pi} \sum_q \text{jump}[\arg F(z_q)] = \frac{1}{2\pi} \sum_m \text{jump}\left[\arg z_q^r \Phi_*^{(s)}(z_q) + \right.$$

$$\left. + \arg\left(1 + \sum_{\substack{m<r\\n\leq s}} z_q^{m-r} \frac{\Phi_m^{(n)}(z)}{\Phi_*^{(s)}(z)}\right)\right]$$

where N_b is the total number of zeros of F in $\text{Int} P_{kb}$ (counting their multiplicities), z_q a discontinuity of the function $z \mapsto \arg F(z)$.

By (6.4.16) the image of the set ∂P_{kb} through the mapping $z \mapsto 1 + \sum_{\substack{m<r\\n\leq s}} z^{m-r} \frac{\Phi_m^{(n)}(z)}{\Phi_*^{(s)}(z)}$ is a curve contained in the circle $|z-1| < 1$, hence N_b

$$= \frac{1}{2\pi} \sum_q \text{jump}[\arg z_q^r \Phi_*^{(s)}(z_q)].$$ Letting b tend to infinity we find that this expression is equal to the number of zeros of the function $z \mapsto z^r \Phi_*^s(z)$ in the strip $-2k\pi + \varepsilon \leq \text{Re} z \leq 2k\pi + \varepsilon$, which in turn by Lemma 6.4.1, is equal to $r + 2k(2s) = 4ks + r$. This completes the proof. □

Theorems 6.4.4 and 6.4.5 give the necessary and sufficient condition for the function $z \mapsto F(z) = f(z, \cos z, \sin z)$ to have real zeros only.

Now we derive a lemma similar to Lemma 6.4.1, related this time to the principal term of the quasipolynomial $z \mapsto h(z, e^z)$. The polynomial (6.4.7) with the principal term $a_{rs} z^r t^s$ may be represented in the form (see Figure 6.4.6)

$$h(z, t) = \sum_{m,n} a_{mn} z^m t^n = z^r \mathcal{X}_*^{(s)}(t) + \sum_{\substack{m<r\\n\leq s}} a_{mn} z^m t^n$$

$$\mathcal{X}_*^{(s)}(t) = \sum_{n\leq s} a_{rn} t^n. \qquad (6.4.20)$$

Lemma 6.4.2 *The function $C \ni z \mapsto \mathcal{X}_*^{(s)}(e^z) \in C$ has the following properties*

(i) *It is $2\pi i$-periodic,*

(ii) *$\forall b \in R$ in the strip $b \leq \text{Im} z < b + 2\pi$ it has at most s zeros, counting their multipliticies,*

(iii) *There exists an uncountable set of real numbers ε, such that $\mathcal{X}_*^{(s)}(e^{x+i\varepsilon}) \neq 0 \; \forall x \in R$.*

Proof As the function $z \mapsto e^z$ is $2\pi i$-periodic, by (6.4.20) the property is clear.

6.4 Criteria for the Location of Characteristic Function Zeros

Let $t_1, \ldots, t_s$ be the roots of the polynomial (of s-th degree) $t \mapsto \mathcal{X}_*^{(s)}(t)$. Then the zeros of the function $z \mapsto \mathcal{X}_*^{(s)}(e^z)$ are solutions to the equations $e^z = t_j$, $j = 1, 2, \ldots, s$. For each fixed $j = 1, 2, \ldots, s$ the last equation has a solution $z_j = \ln|t_j| + i \arg t_j + 2k\pi i$, provided $t_j \neq 0$. Hence, $\forall b \in R$ in the strip $b \leq \operatorname{Im} z < b + 2\pi$ the function $z \mapsto \mathcal{X}_*^{(s)}(e^z)$ has at most s zeros counting their multiplicities. Therefore (ii) holds. Property (iii) is an evident consequence derived from the previous two properties. □

The next result is somewhat similar to Theorem 6.4.5 and refers to the zeros of the quasipolynomial $h(z, e^z)$ in a semi-infinite strip of sufficient width.

Theorem 6.4.6 (i) Let h be a polynomial of the form (6.4.7) with the principal term $a_{rs} z^r t^s$,

(ii) Let $\varepsilon \in R$ be a number such that $\mathcal{X}_*^{(s)}(e^{x+i\varepsilon}) \neq 0$ $\forall x \in R$, and $\mathcal{X}_*^{(s)}$ is the function defined by (6.4.20) (by virtue of Lemma 6.4.2 it always exists),

(iii) The quasipolynomial $z \mapsto H(z) = h(z, e^z)$ has no zeros on the imaginary axis.

Then

$$V(-2k\pi + \varepsilon, 2k\pi + \varepsilon) = 2\pi\left(2ks - N_k + \frac{r}{2}\right) + \delta_k, \quad \delta_k \xrightarrow[k \to \infty]{} 0 \quad (6.4.21)$$

where $V(-2k\pi + \varepsilon, 2k\pi + \varepsilon)$ denotes the angle described by the vector $z = H(iy)$ around the origin when y ranges through the interval $-2k\pi + \varepsilon \leq y \leq 2k\pi + \varepsilon$; N_k is the total number of zeros of the quasipolynomial H in the semi-infinite strip $\{z \in C: \operatorname{Re} z \geq 0, -2k\pi + \varepsilon \leq \operatorname{Im} z \leq 2k\pi + \varepsilon\}$.

Proof Let us consider the rectangular region $P_{ka} = \{z \in C: 0 < \operatorname{Re} z < a, -2k\pi + \varepsilon < \operatorname{Im} z < 2k\pi + \varepsilon\}$. Note that by (6.4.20) $H(z) = h(z, e^z) = z^r \mathcal{X}_*^{(s)}(e^z) + \sum_{\substack{n \leq s \\ m < r}} a_{mn} z^m e^{nz}$, $\mathcal{X}_*^{(s)}(e^z) = \sum_{n \leq s} a_{rn} e^{zn}$. It follows from properties ii) and iii) on Lemma 6.4.2 that off the part of the boundary ∂P_{ka} of the rectangle P_{ka}, which is not contained in the imaginary axis we have

$$H(z) = z^r \mathcal{X}_*^{(s)}(e^z)\left(1 + \sum_{\substack{n \leq s \\ m < r}} a_{mn} z^{m-r} \frac{e^{nz}}{\mathcal{X}_*^{(s)}(e^z)}\right). \quad (6.4.22)$$

Each of the expressions $\dfrac{e^{nz}}{\mathcal{X}_*^{(s)}(e^z)}$ $n = 0, 1, 2, \ldots, s$ is a rational function of e^z and hence they all are bounded on $\partial P_{ka} \setminus \{\text{imaginary axis}\}$. As $m < r$, by choosing sufficiently large a and k we obtain

$$\left|\frac{a_{mn} z^m e^{nz}}{z^r \mathcal{X}_*^{(s)}(e^z)}\right| < \delta \quad \forall z \in \partial P_{ka} \setminus \{\text{the imaginary axis}\} \quad (6.4.23)$$

with δ sufficiently small. On this part of ∂P_{ka} the quasipolynomial has no zeros as we have therein

$$|H(z)| \geq |z^r \mathcal{X}_*^{(s)}(e^z)| - \left|\sum a_{mn} z^m e^{nz}\right| > 0.$$

But together with (iii) this means that the principle of argument (Theorem 6.4.1) can be applied to $\Omega = P_{ka}$ and $f = H$, thus

$$2\pi N_k = -V(-2k\pi+\varepsilon, 2k\pi+\varepsilon) + \lim_{a \to \infty} \{\text{angle described by the vector}$$

$H(z)$ when z moves on $\partial P_{ka} \setminus$ [the imaginary axis] in counter-clockwise direction starting from the point $(0, -2k\pi i + \varepsilon i)\}$. (6.4.24)

By (6.4.22) and (6.4.23), the last term is equal to the sum of similar angles but for the vectors z^r and $\mathcal{X}_*^{(s)}(e^z)$. Assuming sufficiently large a, we can make the angle described by z^r along the horizontal segments arbitrarily close to $\pi r/2$ for each respective segment. At the same time the angle described by z^r along the vertical side is arbitrarily close to 0. As the function $z \mapsto \mathcal{X}_*^{(s)}(e^z)$ is $(2\pi i)$-periodic, the angle described by the vector $\mathcal{X}_*^{(s)}(e^z)$ along the horizontal sides is equal to zero, since they are run over in opposite directions. On the other hand, taking sufficiently large a we can make the angle described by $\mathcal{X}_*^{(s)}(e^z)$ along the vertical side arbitrarily close to the same angle for e^{sz} but the latter is equal to $4ks\pi$. Taking this into account we obtain the result by (6.2.24). □

From Theorem 6.4.6 we conclude that the derivation of the sufficient and necessary condition for the quasipolynomial $z \mapsto H(z) = h(z, e^z)$ to have all roots with negative real parts is based on studying the behaviour of H on the imaginary axis. Here h denotes a polynomial of the form (6.4.7) with a principal term $a_{rs} z^r t^s$.

Now we shall give an analytic and geometric characterization of the behaviour of the function H on the imaginary axis. The analytic characterization is given by the following lemma.

Lemma 6.4.3 *The functions* $(y, u, v) \mapsto f(y, u, v)$, $(y, u, v) \mapsto g(y, u, v)$ *appearing in the following resolution of H*:

$$H(iy) = F(y) + iG(y) = h(iy, e^{iy}) = f(y, \cos y, \sin y) + ig(y, \cos y, \sin y) \quad (6.4.25)$$

have the following properties:

(i) $\forall \lambda, \mu \in R$; $\lambda^2 + \mu^2 \neq 0$ *the function* $\lambda f + \mu g$ *is a polynomial of the form* (6.4.12).

6.4 Criteria for the Location of Characteristic Function Zeros

(ii) Let $y^r\varphi_*^{(s)}(u,v)$, $y^r\psi_*^{(s)}(u,v)$ be the principal terms of the polynomials f and g, respectively, and according to (6.4.13), (6.4.20), let $\Phi_*^{(s)}(y) = \varphi_*^{(s)}(\cos y, \sin y)$, $\Psi_*^{(s)}(y) = \psi_*^{(s)}(\cos y, \sin y)$, $\mathscr{X}_*^{(s)}(t) = \sum_{n \leqslant s} a_{rn} t^n$.

Then

$$i^r \mathscr{X}_*^{(s)}(e^{iy}) = \Phi_*^{(s)}(y) + i\Psi_*^{(s)}(y). \tag{6.4.26}$$

Proof If the polynomial h, $h(z,t) = \sum_{m,n} a_{mn} z^m t^n$ has principal term $a_{rs} z^r t^s$ then $H(iy) = h(iy, e^{iy}) = \sum_{m,n} a_{mn}(iy)^m e^{iny} = \sum_{m,n} a_{mn}(iy)^m (\cos y + i \sin y)^n$. For the reconstruction of f and g from $H(iy)$ we substitute $u = \cos y$, $v = \sin y$ according to (6.4.25), and decompose the last term into the real and imaginary parts. Evidently

$$(u+iv)^n = \alpha^{(n)}(u,v) + i\beta^{(n)}(u,v)$$

where

$$\alpha^{(n)}(u,v) = \frac{(u+iv)^n + (u-iv)^n}{2}$$

$$\beta^{(n)}(u,v) = \frac{(u+iv)^n - (u-iv)^n}{2i}. \tag{6.4.27}$$

Now we have

$$H(iy) = \sum_{m,n} (iy)^m \{[\operatorname{Re} a_{mn} \alpha^{(n)}(u,v) - \operatorname{Im} a_{mn} \beta^{(n)}(u,v)] + \\ + i[\operatorname{Re} a_{mn} \beta^{(n)}(u,v) + \operatorname{Im} a_{mn} \alpha^{(n)}(u,v)]\}$$

therefore

$$f(y,u,v) = \sum_{m,n} y^m \varphi_m^{(n)}(u,v) \quad g(y,u,v) = \sum_{m,n} y^m \psi_m^{(n)}(u,v)$$

where $\varphi_m^{(n)}$, $\psi_m^{(n)}$ are polynomials of degree n which have one of the following forms, depending on m:

$$\begin{aligned} &\pm [\operatorname{Re} a_{mn} \alpha^{(n)}(u,v) - \operatorname{Im} a_{mn} \beta^{(n)}(u,v)] \\ &\pm [\operatorname{Re} a_{mn} \beta^{(n)}(u,v) + \operatorname{Im} a_{mn} \alpha^{(n)}(u,v)]. \end{aligned} \tag{6.4.28}$$

To complete the proof of (i) it suffices to show according to (6.4.12), that

$$\lambda \varphi_m^{(n)}(1, \pm i) + \mu \psi_m^{(n)}(1, \pm i) \neq 0 \quad \forall m \geqslant 0 \quad \forall n \geqslant 1.$$

Let us take for the proof the first forms of $\varphi_m^{(n)}$, $\psi_m^{(n)}$, corresponding to $m = 0, 4, 8, \ldots$ (the proof for other forms is almost the same)

$$\lambda \varphi_m^{(n)}(1, \pm i) + \mu \psi_m^{(n)}(1, \pm i) = [\lambda \operatorname{Re} a_{mn} + \mu \operatorname{Im} a_{mn}] \alpha^{(n)}(1, \pm i) + \\ + [-\lambda \operatorname{Im} a_{mn} + \mu \operatorname{Re} a_{mn}] \beta^{(n)}(1, \pm i).$$

Suppose that for some λ_0, μ_0; $\lambda_0^2 + \mu_0^2 \neq 0$ this expression vanishes. From (6.4.27) for $n \geq 1$ we easily find that

$$[\alpha^{(n)}(1, \pm i)]^2 + [\beta^{(n)}(1, \pm i)]^2 = 0.$$

Thus

$$\begin{bmatrix} \operatorname{Re} a_{mn} & \operatorname{Im} a_{mn} \\ -\operatorname{Im} a_{mn} & \operatorname{Re} a_{mn} \end{bmatrix} \begin{bmatrix} \lambda_0 \\ \mu_0 \end{bmatrix} = \begin{bmatrix} 0 \\ 0 \end{bmatrix}.$$

But this means that $|a_{mn}| = 0$. According to (6.4.28) it follows that $\varphi_m^{(n)} = \psi_m^{(n)} = 0$, in contradiction to the fact that $\varphi_m^{(n)}, \psi_m^{(n)}$ are polynomials of degree $n \geq 1$. The proof of (i) is completed. For the proof of (ii) notice that

$$H(iy) = (iy)^r \mathcal{X}_*^{(s)}(e^{iy}) + \ldots = y^r [\varphi_*^{(s)}(\cos y, \sin y) + i\psi_*^{(s)}(\cos y, \sin y) + \ldots] = y^r [\Phi_*^{(s)}(y) + i\Psi_*^{(s)}(y)]. \quad \square$$

Lemma 6.4.4 *Assume that* $F^2(y) + G^2(y) \neq 0$, $\forall y \in R$ *(geometrically this means that the curve* $\{H(iy) \in C: y \in R\}$ *does not pass through the point* $0 \in C$*) The angle of rotation* $V(a, b)$, *described by the vector* $H(iy)$ *when* y *ranges from* a *to* b, *has the following properties*:

(i) *The function* $y \mapsto V(0, y)$ *is analytic, and its derivative at* y, $V'(0, y)$ *satisfies*

$$V'(0, y) = \frac{G'(y)F(y) - F'(y)G(y)}{F^2(y) + G^2(y)} \qquad (6.4.29)$$

where F', G' *are the derivatives of* F *and* G

$$\lim_{|y| \to \infty} V'(0, y) = c < \infty, \qquad (6.4.30)$$

(ii) $V(a+\varepsilon, b+\varepsilon) = V(a, b) + [V'(0, \bar{a}) - V'(0, \bar{b})]\varepsilon \qquad (6.4.31)$
for some $\bar{a} \in [a, a+\varepsilon]$, $\bar{b} \in [b, b+\varepsilon]$,

(iii) *The following implication holds*:

$$V(-2k\pi, 2k\pi) = \tau(4k\pi s + r\pi) + \delta_k, \underset{k \to \infty}{\delta_k \to 0}, \quad \tau = \pm 1 \qquad (6.4.32)$$

$\Rightarrow \forall \lambda, \mu \in R$; $\lambda^2 + \mu^2 \neq 0$ *the function* $y \mapsto \lambda F(y) + \mu G(y)$ *has only real and simple zeros and*

$$\tau[G'(y)F(y) - F'(y)G(y)] > 0 \quad \forall y \in R. \qquad (6.4.33)$$

Proof It is obvious that the function $y \mapsto V(0, y)$ is analytic. Since then $V(0, y)$

$= \dfrac{G(y)}{F(y)}$,

$$\frac{F^2(y) + G^2(y)}{F^2(y)} = [1 + \tan^2 V(0, y)]V'(0, y) = \frac{G'(y)F(y) - F'(y)G(y)}{F^2(y)}$$

but this implies (6.4.29).

6.4 Criteria for the Location of Characteristic Function Zeros 197

Further, $F(y) = f(y, \cos y, \sin y)$, $G(y) = g(y, \cos y, \sin y)$ and f, g are polynomials of the form (6.4.12), and so each of these functions is estimated from above and below by a polynomial in y of degree r which, together with (6.4.29), means that $V'(0, y)$ is estimated by the quotient of two polynomials in y of the same degree r. Thus we have proved that (6.4.30) holds. Applying the mean-value theorem, and noting that $V(0, b) - V(0, a) = V(a, b)$ we obtain

$$V(a+\varepsilon, b+\varepsilon) = V(0, b+\varepsilon) - V(0, a+\varepsilon)$$
$$= V(0, b) + V'(0, \bar{b})\varepsilon - V(0, a) - V'(0, \bar{a})\varepsilon = V(a, b) + \varepsilon[V'(0, \bar{b}) - V'(0, \bar{a})]$$

where $\bar{a} \in [a, a+\varepsilon]$, $\bar{b} \in [b, b+\varepsilon]$.

Hence (6.4.31) is true. Let us fix arbitrary $\lambda, \mu \in R$; $\lambda^2 + \mu^2 \neq 0$. By Lemmas 6.4.3, 6.4.1 there exists $\varepsilon \in R$ such that

$$\lambda \Phi_*^{(s)}(\varepsilon + iy) + \mu \Psi_*^{(s)}(\varepsilon + iy) \neq 0 \quad \forall y \in R. \tag{6.4.34}$$

Now, (6.4.32) and (6.4.31) yield

$$V(-2k\pi + \varepsilon, 2k\pi + \varepsilon) = V(-2k\pi, 2k\pi) + \varepsilon[V'(0, \bar{b}) - V'(0, \bar{a})]$$
$$= \tau(4k\pi s + \pi r) + \delta_k + \varepsilon[V'(0, \bar{b}) - V'(0, \bar{a})]$$
$$\bar{a} \in [-2k\pi, 2k\pi + \varepsilon], \bar{b} \in [2k\pi, 2k\pi + \varepsilon].$$

Since $\delta_k \to 0$ and $[V'(0, \bar{b}) - V'[0, \bar{a}]] \to 0$ as $k \to \infty$, by (6.4.30) it follows that starting from a sufficiently large $k \in N$, $V(-2k\pi + \varepsilon, 2k\pi + \varepsilon) = \tau(4k\pi s + \pi r)$. Geometrically, this means that the curve $\{H(iy) = F(y) + iG(y): y \in [-2k\pi + \varepsilon, 2k\pi + \varepsilon]\}$ has exactly $4ks + r$ common points with the half-line starting from $0 \in C$ and contained in the straight line $\lambda F + \mu G = 0$. This implies, however, that the function $y \mapsto \lambda F(y) + \mu G(y)$ has exactly $4sk + r$ real zeros in the interval $[-2k\pi + \varepsilon, 2k\pi + \varepsilon]$. By virtue of (6.4.34) and Theorem 6.4.5 we conclude that this function has only real zeros. Suppose that one of them is multiple. The tangent to the curve $\{H(iy) \in C: y \in R\}$ at the point $H(iy)$ corresponding to it would be then contained in the straight line $\lambda F + \mu G = 0$. In this case a small variation of λ and μ would cause a change in the number of intersections of the plot $H(iy)$ with the straight line $\lambda F + \mu G = 0$, in contradiction to the above results, where λ, μ were arbitrary.

It thus turns out that our supposition leads to contradictions, which implies that the zeros of the function $y \mapsto \lambda F(y) + \mu G(y)$ are simple. The same geometrical arguments also lead to the conclusion that $\tau V'(0, y) > 0$. Hence by (6.4.29) this yields (6.4.33). □

To formulate the final result giving the necessary and sufficient conditions for all roots of a quasipolynomial with a principal term to have negative real parts, we need the following definition.

Definition 6.4.4 Let $R \ni y \mapsto p(y) \in R$, $R \ni y \mapsto q(y) \in R$ be two arbitrary functions. We say that their roots alternate if and only if
  (i) *Neither of the functions p, q has multiple roots*;
  (ii) *Between each two roots of p there is exactly one root of q*;
  (iii) $p^2(y) + q^2(y) \neq 0$ $\forall y \in R$.

The following theorem is the main result of Pontryagin's theory.

Theorem 6.4.7 Let $z \mapsto H(z) = h(z, e^z)$ be the quasipolynomial corresponding to the polynomial h of the form (6.4.7) with principal term $a_{rs}z^r t^s$. Consider the resolution (6.4.25) for $H(iy)$. If all zeros of H have negative real parts, then the zeros of the functions F and G are real, alternate and for each $y \in R$

$$G'(y)F(y) - F'(y)G(y) > 0. \qquad (6.4.35)$$

Each of the conditions given below is sufficient for all zeros of H to have negative real parts:
  (i) *All zeros of the functions F and G are real, alternate and (6.4.35) holds at some* $y \in R$;
  (ii) *All zeros of the function F are real and at each zero* y_0 *(6.4.35) holds* i.e. $F'(y_0)G(y_0) < 0$;
  (iii) *All zeros of the function G are real and at each zero* y_0 *(6.4.35) holds* i.e. $G'(y_0)F(y_0) > 0$.

Proof. Necessity Let $\varepsilon \in R$ be such that $\mathcal{X}_*^{(s)}(e^{\varepsilon+ix}) \neq 0$ $\forall x \in R$. It follows from Theorem 6.4.6 that $V(-2k\pi + \varepsilon, 2k\pi + \varepsilon) = 4k\pi s + \pi r + \delta_k$, $\delta_k \underset{k \to \infty}{\to} 0$
As $F^2(y) + G^2(y) \neq 0$ $\forall y \in R$, by (6.4.31):
$$V(-2k\pi, 2k\pi) = 4k\pi s + \pi r + \delta'_k, \quad \delta'_k \underset{k \to \infty}{\to} 0.$$

In virtue of Lemma 6.4.4 all zeros of F and G are real and simple. Moreover for every $y \in R$ (6.4.35) is fulfilled. Interpreting geometrically all the information obtained about F and G up to now, we conclude that the vector $H(iy)$ rotates in counter-clockwise direction intersecting alternately the real and imaginary axes, hence the roots of F and G alternate.

Sufficiency As the roots of F and G alternate it follows that $F^2(y) + G^2(y) \neq 0$ (the quasipolynomial has no zeros on the imaginary axis), which together with Theorem 6.4.6 yields

$$V(-2k\pi + \varepsilon, 2k\pi + \varepsilon) = 2\pi(2ks + \tfrac{1}{2}r - N_k) + \delta_k, \quad \delta_k \underset{k \to \infty}{\to} 0,$$

with $\varepsilon \in R$ such that $\mathcal{X}_*^{(s)}(e^{x+i\varepsilon}) \neq 0$ $\forall x \in R$. Therefore, due to (6.4.31) for every sufficiently large k

$$2\pi N_k = 4k\pi s + r\pi - V(-2k\pi, 2k\pi) \qquad (6.4.36)$$

6.4 Criteria for the Location of Characteristic Function Zeros 199

where N_k is the number of zeros of H in the semi-strip described in Theorem 6.4.6. Notice that by (6.4.26) where we may substitute $y = \varepsilon - ix$ we also have $\Phi_*^{(s)}(\varepsilon - ix) \neq 0$, $\Psi_*^{(s)}(\varepsilon - ix) \neq 0$ $\forall x \in R$. As we have assumed that all zeros of F and G are real, by Theorem 6.4.5 we establish that in the interval $[-2k\pi + \varepsilon, 2k\pi + \varepsilon]$ each of these functions has exactly $4ks + r$ zeros. However, taken with the fact that (6.4.35) holds at one point at least, this means that $N_k = 0$ (6.4.36).
Since k is arbitrarily large the theorem holds. □

Examples of practical applications of Theorem 6.4.7 to the stability analysis of classical automatic control systems will be given in Chapter 7. Here we note only that Theorem 6.4.5 plays an essential, auxiliary role with respect to Theorem 6.4.7, because it allows the conditions under which the appropriate functions have only real roots to be effectively established.

7

Conventional Regulation Problems

7.1 Stability Region of a Continuous-time Conventional Regulation System

We shall use the word 'regulation' for feedback control if the control signal is produced by a conventional controller or 'regulator'. A simple one-loop regulation system is shown in Figure 7.1.1. O represents the controlled plant with input u and output y. R is the conventional controller (regulator). Its input is denoted by ε and output by v. The disturbances are represented by z; w is a given reference input. All these quantities are scalar-valued functions of time.

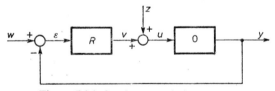

Figure 7.1.1 One-loop regulation system

In this chapter we consider linear regulators, characterized in the most general case by the transfer function

$$G_R(s) = \frac{\hat{v}(s)}{\hat{\varepsilon}(s)} = K_R\left(1 + \frac{1}{T_i s} + T_d s\right) \qquad (7.1.1)$$

or, in a more realistic description,

$$G_R(s) = K_R\left(1 + \frac{1}{T_i s} + \frac{T_d s}{T s + 1}\right). \qquad (7.1.2)$$

7.1 Continuous-time Conventional Regulation System

K_R, T_i and T_d are adjustable parameters to be set according to control task, T is a given constant. A regulator of this form is called a PID-regulator. The following regulators may be regarded as special cases of (7.1.1), (7.1.2):

— the proportional action (or P) regulator

$$G_R(s) = K_R \qquad (7.1.3)$$

— the astatic (integral action or I) regulator

$$G_R(s) = \frac{1}{T_i s} \qquad (7.1.4)$$

— the proportional plus reset (or PI) regulator

$$G_R(s) = K_R\left(1 + \frac{1}{T_i s}\right) \qquad (7.1.5)$$

— the proportional and differential action (or PD) regulator

$$G_R(s) = K_R(1 + T_d s) \qquad (7.1.6)$$

or

$$G_R(s) = K_R\left(1 + \frac{T_d s}{Ts+1}\right). \qquad (7.1.7)$$

We shall develop methods for determining the appropriate type of regulator and the optimal parameter settings. Consider the simple but typical regulation system in Figure 7.1.1. It has two basic tasks:
(a) the output signal y should reproduce the shape of the reference input w as far as possible;
(b) the output y should be insensitive to disturbances z, i.e. the deviation resulting from the action of the disturbances should be as low as possible, and if deviations appear, they should be cancelled as quickly as possible.

Depending on the destination of the system, tasks (a) and (b) may have different degrees of importance. In the stabilization systems the input $w(t)$ = const and task (b) is more important. In programme control systems, task (a) is crucial, since such systems should precisely reproduce the standard control signal w. In follow-up systems task (a) and (b) are of equal importance.

The performance of the control system is evaluated by means of certain functions (functionals) of the adjustable parameters, called performance criteria. If the optimization framework is used, the regulator parameter settings are chosen so that the main performance criterion reaches its extreme value (maximum or minimum depending on whether the criterion represents profits or losses), and the others assume admissible values. Often, however, the settings are determined by means of a simplified procedure which only requires that all criteria should assume 'satisfactory' values. They are not strictly optimal, but guarantee satisfactory performance of the system for different disturbances z and reference inputs w.

The very general control tasks (a) and (b) implicate several more specific requirements with respect to the regulation system. These are analogous to those formulated for control systems without delays. A fundamental requirement which must be fulfilled in every regulation system is the stability (asymptotic stability) of the whole system no matter whether the controlled plant is stable or not.

We shall analyse the asymptotic stability region in the space of parameters of the conventional regulation system (Figure 7.1.1). We assume a linear controlled plant with delay described by the transfer function

$$G_0(s) = \frac{K_0 e^{-sh}}{T_0 s + q} \qquad (7.1.8)$$

$$T_0 > 0 \quad h > 0 \quad q = 0 \text{ or } 1.$$

If $q = 1$, the plant is a first-order inertial system, and if $q = 0$, it is a first-order astatic system with delay.

We shall write the transfer function of the regulator in the form

$$G_R(s) = K_R \left(p_0 + \frac{p_1}{T_i s} + T_d s \right) \qquad (7.1.9)$$

where $T_i \neq 0$, p_0 and p_1 are equal to zero or one. Without loss of generality we assume $h = 1$ and denote

$$a = \frac{q}{T_0} \quad b = K_0 K_R \frac{T_d}{T_0} \quad c = K_0 K_R \frac{p_0}{T_0} \quad d = \frac{K_0 K_R}{T_i T_0} p_1. \qquad (7.1.10)$$

The transfer function of the whole regulation system is

$$G(s) = \frac{G_0(s) G_R(s)}{1 + G_0(s) G_R(s)} = \frac{bs^2 + cs + d}{(s^2 + as) e^s + bs^2 + cs + d}. \qquad (7.1.11)$$

Since basically stability is a property of solutions of differential equations, it will be helpful to discuss it in this framework. Of course, we put $w = 0$ and $z = 0$. For $t \in [0, 1]$ we have

$$\dot{y}(t) + ay(t) = \frac{K_0}{T_0} u(t-1) \quad y(0) = y_0. \qquad (7.1.12)$$

For $t > 1$ we distinguish two cases. If $p_1 = 0$ we have

$$\dot{y}(t) + ay(t) = -cy(t-1) - b\dot{y}(t-1) \qquad (7.1.13)$$

and if $p_1 = 1$

$$\dot{y}(t) + ay(t) = -cy(t-1) - b\dot{y}(t-1) - d \int_0^{t-1} y(s) ds + \frac{K_0 K_R}{T_0} \varepsilon_0 \qquad (7.1.14)$$

7.1 Continuous-time Conventional Regulation System

where ε_0 is a constant. Thus the problem of stability of the regulation system is reduced to the problem of stability of the neutral equation (7.1.13) or (7.1.14), whose initial conditions are determined by equation (7.1.12).

The region of stability in the space of parameters depends on the choice of the space of initial conditions. Generally speaking, the more elements in the space of initial conditions, the smaller the stability region. This choice may be critical when we want to decide if a boundary point belongs to the region of asymptotic stability. In order to obtain simple sufficient and necessary conditions of asymptotic stability, we shall assume a space of initial conditions which may be too large for some practical applications, i.e. some boundary points excluded from the region of asymptotic stability will give trajectories vanishing in time for all practical initial conditions.

Substituting $t \leftarrow t-1$ and $x(t) \leftarrow y(t+1)$ we can rewrite equation (7.1.13) in the form

$$\frac{\mathrm{d}}{\mathrm{d}t}[x(t)+bx(t-1)]+ax(t)+cx(t-1) = 0 \quad t \geqslant 0 \quad (7.1.15)$$

$$x(t) = \varphi(t) \quad (t \in [-1, 0])$$

where φ is a given integrable and bounded function.

For equation (7.1.14) define

$$x_1(t) = \int_{-1}^{t} y(s)\mathrm{d}s - T_t \varepsilon_0 \quad x_2(t) = \dot{x}_1(t). \quad (7.1.16)$$

Hence for $t \geqslant 0$

$$\dot{x}_1(t) = x_2(t)$$

$$\frac{\mathrm{d}}{\mathrm{d}t}[x_2(t)+bx_2(t-1)]+ax_2(t)+dx_1(t-1)+cx_2(t-1) = 0 \quad (7.1.17)$$

$$x_1(t) = \varphi_1(t) \quad x_2(t) = \varphi_2(t) \quad t \in [-1, 0]$$

φ_1 and φ_2 are integrable and bounded functions.

System (7.1.15) (resp. (7.1.17)) is asymptotically stable if and only if for every bounded and integrable initial function φ (resp. φ_1, φ_2) the solution tends to zero, $x(t) \to 0$, $t \to \infty$ (resp. $x_1(t) \to 0$, $x_2(t) \to 0$, $t \to \infty$).

The characteristic quasipolynominal of (7.1.15) is

$$g(s) = \mathrm{e}^{-s}H(s)$$

where

$$H(s) = (s+a)\mathrm{e}^s+bs+c. \quad (7.1.18)$$

The characteristic quasipolynominal of (7.1.17) is

$$g(s) = \mathrm{e}^{-s}H(s)$$

$$H(s) = (s^2+as)\mathrm{e}^s+bs^2+cs+d. \quad (7.1.19)$$

It is noteworthy that in both cases $H(s)$ is equal to the denominator of the transfer function (7.1.11).

From the results of the previous chapter it follows immediately that the regulation system is asymptotically stable if and only if $|b| < 1$ and all roots of H ((7.1.18) for $p_1 = 0$, and (7.1.19) for $p_1 = 1$) have negative real parts. We shall use Pontryagin's theory to derive explicit conditions of asymptotic stability.

I. Determination of the asymptotic stability region in the case $p_1 = 0$

We have $H(i\omega) = E(\omega) + iG(\omega)$,

$$E(\omega) = a\cos\omega - \omega\sin\omega + c \qquad (7.1.20)$$
$$G(\omega) = \omega\cos\omega + a\sin\omega + b\omega. \qquad (7.1.21)$$

First we shall determine conditions for G to have only real roots. The principal term of G is $\omega\cos\omega$, and so $r = 1, s = 1$,

$$\varphi^*(\omega) = \cos\omega + b.$$

Notice that $y \mapsto \varphi^*(\varepsilon + iy)$ has no real roots for $b > -1$ provided $\varepsilon = 0$. Define

$$f(\omega) = \cos\omega + a\frac{\sin\omega}{\omega} + b \quad \omega \neq 0, \quad f(0) = 1+a+b. \qquad (7.1.22)$$

All roots of f are roots of G, and the only root of G which may not be a root of f is 0. We examine the extrema of f:

$$f'(\omega) = -\sin\omega + a\frac{\omega\cos\omega - \sin\omega}{\omega^2} = 0.$$

Hence

$$a\omega\cos\omega = (a+\omega^2)\sin\omega. \qquad (7.1.23)$$

Notice that if ω is a root of (7.1.23), then $-\omega$ is also a root; $\omega_0 = 0$ is a root, and there is exactly one root ω_k in each interval $[k\pi, k\pi + \pi/2)$, $k = 1, 2, \ldots$ If $a = 0$, we have $\omega_k = k\pi$. If $a > 0$, then $\omega_k > k\pi$ and $\omega_k - k\pi \to 0, k \to \infty$. There is a k such that

$$\omega_{j+1} - (j+1)\pi < \omega_j - j\pi \quad \forall j \geq k.$$

Function f has maxima at points $\pm\omega_k$, $k = 0, 2, 4, \ldots$, and minima at points $\pm\omega_k$, $k = 1, 3, 5, \ldots$ It is easy to show that

$$f(\omega_k) = (-1)^k \frac{a+a^2+\omega_k^2}{\sqrt{a^2\omega_k^2 + (a+\omega_k^2)^2}} + b \quad (k = 1, 2, 3, \ldots). \qquad (7.1.24)$$

7.1 Continuous-time Conventional Regulation System

The function $x \mapsto \dfrac{a+a^2+x}{\sqrt{a^2x+(a+x)^2}}$ is identically equal to one if $a = 0$, and is strictly decreasing to one if $a > 0$. Therefore we obtain the result that all maxima of f are positive and all minima negative, if

$$b \in (-1, 1). \tag{7.1.25}$$

Now assume (7.1.25). There is then exactly one root of f in each interval (ω_{k-1}, ω_k). We denote this by v_k. From (7.1.23) it easily follows that for sufficiently large k

$$\omega_k - k\pi \leqslant \frac{ak\pi}{a+(k\pi)^2} \leqslant \frac{a}{k\pi}, \tag{7.1.26}$$

and there is a constant $C > 0$, independent of k, such that

$$\omega_k - v_k > C\sqrt{|f(\omega_k)|}. \tag{7.1.27}$$

Since

$$|f(\omega_k)| > 1 - |b|,$$

we obtain

$$v_k < \omega_k - C\sqrt{1-|b|} \leqslant k\pi + \frac{a}{k\pi} - C\sqrt{1-|b|}. \tag{7.1.28}$$

Therefore, for sufficiently large k

$$v_k < k\pi. \tag{7.1.29}$$

Thus function f has $4k$, and G has $4k+1$ real roots in the interval $[-2k\pi, 2k\pi]$. We conclude that if $b \in (-1, 1)$ and $a \geqslant 0$, then G has only real roots.

Now we have to find out when $G'(v_k)E(v_k) > 0$, $k = 0, 1, 2, \ldots$ Since $G'(v_0)E(v_0) = (1+a+b)(a+c)$, $a \geqslant 0$ and $b \in (-1, 1)$, we obtain the condition

$$a + c > 0. \tag{7.1.30}$$

Obviously $G'(v_k) \neq 0$, and $\text{sign}\, G'(v_k) = (-1)^k$. Then $G'(v_k)E(v_k) > 0$ if and only if

$$\text{sign}\, E(v_k) = (-1)^k \quad (k = 1, 2, \ldots). \tag{7.1.31}$$

For every positive integer k, the mapping $[0, \infty) \times (-1, 1) \ni (a, b) \mapsto v_k$ is single-valued and continuous, and v_k does not depend on c. Therefore the equation $E(v_k) = 0$, or

$$c = v_k \sin v_k - a \cos v_k \tag{7.1.32}$$

determines a continuous surface in the space of parameters a, b, c on each side of which E is of constant sign. From the defining equation of v_k,

$$v_k \cos v_k + a \sin v_k + b v_k = 0, \tag{7.1.33}$$

one obtains

$$\sin v_k = \frac{v_k}{a^2+v_k^2}\left[-ab \pm \sqrt{a^2b^2+(1-b^2)(a^2+v_k^2)}\right]. \quad (7.1.34)$$

Since v_k depends continuously on a, b, for each $k > 0$ only one sign, plus or minus, may be in force in the whole domain of v_k. It is easy to verify that

$$\sin v_k = \frac{-v_k}{a_2+v_k^2}\left[ab+(-1)^k\sqrt{a^2+(1-b^2)v_k^2}\right]. \quad (7.1.35)$$

Now we shall derive expressions for $E(v_k)$. From (7.1.33) for $k > 0$

$$\cos v_k = -\frac{a}{v_k}\sin v_k - b \quad (7.1.36)$$

$$E(v_k) = \left(-\frac{a^2}{v_k}-v_k\right)\sin v_k - ab + c. \quad (7.1.37)$$

Substituting (7.1.34) we get

$$E(v_k) = c+(-1)^k\sqrt{a^2+(1-b^2)v_k^2}. \quad (7.1.38)$$

Condition (7.1.31) takes the form

$$(-1)^k c+\sqrt{a^2+(1-b^2)v_k^2} > 0 \quad (k = 1, 2, 3, \ldots). \quad (7.1.39)$$

As the sequence v_k is increasing and $c+a > 0$, this inequality is satisfied for every even k. For odd k, it is equivalent to

$$c < \sqrt{a^2+(1-b^2)v_1^2} \quad (7.1.40)$$

where v_1 is the unique solution of $v_1\cos v_1 + a\sin v_1 + bv_1 = 0$ in $(0, \pi)$.

Thus the region of asymptotic stability is determined by (7.1.40) and $c+a > 0$, $b \in (-1, 1)$. Recall that it has been assumed that $a \geq 0$. We shall rewrite these conditions in a more explicit form. On the boundary of the region

$$c = \sqrt{a^2+(1-b^2)v_1^2} \quad (7.1.41)$$

or

$$v_1 = \sqrt{\frac{c^2-a^2}{1-b^2}}. \quad (7.1.42)$$

Substituting this into (7.1.35) we obtain

$$\sin v_1 = \frac{\sqrt{(c^2-a^2)(1-b^2)}}{ab+c} \quad (7.1.43)$$

and from (7.1.36) we obtain further

$$\cos v_1 = -\frac{a+bc}{ab+c}. \quad (7.1.44)$$

7.1 Continuous-time Conventional Regulation System

The boundary equation (7.1.41) can then be written in the form

$$\sqrt{\frac{c^2-a^2}{1-b^2}} = \arccos\left(-\frac{a+bc}{ab+c}\right). \tag{7.1.45}$$

Notice that if $c-a > 0$, the left-hand side decreases with decreasing c, and the right-hand side increases. Therefore in the region of asymptotic stability the left-hand side is less than the right-hand side.

Our results so far may be summarized as follows. It is assumed that $a \geqslant 0$. If $c-a \leqslant 0$, the asymptotic stability region is determined by the inequalities: $c+a > 0$, $-1 < b < 1$. If $c-a > 0$, the system is asymptotically stable in the region determined by $c+a > 0$, $-1 < b < 1$

$$\sqrt{\frac{c^2-a^2}{1-b^2}} < \arccos\left(-\frac{a+bc}{ab+c}\right). \tag{7.1.46}$$

Our results are also valid for any arbitrary positive time delay $h > 0$. One has only to replace in all the formulas a by ah, and c by ch. After this substitution, all the conditions determining the region of asymptotic stability remain unchanged, except for (7.1.46) which takes the form

$$h\sqrt{\frac{c^2-a^2}{1-b^2}} < \arccos\left(-\frac{a+bc}{ab+c}\right).$$

We come to the conclusion that if the regulation system is asymptotically stable for some time delay $h' > 0$, then it is asymptotically stable for every h, $0 < h \leqslant h'$.

The stability regions for the unit delay $h = 1$ in the space of parameters $K = K_0 K_R$, T_d and T_0 are shown in Figures 7.1.2–7.1.4.

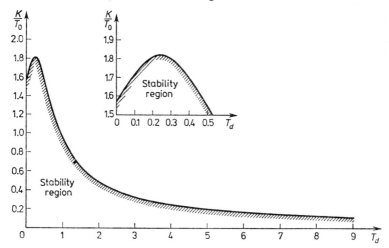

Figure 7.1.2 Stability region for astatic plant ($q = 0$) and P- ($T_d = 0$) or PD-regulator ($T_d > 0$)

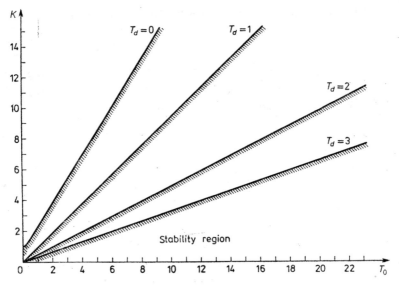

Figure 7.1.3 Stability region for inertial plant ($q = 1$) and P- ($T_d = 0$) or PD-regulator ($T_d > 0$)

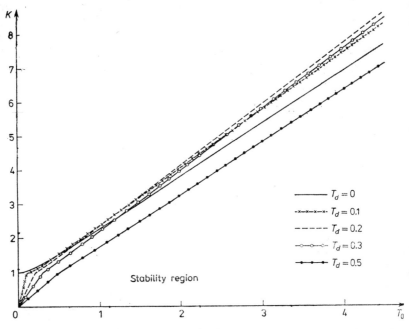

Figure 7.1.4 Stability region for inertial plant ($q = 1$) and P- ($T_d = 0$) or PD- regulator ($T_d > 0$)

7.1 Continuous-time Conventional Regulation System

II. Determination of the asymptotic stability region in the case $p_1 = 1$

In this case
$$H(i\omega) = E(\omega) + iG(\omega)$$
$$E(\omega) = -\omega^2\cos\omega - a\omega\sin\omega - b\omega^2 + d \quad (7.1.47)$$
$$G(\omega) = \omega(a\cos\omega - \omega\sin\omega + c). \quad (7.1.48)$$

We must find out when G has only real roots. The principal term of G is $-\omega^2\sin\omega$, and so $r = 2$, $s = 1$,
$$\varphi^*(\omega) = -\sin\omega.$$

The function $y \mapsto \varphi^*(\varepsilon + iy)$ has no real roots for $\varepsilon = \dfrac{\pi}{2}$. Define
$$f(\omega) = a\cos\omega - \omega\sin\omega + c. \quad (7.1.49)$$

All roots of f are roots of G, and the only root of G that may not be a root of f is 0. We examine the extrema of f:
$$f'(\omega) = -(a+1)\sin\omega - \omega\cos\omega = 0. \quad (7.1.50)$$

If ω is a root of (7.1.50), then $-\omega$ is a root too; $\omega_0 = 0$ is a root, and there is exactly one root ω_k in each interval $\left(k\pi - \dfrac{\pi}{2}, k\pi\right)$, $k = 1, 2, \ldots$ It is easy to see that $\omega_k - k\pi + \dfrac{\pi}{2} \to 0$, $k \to \infty$, $\omega_{k+1} - (k+1)\pi + \dfrac{\pi}{2} < \omega_k - k\pi + \dfrac{\pi}{2}$, $\forall k > 0$, and $\omega_k - k\pi + \dfrac{\pi}{2} < \operatorname{arccot}\left(k\pi + \dfrac{\pi}{2}\right)$.

The maxima of f are at $\pm\omega_k$, $k = 0, 2, 4, \ldots$, and the minima at $\pm\omega_k$, $k = 1, 3, 5, \ldots$ The extremal values of f are
$$f(\omega_k) = c + (-1)^k \frac{a + a^2 + \omega_k^2}{\sqrt{(1+a)^2 + \omega_k^2}} \quad (k = 0, 1, 2, \ldots). \quad (7.1.51)$$

The function $x \mapsto \dfrac{a + a^2 + x}{\sqrt{(1+a)^2 + x}}$ is strictly increasing for $x \geq 0$. Therefore all the maxima of f are positive if and only if
$$f(\omega_0) = c + a > 0 \quad (7.1.52)$$
and the minima are negative if and only if
$$c < \frac{a + a^2 + \omega_1^2}{\sqrt{(1+a)^2 + \omega_1^2}} \quad (7.1.53)$$
where ω_1 is the only solution of
$$\tan\omega_1 = -\frac{\omega_1}{a+1}$$

in $\left(\frac{\pi}{2}, \pi\right)$. Conditions (7.1.52) and (7.1.53) are necessary and sufficient for f to have exactly one root v_k in every interval (ω_{k-1}, ω_k). Taking into account that the sequence $\{f(\omega_{2k})\}$ is increasing, and $\{f(\omega_{2k+1})\}$ decreasing, we come to the conclusion that there exists a real $\delta > 0$ (independent of k) such that $\omega_k - v_k > \delta$. Thus for sufficiently large k

$$v_k < k\pi - \frac{\pi}{2}.$$

Hence for sufficiently large k, in every interval $\left[-2k\pi + \frac{\pi}{2}, 2k\pi + \frac{\pi}{2}\right]$ f has exactly $4k+1$ roots, and G exactly $4k+2$ roots. Thus we have proved that G has only real roots if and only if inequalities (7.1.52) and (7.1.53) hold.

Condition (7.1.53) can be written in a more explicit form. Denote

$$Z = \sqrt{c^2 + 4a + 4}.$$

After some calculations which are omitted here we obtain that c is equal to the right-hand side of (7.1.53) if and only if

$$\arccos \frac{1}{2}(c - Z) = \frac{1}{2}\sqrt{(c+Z)^2 - 4(a+1)^2}. \qquad (7.1.54)$$

Now we shall determine the region in the space of parameters where $G'(v_k)E(v_k) > 0$, $k = 0, 1, 2, \ldots$ For $k = 0$ we have

$$G'(v_0)E(v_0) = (a+c)d.$$

Since $a + c > 0$, we obtain the condition

$$d > 0. \qquad (7.1.55)$$

Of course $G'(v_k) \neq 0$ and $\operatorname{sign} G'(v_k) = (-1)^k$, hence the condition $G'(v_k)E(v_k) > 0$ takes the form

$$\operatorname{sign} E(v_k) = (-1)^k \quad (k = 1, 2, 3, \ldots). \qquad (7.1.56)$$

For every positive integer k, the mapping $(a, c) \mapsto v_k$ is single-valued and continuous, provided a and c satisfy conditions (7.1.52), (7.1.53) and $a \geqslant 0$. Notice that v_k does not depend on b or d. Therefore the equation $E(v_k) = 0$, or

$$d = v_k(v_k \cos v_k + a \sin v_k + b v_k) \qquad (7.1.57)$$

determines a continuous surface in the space of parameters a, b, c, d on each side of which E is of constant sign. From the equation defining v_k, $k > 0$

$$a \cos v_k - v_k \sin v_k + c = 0 \qquad (7.1.58)$$

7.1 Continuous-time Conventional Regulation System

we obtain

$$\sin v_k = \frac{cv_k \pm a\sqrt{a^2 - c^2 + v_k^2}}{a^2 + v_k^2}. \tag{7.1.59}$$

It can be easily shown that the radicand is nonnegative, and strictly positive for $k > 1$. Using condition (7.1.53) it can be proved following a lengthy analysis that

$$\sin v_k = \frac{cv_k - \varepsilon_k a\sqrt{a^2 - c^2 + v_k^2}}{a^2 + v_k^2} \tag{7.1.60}$$

where

$$\varepsilon_k = (-1)^k \quad (k = 2, 3, 4, \ldots)$$

$$\varepsilon_1 = \begin{cases} -1 & (c \leq \sqrt{a^2 + V_1^2}) \\ +1 & (c > \sqrt{a^2 + V_1^2}) \end{cases} \tag{7.1.61}$$

V_1 is the smallest positive root of the equation

$$a\sin x + x\cos x = 0. \tag{7.1.62}$$

Now we shall calculate $E(v_k)$. From (7.1.58), if $a > 0$,

$$\cos v_k = \frac{v_k}{a}\sin v_k - \frac{c}{a} \tag{7.1.63}$$

and

$$E(v_k) = -\frac{v_k}{a}(v_k^2 + a^2)\sin v_k + \frac{c}{a}v_k^2 - bv_k^2 + d. \tag{7.1.64}$$

Substituting (7.1.60) we obtain

$$E(v_k) = d - bv_k^2 + \varepsilon_k v_k \sqrt{a^2 - c^2 + v_k^2}. \tag{7.1.65}$$

Because of the continuity of E, this formula is also valid for $a = 0$. Condition (7.1.56) takes the form

$$d < bv_k^2 - \varepsilon_k v_k \sqrt{a^2 - c^2 + v_k^2} \quad (k = 1, 3, 5, \ldots) \tag{7.1.66}$$

$$d > bv_k^2 - \varepsilon_k v_k \sqrt{a^2 - c^2 + v_k^2} \quad (k = 2, 4, 6, \ldots). \tag{7.1.67}$$

Simple analysis of the dependence of the right-hand sides of these inequalities on k allows us to rewrite them in the form

$$d < bv_1^2 - \varepsilon_1 v_1\sqrt{a^2 - c^2 + v_1^2} \tag{7.1.68}$$

$$d > bv_2^2 - v_2\sqrt{a^2 - c^2 + v_2^2}. \tag{7.1.69}$$

It can be shown that if the interval for d determined by these two inequalities is nonempty, then condition (7.1.53) is always satisfied. Therefore (7.1.53) is redundant and may be replaced by the requirement that all expressions must make sense (e.g. all radicands must be nonnegative).

Let us summarize our results. By virtue of our assumptions, $a \geq 0$. The regulation system is asymptotically stable if and only if $-1 < b < 1$, $c+d > 0$, $d > 0$, and

$$d > bv_2^2 - v_2\sqrt{a^2 - c^2 + v_2^2} \qquad (7.1.70)$$

$$d < \begin{cases} bv_1^2 + v_1\sqrt{a^2 - c^2 + v_1^2} & \text{for} \quad c \leq \sqrt{a^2 + V_1^2} \\ bv_1^2 - v_1\sqrt{a^2 - c^2 + v_1^2} & \text{for} \quad c > \sqrt{a^2 + V_1^2}. \end{cases} \qquad (7.1.71)$$

Here V_1 is the smallest positive root of the equation

$$a\sin x + x\cos x = 0.$$

v_1 and v_2, $v_1 < v_2$, are the two smallest positive roots of

$$a\cos x - x\sin x + c = 0.$$

It is implicitly understood that if any radicand is negative, the system is not asymptotically stable.

To end with, we shall transform conditions (7.1.70) and (7.1.71) to a more explicit form (though less convenient for computations). We derive only the equation of the boundary surface of the stability region corresponding to (7.1.70), (7.1.71). Denote

$$Z = \sqrt{(a^2 - c^2)(a^2 - c^2 + 4bd) + 4d^2}.$$

It is easy to show that on this surface both v_1 and v_2 satisfy the equalities

$$v_{1,2}^2 = \frac{-a^2 + c^2 - 2bd + Z}{2(1 - b^2)}$$

and

$$\cos v_{1,2} = \frac{d - ac + a^2 b}{a^2 + v_{1,2}^2} - b.$$

Hence we obtain

$$\sqrt{\frac{-a^2 + c^2 - 2bd + Z}{2(1 - b^2)}} = \arccos \frac{-a^2 - 2abc - c^2 + Z}{2(ac + a^2 b + d)}. \qquad (7.1.72)$$

Since $v_1 < \omega_1 < \pi$, this equality is satisfied on the boundary of (7.1.71) (i.e. if we replace $<$ by $=$). If $v_2 \leq \pi$, then (7.1.72) is satisfied on the boundary of (7.1.70) (after replacement of $>$ by $=$).

7.1 Continuous-time Conventional Regulation System

By virtue of equation (7.1.58) $v_2 = \pi$ if and only if $a = c$. From the continuity of v_2 it follows that $v_2 \geq \pi$ if and only if $a-c \geq 0$, i.e. $a^2 - c^2 \geq 0$. In this case, however, the constraint (7.1.70) is redundant. (7.1.70) may then be non-redundant only if $v_2 < \pi$, and we can conclude that (7.1.72) describes all of the boundary surface determined by (7.1.70), (7.1.71).

On the basis of the analysis of the asymptotic stability region and (7.1.72), we can verify by straightforward calculation that conditions (7.1.70) and (7.1.71) are equivalent to

$$\sqrt{\frac{-a^2+c^2-2bd+Z}{2(1-b^2)}} < \arccos \frac{-a^2-2abc-c^2+Z}{2(ac+a^2b+d)}. \qquad (7.1.73)$$

It is not difficult to generalize our results to the case of an arbitrary positive delay $h > 0$. All we have to do is to replace a by ah, c by ch, and d by dh^2 in the above formulas. No condition will be changed, except (7.1.73) (or (7.1.70), (7.1.71)), which will take the form

$$h\sqrt{\frac{-a^2+c^2-2bd+Z}{2(1-b^2)}} < \arccos \frac{-a^2-2abc-c^2+Z}{2(ac+a^2b+d)} \qquad (7.1.74)$$

where Z is as in (7.1.73).

We see that if the regulation system is asymptotically stable for some delay $h' > 0$, then it is asymptotically stable for every delay h, $0 < h \leq h'$. The stability regions for the delay $h = 1$ in the space of parameters $K = K_0 K_R$, T_i, T_d and T_0 are shown in Figures 7.1.5–7.1.9.

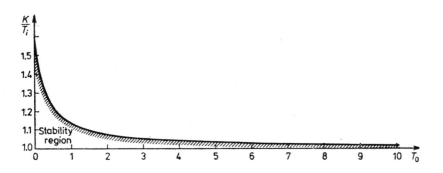

Figure 7.1.5 Stability region for inertial plant ($q = 1$) and I - regulator

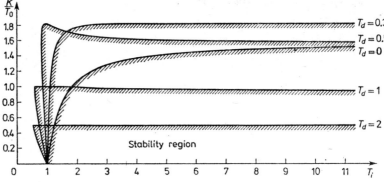

Figure 7.1.6 Stability region for astatic plant ($q = 0$) and PI- ($T_d = 0$) or PID-regulator ($T_d > 0$)

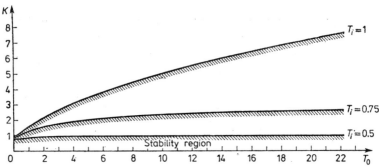

Figure 7.1.7 Stability region for inertial plant ($q = 1$) and PI-regulator; small values of T_i

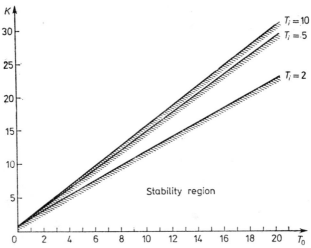

Figure 7.1.8 Stability region for inertial plant ($q = 1$) and PI-regulator; large values of T_i

7.2 Parametric Synthesis of Continuous Regulators

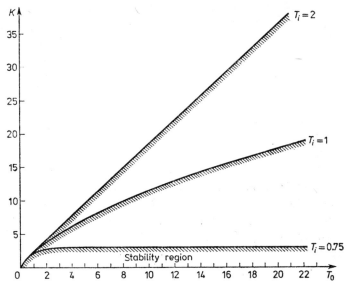

Figure 7.1.9 Stability region for inertial plant ($q = 1$) and PID-regulator; $T_d = 0.3$

7.2 Parametric Synthesis of Continuous Regulators

In this section we give a review of simple criteria which allow us to choose the parameter settings of continuous regulators. Indications are also given referring to the choice of the regulator type most suitable for a given controlled plant. Since we are dealing here with simple control structures, it is reasonable to consider in the first place those methods which do not require sophisticated computations and measurements.

7.2.1 Determination of the regulator type

It is sometimes convenient to divide the process of determining regulator settings into two parts: first, the determination of the regulator type and second, the calculation of its settings.

Let us consider a controlled plant which can be described by the transfer function (7.1.8) with $q = 1$

$$G_0(s) = \frac{K_0 e^{-sh}}{T_0 s + 1}. \qquad (7.2.1)$$

The disturbance action onto the object is characterized by the ratio

$$m = \frac{y_{zm}}{y_{wm}} \qquad (7.2.2)$$

where y_{zm} is the steady state-output deviation resultant from the action of the maximum constant disturbance onto the plant without a controller, and y_{wm} is the steady state-output deviation resultant from the maximum variation of the set value w_m acting onto the plant without a controller. If the ratio (7.2.2) cannot be determined, we replace it with the ratio of respective derivatives after a sufficiently long time

$$m = \frac{\dot{y}_{zm}}{\dot{y}_{wm}}. \tag{7.2.3}$$

We denote by t_r the setting time, i.e. the time period after which the output deviation does not exceed 5% of its maximal (or initial) value. The output is the response of the plant to a step input. By $\varepsilon_{u\%}$ we denote the proportional admissible steady-state deviation, or static error (in proportion to the set value). This deviation is taken into consideration also in the systems with I-, PI- and PID-regulators where it results from the existence of insensitivity zones. We define

$$T = T_0/h \quad T_r = t_r/h.$$

The ratio $\varepsilon_{u\%}/m$ is a characteristic quantity, helpful in determining the suitable type of regulator.

The selection of regulator type is carried out using Lerner's diagram (Lerner, 1958), see Figure 7.2.1.

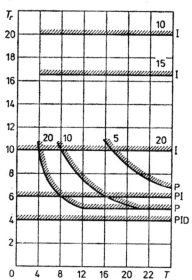

Figure 7.2.1 Lerner's diagram; for each regulator type its region of applicability is shaded; the numbers denote $\varepsilon_{u\%}/m$

7.2 Parametric Synthesis of Continuous Regulators

As can be seen, the application of I-regulators is limited to control plants that permit high setting times and high steady-state deviations resultant from the existence of the zones of insensitivity. The proportional action regulators may be used when either a high setting time or a high value of $\varepsilon_{u\%}/m$ is acceptable. A low setting time, about $5h$, and $\varepsilon_{u\%}/m \leqslant 10\%$ may be obtained in the case of small delays, $h < 20T_0$.

The PI-regulators may be used for any disturbances and fulfil practically any requirements regarding steady state, unless setting time is required to be less than $6h$. Lower setting times, between $4h$ and $6h$, may be obtained using PID-regulators. Further shortening of the setting time to the theoretical limit of $2h$ is possible with Smith-type special controllers which will be discussed later.

In conclusion, the PI-regulator is most suitable for the plant (7.2.1) and the proportional action component is most important. It is advisable to introduce the integration action component together with the P-component because this helps to avoid high gains. It is not advisable to use I-regulators since the transient processes are then very slow.

We shall give some pointers concerning choice of regulator type for controlled plants which can be described by the transfer function

$$G_0(s) = \frac{K_0 e^{-sh}}{(1+sT_{01})(1+sT_{02})}. \tag{7.2.4}$$

The selection of regulator type may be carried out on the basis of the diagram (Figure 7.2.2) given by Rijnsdorp (Grabbe et al., 1961). This diagram was obtained by means of simulation experiments on analogue computers. The coordinates are $T_1 = T_{01}/h$ and $T_2 = T_{02}/h$.

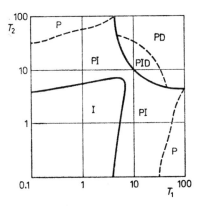

Figure 7.2.2 Diagram for selecting regulator type

If the controlled plant is described by the transfer function

$$G_0(s) = \frac{K_0 e^{-sh}}{(1+sT_0)^n} \quad n > 2 \tag{7.2.5}$$

the P-regulator is not advisable, and the addition of the D-component does not help very much. In this case I- or PI-regulators are most suitable.

7.2.2 Criteria for regulator setting selection based on characteristic roots

A simple criterion is $\sup_i \operatorname{Re} \lambda_i$, where λ_i are the characteristic roots of the regulation system. Of course they are also the roots of the denominator of the transfer function. According to this criterion the adjustable parameters of the regulator are chosen so that the supremum of real parts of the characteristic roots takes as low a value as possible. This leads to quick vanishing of the transients, therefore the input signals (reference signals) are correctly reproduced.

We shall now go on to consider in detail a generalization of the aperiodic stability criterion, well known in the theory of linear finite-dimensional control systems. The regulator settings are chosen in such a way that there exists a real characteristic root of the regulation system and this root has maximal attainable multiplicity. Unlike the case of finite-dimensional systems, this criterion does not ensure that the transients are non-oscillatory. However one can expect, especially in delayed (non-neutral) systems with small delays, that the transient components corresponding to the complex characteristic roots will decay much quicker than those corresponding to the multiple real characteristic root. In every case these results should be applied carefully, and the asymptotic stability of the regulation system should be verified.

We now consider the conventional regulation system of Figure 7.1.1 with the controlled plant (7.1.8). For simplicity of calculation we shall define in this section the characteristic polynomial as the denominator of the transfer function (7.1.11). Of course this does not change the results.

For P- and PD-regulators the characteristic quasipolynomial is

$$g(s) = (s+a)e^s + bs + c \tag{7.2.6}$$

and for all other regulators considered

$$g(s) = (s^2 + as)e^s + bs^2 + cs + d \tag{7.2.7}$$

where a, b, c, d are given by (7.1.10). By definition, g has a root s of multiplicity k if and only if $g^{(j)}(s) = 0$, $j = 0, \ldots, k-1$, and $g^{(k)}(s) \neq 0$.

First we consider the case of P- and PD-regulators, i.e. $p_1 = 0$ and g is given by (7.2.6). We have

7.2 Parametric Synthesis of Continuous Regulators

$$\dot{g}(s) = (s+a+1)e^s + b \tag{7.2.8}$$

$$\ddot{g}(s) = (s+a+2)e^s \tag{7.2.9}$$

$$\dddot{g}(s) = (s+a+3)e^s. \tag{7.2.10}$$

For all s, $\ddot{g}(s) \neq \dddot{g}(s)$ and therefore the multiplicity of characteristic roots cannot exceed three. For further discussion we must distinguish two cases: $b = 0$ (P-regulator) and $b \geqslant 0$ (PD-regulator). If $b = 0$, we can see from (7.2.8) and (7.2.9) that the maximal multiplicity of characteristic roots is two. If we denote a double root by r, we have

$$(r+a)e^r + c = 0$$

and

$$(r+a+1)e^r = 0.$$

Hence

$$r = -a-1 \tag{7.2.11}$$

$$c = \exp(-a-1). \tag{7.2.12}$$

For the PD-regulator and a triple root r we now have

$$(r+a)e^r + br + c = 0$$
$$(r+a+1)e^r + b = 0$$
$$(r+a+2)e^r = 0.$$

Hence

$$r = -a-2 \tag{7.2.13}$$

$$b = \exp(-a-2) \quad c = (a+4)\exp(-a-2). \tag{7.2.14}$$

Now let us consider the regulators with an integral action component. Then $p_1 = 1$ and the characteristic quasipolynomial is given by (7.2.7). We calculate its derivatives

$$\dot{g}(s) = [s^2 + (a+2)s + a]e^s + 2bs + c \tag{7.2.15}$$

$$\ddot{g}(s) = [s^2 + (a+4)s + 2a + 2]e^s + 2b \tag{7.2.16}$$

$$g^{(3)}(s) = [s^2 + (a+6)s + 3a + 6]e^s \tag{7.2.17}$$

$$g^{(4)}(s) = [s^2 + (a+8)s + 4a + 12]e^s. \tag{7.2.18}$$

It is easy to see that the multiplicity of characteristic roots cannot be greater than four. Let us begin with the case $b = c = 0$ (I-regulator). Maximal multiplicity is then two and a double root r satisfies

$$(r^2 + ar)e^r + d = 0$$
$$[r^2 + (a+2)r + a]e^r = 0$$

hence

$$r = -\frac{a}{2} - 1 + \sqrt{\left(\frac{a}{2}\right)^2 + 1} \qquad (7.2.19)$$

$$d = \left(-2 + \sqrt{a^2 + 4}\right) e^r. \qquad (7.2.20)$$

The negative value of the square root in (7.2.19) is rejected because it yields a negative value for d.

For the PI-regulator we have $b = 0$, $c \geqslant 0$. The maximal multiplicity of characteristic roots is three and a triple root r satisfies

$$(r^2 + ar)e^r + cr + d = 0$$
$$[r^2 + (a+2)r + a]e^r + c = 0$$
$$[r^2 + (a+4)r + 2a + 2]e^r = 0,$$

hence

$$r = -\frac{a}{2} - 2 + \sqrt{\left(\frac{a}{2}\right)^2 + 2} \qquad (7.2.21)$$

$$c = \left(-2 + \sqrt{a^2 + 8}\right) e^r_{\,} \qquad (7.2.22)$$

$$d = \frac{1}{2}\left[(a+10)\sqrt{a^2+8} - (a+1)^2 - 27\right] e^r. \qquad (7.2.23)$$

In the case of the PID-regulator the maximal multiplicity is four and a root r of multiplicity four satisfies

$$(r^2 + ar)e^r + br^2 + cr + d = 0$$
$$[r^2 + (a+2)r + a]e^r + 2br + c = 0$$
$$[r^2 + (a+4)r + 2a + 2]e^r + 2b = 0$$
$$[r^2 + (a+6)r + 3a + 6]e^r = 0.$$

Hence

$$r = -\frac{a}{2} - 3 + \sqrt{\left(\frac{a}{2}\right)^2 + 3} \qquad (7.2.24)$$

$$b = \left[-1 + \sqrt{\left(\frac{a}{2}\right)^2 + 3}\right] e^r \qquad (7.2.25)$$

$$c = \left[\left(6 + \frac{a}{2}\right)\sqrt{a^2+12} - \frac{a^2}{2} - a - 18\right] e^r \qquad (7.2.26)$$

$$d = \left[\left(\frac{a^2}{4} + 3a + 21\right)\sqrt{a^2+12} - \frac{a^3}{4} - 3a^2 - 9a - 72\right] e^r. \qquad (7.2.27)$$

It is evident that our criterion may be regarded as an analogue of the aperiodic stability criterion only if all complex characteristic roots have real

7.2 Parametric Synthesis of Continuous Regulators

parts less than r. We shall only give a necessary condition for this. If the regulator contains the D-component, then we have for the characteristic roots the asymptotic formula (3.2.49) with a constant real part equal to $\ln b$ ($b > 0$). Therefore we require

$$\ln b < r. \tag{7.2.28}$$

It is easy to check that this condition cannot be satisfied in a system with the PD-regulator, and in a system with the PID-regulator it implies that $a < 2$.

Finally, we gather all formulas defining the adjustable parameters as functions of T_0 or $A = \frac{1}{2}a = \frac{1}{2}q/T_0$. We denote $K = K_0 K_R$:

P-regulator:
$$r = -\frac{1+T_0}{T_0}, \quad K = T_0 e^r \tag{7.2.29}$$

PD-regulator:
$$r = -\frac{1+2T_0}{T_0}$$

$$T_d = \frac{T_0}{1+4T_0}, \quad K = (1+4T_0)e^r \tag{7.2.30}$$

PI-regulator:
$$r = \sqrt{1+A^2} - 1 - A$$

$$\frac{K}{T_i} = 2(\sqrt{1+A^2}-1)e^r \tag{7.2.31}$$

PI-regulator:
$$r = \sqrt{2+A^2} - 2 - A$$

$$T_i = \frac{1+A^2}{3+A+4A^2+A^3-(2+A^2)\sqrt{2+A^2}}$$

$$K = 2T_0(\sqrt{2+A^2}-1)e^r \tag{7.2.32}$$

PID-regulator: $r = \sqrt{3+A^2} - 3 - A$

$$T_i = \frac{(6+A)\sqrt{3+A^2}-9-A-A^2}{(21+6A+A^2)\sqrt{3+A^2}-36-9A-6A^2-A^3}$$

$$T_d = \frac{1}{2}\frac{\sqrt{3+A^2}-1}{(6+A)\sqrt{3+A^2}-9-A-A^2}$$

$$K = 2T_0\left[(6+A)\sqrt{3+A^2}-9-A-A^2\right]e^r. \tag{7.2.33}$$

These formulas are valid also for the delay $h \neq 1$. One has only to replace T_0, T_d and T_i by T_0/h, T_d/h and T_i/h, respectively.

7.2.3 Criteria based on frequency characteristics

In principle, for conventional regulator synthesis in frequency domain we should know the transfer function of the controlled plant, together with the frequency characteristics of the reference input and of disturbances, e.g.

their spectra. The regulator settings are chosen so as to obtain a frequency characteristic of the regulation system which would ensure the reproduction of the reference input with minimal distortions and maximal suppression of disturbances. These two requirements are usually contradictory. Synthesis is possible due to the fact that in most cases the dangerous disturbances are high-frequency noise, whereas the reference input consists of low-frequency components.

In practice, the disturbances are often negligible and if not, it might be very difficult to find their frequency characteristics. In such situations one chooses the settings so that the low-frequency part of the attenuation diagram of the regulation system is as flat as possible, and then 'hopes for the best'. Of course, this only ensures that low-frequency reference inputs are reproduced with small distortions.

We present here (after Maršik, 1958) a computationally simple way to make the low-frequency part of the attenuation diagram flat. The regulator settings are chosen so that

$$\frac{d^k}{d\omega^k}|G(i\omega)|^2_{\omega=0} = 0 \quad (k = 1, 2, ..., l) \tag{7.2.34}$$

where G is the transfer function of the closed-loop regulator system and is as large as possible. Of course all odd derivatives of the function $\gamma(\omega) = |G(i\omega)|^2$ vanish at zero, and so we may assume l to be even in (7.2.34). In our case $\gamma(\omega) = L(\omega)/M(\omega)$ where L and M are symmetric real functions. It is easy to show that condition (7.2.34) is equivalent to

$$L(0)M^{(2k)}(0) = M(0)L^{(2k)}(0) \quad (k = 1, ..., l/2). \tag{7.2.35}$$

We take for $l/2$ the number of adjustable parameters to be determined.

Criterion (7.2.34) (or (7.2.35)) is called the optimal absolute value criterion. We shall apply this to determine the regulator settings for a controlled plant

$$G_0(s) = \frac{K_0 e^{-sh}}{T_0 s + q} \quad h > 0 \quad q = 0 \text{ or } 1. \tag{7.2.36}$$

We assume $h = 1$, but the results may be readily extended to any positive delay in the same fashion as in the preceding section. For the P- and PI regulators we have

$$\gamma(\omega) = \frac{c^2 + b^2\omega^2}{(a\cos\omega - \omega\sin\omega + c)^2 + (\omega\cos\omega + a\sin\omega + b\omega)^2}. \tag{7.2.37}$$

Condition (7.2.35) yields for $k = 1$

$$c^2[(1+b)^2 + 2ab - ac - 2c] = (a+c)^2 b^2 \tag{7.2.38}$$

7.2 Parametric Synthesis of Continuous Regulators

and for $k = 2$

$$c(a+4) = 4b(a+3). \tag{7.2.39}$$

For the P-regulator $b = 0$, and the first condition yields

$$K = \frac{T_0^9}{2T_0+q}. \tag{7.2.40}$$

For the PD-regulator we use both equalities (7.2.38) and (7.2.39) to obtain

$$K = \frac{(6T_0+q)(12T_0^2-q)}{2(48T_0^2+30qT_0+5q)} \tag{7.2.41}$$

$$T_d = \frac{4T_0+q}{4(3T_0+q)}. \tag{7.2.42}$$

Let us note that for small values of T_0 formula (7.2.41) yields negative gains, which is not admissible. As a rule, the settings resulting from criterion (7.2.34) must be applied with great care; in particular, the stability of the regulation system should always be verified.

If the regulator has an I-component, we have

$$\gamma(\omega) = \frac{c^2\omega^2+(d-b\omega^2)^2}{(d-b\omega^2-\omega^2\cos\omega-a\omega\sin\omega)^2+(c\omega+a\omega\cos\omega-\omega^2\sin\omega)^2} \tag{7.2.43}$$

and from (7.2.35) we obtain for $k = 1, 2, 3$

$$2ac-2d(a+1) = -a^2 \tag{7.2.44}$$

$$6b(a+1)-3c(a+2)+d(a+3) = -3 \tag{7.2.45}$$

$$20b(a+3)-5c(a+4)+d(a+5) = 0. \tag{7.2.46}$$

From (7.2.44)

$$K = \frac{1}{2} \frac{q}{\frac{1}{T_i}(T_0+q)-q}. \tag{7.2.47}$$

It is evident from this formula that if $q = 0$, criterion (7.2.35) cannot be used to determine the settings of a regulator with I-component. Let us then assume $q = 1$. For the I-regulator we calculate from (7.2.44)

$$\frac{K}{T_i} = \frac{1}{2(T_0+1)}. \tag{7.2.48}$$

For the PI-regulator we calculate from (7.2.44) and (7.2.45)

$$K = \frac{6T_0^3+6T_0^2+3T_0+1}{4(3T_0^2+3T_0+1)} \tag{7.2.49}$$

$$\frac{1}{T_i} = \frac{6T_0^2+6T_0+3}{6T_0^3+6T_0^2+3T_0+1}. \tag{7.2.50}$$

Finally, to obtain the settings for the PID-regulator we use all three equalities (7.2.44), (7.2.45), (7.2.46):

$$K = \frac{1}{\frac{2}{T_i}(T_0+1)-2} \qquad (7.2.51)$$

$$\frac{1}{T_i} = \frac{15(2T_0+1)(6T_0^2+3T_0+1)}{180T_0^4+240T_0^3+135T_0^2+42T_0+7} \qquad (7.2.52)$$

$$T_d = \frac{60T_0^4+60T_0^3+27T_0^2+7T_0+1}{180T_0^4+240T_0^3+135T_0^2+42T_0+7}. \qquad (7.2.53)$$

7.2.4 Integral criteria based on step response

It is generally assumed that a good regulation system should reproduce a step change of reference input as exactly as possible on its output. Therefore regulator settings should be chosen so as to minimize the error of reproduction. Similar criteria result from the requirement that the error caused by a step or delta-type disturbance should be as small as possible. The error is defined in an integral form

$$E = \int_0^\infty e(t)\,dt. \qquad (7.2.54)$$

Of course, E depends on the initial conditions. We shall follow the standard approach here and assume that these are equal to zero. Depending on the integrand, different criteria are obtained:

Integral error (IE)

$$e(t) = y(t)-y_\infty \qquad y_\infty = \lim y(t) \qquad (t \to \infty).$$

This criterion is easy to express analytically (E is equal to $dG(s)/ds$ at zero if this derivative exists), but its application is very limited—to systems without overshoots.

Integral of squared error (ISE)

$$e(t) = \bigl(y(t)-y_\infty\bigr)^2.$$

This criterion is often used since it leads to systems least sensitive to disturbances. Its application results in strongly oscillatory transients and strong control action (large values of gain K and advance time T_d, small values of T_i). The overshoots may reach 40–50%.

Integral of absolute error (IAE)

$$e(t) = |y(t)-y_\infty|.$$

7.2 Parametric Synthesis of Continuous Regulators 225

This criterion gives good transient properties of the regulation system if the transient amplitude is not too high.

Integral of absolute error multiplied by time (ITAE)

$$e(t) = t|y(t) - y_\infty|.$$

As compared with the previous one, this criterion gives improved transients for large values of t.

Miller et al. (1967) have determined optimal regulator settings by means of computer calculations for different criteria and for all types of continuous conventional regulators. The controlled plant is described by the transfer function (7.1.8) with $h = 1$, $q = 1$. The reference input is equal to zero and the system is stimulated by a unit step function on the input of the controlled plant. The optimal values of regulator parameters depend on T_0 according to the formulas:

$$K = \alpha T_0^\beta \quad T_i = \gamma T_0^\delta \quad T_d = \mu T_0^\lambda \quad K = K_0 K_R.$$

Table 7.1 also shows optimal parameter values according to the Ziegler–Nichols (ZN) criterion and the 'three constraints' (3C) criterion, which are basically similar to those in systems without delays.

Table 7.1 Optimal regulator settings according to various performance criteria

Criterion	Regulator	Optimal values of parameters					
		α	β	γ	δ	μ	λ
ZN	P	1.000	1.000				
3C	P	1.208	0.936				
IAE	P	0.902	0.985				
ISE	P	1.411	0.917				
ITAE	P	0.490	1.084				
ZN	PI	0.900	1.000	3.333	0.000		
3C	PI	0.928	0.946	0.928	0.417		
IAE	PI	0.984	0.986	1.644	0.293		
ISE	PI	1.305	0.959	2.033	0.261		
ITAE	PI	0.859	0.977	1.484	0.320		
ZN	PID	1.200	1.000	2.000	0.000	0.500	0.000
3C	PID	1.370	0.950	0.740	0.262	0.365	0.050
IAE	PID	1.435	0.921	1.139	0.251	0.482	−0.137
ISE	PID	1.495	0.945	0.917	0.229	0.560	−0.006
ITAE	PID	1.357	0.947	1.176	0.262	0.381	0.005

An analysis of transient processes, given also by Miller et al. (1967), shows that the best processes result from the ITAE criterion; other integral criteria yield slightly worse transients, and the ZN and 3C criteria give transients with much worse properties.

We shall now discuss the ISE criterion in more detail. Though its analysis is much more difficult than in case of systems without delays, first analytic results in this direction were obtained as early as 1965 (Repin, 1965). Here we shall follow the Lyapunov equation method. The same results can be obtained by means of Laplace transform techniques (Górecki and Popek 1982). Let us consider the regulation system shown in Figure 7.2.3.

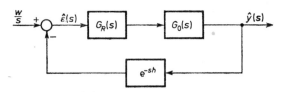

Figure 7.2.3 Regulation system with a delay in the feedback loop

The controlled plant is described by the transfer function

$$G_0(s) = \frac{K_0}{T_0 s + q} \quad T_0, K_0 > 0 \quad q = 0 \text{ or } 1.$$

The regulator is of I- or PI-type, $G_R(s) = K_R(p_0 + 1/T_i s)$, $p_0 = 0$ or 1. The delay is in the feedback loop, not in the plant. We assume that the reference input w is constant and the system is in the state of equilibrium until $t = 0$ and that at $t = 0$ a delta-type disturbance appears and changes the value of y from y_∞ to some y_0. Our purpose is to find an analytic expression for the ISE criterion

$$E = \int_0^\infty [y(t) - y_\infty]^2 dt$$

which could be helpful in determining regulator settings.

Denoting $\varepsilon(t) = w - y(t-h)$ we obtain the following equation for the system's dynamics:

$$T_0 \dot{y}(t) + q y(t) = p_0 K \varepsilon(t) + \frac{K}{T_i} \int_{t_0}^{t} \varepsilon(s) ds + C \quad t \geq t_0 \quad (7.2.55)$$

7.2 Parametric Synthesis of Continuous Regulators

where t_0 is the initial time, C is a constant and $K = K_0 K_R$. From this equation it is evident that $y_\infty = w$ and $C = qw$. Define

$$x_2(t) = y(t) - w \quad x_1(t) = \int_0^t x_2(s)\,ds. \tag{7.2.56}$$

The equations of the regulation system then take the form

$$\begin{aligned} \dot{x}_1(t) &= x_2(t) \\ \dot{x}_2(t) + a x_2(t) + d x_1(t-h) + c x_2(t-h) &= 0 \end{aligned} \tag{7.2.57}$$

with initial conditions

$$\begin{aligned} x_1(t) &= 0 \quad x_2(t) = 0 \quad t < 0 \\ x_1(0) &= 0 \quad x_2(0) = x_{20} = y_0 - w. \end{aligned} \tag{7.2.58}$$

As before, we denote $a = q/T_0$, $c = p_0 K/T_0$, $d = K/(T_i T_0)$. Now the ISE criterion is

$$E = \int_0^\infty x_2(t)^2\,dt. \tag{7.2.59}$$

We assume further that this integral exists.

It will be helpful to study our problem in a more general form. Let the system equation be

$$\dot{x}(t) = Ax(t) + Bx(t-h) \quad t \geq 0 \tag{7.2.60}$$

$$x(t) \in R^n \quad h > 0$$

$$x(0) = x_0 \quad x(t) = 0 \quad t < 0.$$

We want to compute

$$E = \int_0^\infty x(t)^T V x(t)\,dt \quad V = V^T \geq 0. \tag{7.2.61}$$

Using the matrix-valued fundamental solution Φ defined in Section 3.2 (formulas (3.2.11), (3.2.12)), for the solution of (7.2.60) we have

$$x(t) = \Phi(t) x_0.$$

Hence $E = x_0^T M x_0$,

$$M = \int_0^\infty \Phi(t)^T V \Phi(t)\,dt. \tag{7.2.62}$$

Of course, for every s

$$M = \int_s^\infty \Phi(t-s)^T V \Phi(t-s)\,dt.$$

Thus, by differentiating with respect to s one readily obtains
$$0 = V + A^T M + MA + L(h)^T + L(h) \tag{7.2.6}$$
where
$$L(h) = \int_0^\infty \Phi(t)^T V \Phi(t-h) B \, dt = \int_0^\infty \Phi(t+h)^T V \Phi(t) B \, dt. \tag{7.2.6}$$
We have here used the following lemma.

Lemma 7.2.1 The following identity holds:
$$A\Phi(t) + B\Phi(t-h) = \Phi(t)A + \Phi(t-h)B. \tag{7.2.6}$$

Proof It is easy to show that the Laplace transform $\hat{\Phi}$ of Φ is
$$\hat{\Phi}(s) = (sI - A - e^{-sh}B)^{-1}.$$
Applying a Laplace transform to both sides of (7.2.65) we obtain
$$A\hat{\Phi}(s) + Be^{-sh}\hat{\Phi}(s) = \hat{\Phi}(s)A + e^{-sh}\hat{\Phi}(s)B \tag{7.2.6}$$
which can be verified by straightforward substitution. □

Let us define
$$L(s) = \int_0^\infty \Phi(t+s)^T V \Phi(t) B \, dt \quad s \in [0, h]. \tag{7.2.6}$$
By virtue of (7.2.65) it follows that
$$\dot{L}(s) = \int_0^\infty [\Phi(t+s)A + \Phi(t+s-h)B]^T V \Phi(t) B \, dt$$
$$= A^T L(s) + \int_{s-h}^\infty B^T \Phi(t)^T V \Phi(t-s+h) B \, dt$$
$$= A^T L(s) + L(h-s)^T B. \tag{7.2.6}$$

In this way we have obtained the so-called Lyapunov set of equations determining M:
$$A^T M + MA + V + L(h)^T + L(h) = 0 \tag{7.2.6}$$
$$\dot{L}(s) = A^T L(s) + L(h-s)^T B \quad s \in [0, h] \tag{7.2.7}$$
$$L(0) = MB. \tag{7.2.7}$$

It can be shown that this set of equations has a unique solution if system (7.2.60) is asymptotically stable (see Chapter 6). Following Castelan and Infante (1977) we shall outline a method for solving this set. Define $Z($ $= L(h-s)$. From (7.2.70), (7.2.71) we then have

7.2 Parametric Synthesis of Continuous Regulators

$$\dot{L}(s) = A^T L(s) + Z(s)^T B, \quad L(0) = MB \tag{7.2.72}$$

$$\dot{Z}(s) = -A^T Z(s) - L(s)^T B, \quad Z(h) = MB. \tag{7.2.73}$$

This is a linear autonomous system with split boundary conditions. If its solution is unique, it can be easily found (at least in principle) using elementary methods.

Let us return to the original problem (7.2.57), (7.2.58), (7.2.59). We denote

$$L = \begin{bmatrix} L_1 & L_2 \\ L_3 & L_4 \end{bmatrix}$$

and similarly for Z and M. Then after obvious transformations system (7.2.72), (7.2.73) takes the form

$$\begin{aligned}
\dot{L}_1 &= -dZ_3 & L_1(0) &= -dM_2 \\
\dot{L}_2 &= -cZ_3 & L_2(0) &= -cM_2 \\
\dot{L}_3 &= L_1 - aL_3 - dZ_4 & L_3(0) &= -dM_4 \\
\dot{L}_4 &= L_2 - aL_4 - cZ_4 & L_4(0) &= -cM_4 \\
\dot{Z}_1 &= dL_3 & Z_1(h) &= -dM_2 \\
\dot{Z}_2 &= cL_3 & Z_2(h) &= -cM_2 \\
\dot{Z}_3 &= -Z_1 + aZ_3 + dL_4 & Z_3(h) &= -dM_4 \\
\dot{Z}_4 &= -Z_2 + aZ_4 + cL_4 & Z_4(h) &= -cM_4.
\end{aligned} \tag{7.2.74}$$

Notice that the solution satisfies the relationship

$$c\,\text{col}(L_1, L_3, Z_1, Z_3) = d\,\text{col}(L_2, L_4, Z_2, Z_4). \tag{7.2.75}$$

To simplify further analysis we make the natural assumption $d > 0$. Let

$$\Delta_{1,2} = \frac{1}{2}\left[\sqrt{(a^2-c^2)^2 + 4d^2} \pm (a^2 - c^2)\right]$$

$$q = \sqrt{\Delta_1} \quad r = \sqrt{\Delta_2}.$$

We then have

$$\begin{aligned}
L_1(s) &= \alpha_1\left\{y_1 \cosh\left[q\left(s - \frac{h}{2}\right)\right] + y_3 \sinh\left[q\left(s - \frac{h}{2}\right)\right]\right\} + \\
&\quad + \alpha_2\left\{z_1 \cos\left[r\left(s - \frac{h}{2}\right)\right] + z_3 \sin\left[r\left(s - \frac{h}{2}\right)\right]\right\} \\
L_3(s) &= \alpha_1\left\{y_2 \cosh\left[q\left(s - \frac{h}{2}\right)\right] + y_4 \sinh\left[q\left(s - \frac{h}{2}\right)\right]\right\} + \\
&\quad + \alpha_2\left\{z_2 \cos\left[r\left(s - \frac{h}{2}\right)\right] + z_4 \sin\left[r\left(s - \frac{h}{2}\right)\right]\right\},
\end{aligned} \tag{7.2.76}$$

and from (7.2.75)

$$L_2(s) = \frac{c}{d} L_1(s) \quad L_4(s) = \frac{c}{d} L_3(s).$$

We have denoted

$$y = \mathrm{col}\left((c+a)d, \quad d-q^2, \quad \frac{d}{q}(q^2-d), \quad (c+a)q\right)$$

$$z = \mathrm{col}\left((c+a)d, \quad d+r^2, \quad -\frac{d}{r}(r^2+d), \quad -(c+a)r\right)$$

for $c+a \neq 0$:

$$y = \mathrm{col}(d, \quad -a, \quad a\sqrt{d}, \quad \sqrt{d})$$
$$z = \mathrm{col}(0, \quad 1, \quad -\sqrt{d}, \quad 0)$$

for $c+a = 0$ and $a \neq 0$; and

$$y = \mathrm{col}(\sqrt{d}, \quad 0, \quad 0, \quad 1)$$
$$z = \mathrm{col}(0, \quad 1, \quad -\sqrt{d}, \quad 0)$$

(7.2.77)

for $a = c = 0$.

Let $Q = \tfrac{1}{2}hq$ and $R = \tfrac{1}{2}hr$. We calculate

$$\begin{aligned}
L_1(0) &= \alpha_1(y_1\cosh Q - y_3\sinh Q) + \alpha_2(z_1\cos R - z_3\sin R) \\
L_3(0) &= \alpha_1(y_2\cosh Q - y_4\sinh Q) + \alpha_2(z_2\cos R - z_4\sin R) \\
L_1(h) &= \alpha_1(y_1\cosh Q + y_3\sinh Q) + \alpha_2(z_1\cos R + z_3\sin R) \\
L_3(h) &= \alpha_1(y_2\cosh Q + y_4\sinh Q) + \alpha_2(z_2\cos R + z_4\sin R).
\end{aligned}$$

(7.2.78)

The initial conditions in (7.2.74) and the matrix equation (7.2.69) yield

$$\begin{aligned}
L_1(0) &= -dM_2 \quad L_3(0) = -dM_4 \\
L_1(h) &= 0 \\
L_2(h) + L_3(h) &= aM_2 - M_1 \\
L_4(h) &= aM_4 - M_2 - \tfrac{1}{2}.
\end{aligned}$$

(7.2.79)

Hence finally we obtain

$$L_1(h) = 0$$

$$L_1(0) - aL_3(0) - cL_3(h) = \frac{d}{2}$$

$$M_2 = -\frac{1}{d} L_1(0)$$

(7.2.80)

$$M_4 = -\frac{1}{d} L_3(0)$$

$$M_1 = aM_2 - L_3(h).$$

7.2 Parametric Synthesis of Continuous Regulators

Using relationships (7.2.78) we determine the constants α_1 and α_2 from the first two equations, and later we compute $M_2 = M_3$, M_4 and M_1 from the three last equations.

Let us consider another example. The controlled plant is described by the transfer function

$$G_0(s) = \frac{K_0 e^{-sh}}{s} \quad (K_0 > 0)$$

and the regulator is of P-type with gain $K_R > 0$. The whole regulation system is shown in Figure 7.2.4.

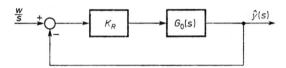

Figure 7.2.4 Astatic plant with delay and P-regulator

Again we assume that the reference input w is constant, the system is in the state of equilibrium until $t = 0$, and at $t = 0$ the value of y is changed from y_∞ to some y_0. We seek an analytic expression for the ISE criterion

$$E = \int_0^\infty [y(t) - y_\infty]^2 dt.$$

In the state of equilibrium $y(t) = y_\infty = w$. The system's dynamics is described by the equation

$$\dot{y}(t) = K[w - y(t-h)] \quad K = K_0 K_R \quad t \geq 0.$$

Denoting $x(t) = y(t) - w$ we come to the following problem. Compute

$$E = \int_0^\infty x(t)^2 dt \qquad (7.2.81)$$

where x is the solution of the initial-value problem

$$\begin{aligned}\dot{x}(t) + Kx(t-h) &= 0 & t \geq 0 \\ x(0) = y_0 - w \quad x(t) &= 0 & t < 0.\end{aligned} \qquad (7.2.82)$$

Putting $A = 0$ and $B = -K$ into equations (7.2.72), (7.2.73) we get

$$\begin{aligned}\dot{L}(s) &= -KZ(s) & L(0) &= -KM \\ \dot{Z}(s) &= KL(s) & Z(h) &= -KM.\end{aligned} \qquad (7.2.83)$$

Hence

$$L(s) = -KM\left(\cos Ks + \frac{\sin Kh - 1}{\cos Kh}\sin Ks\right). \qquad (7.2.84)$$

We may assume here $\cos Kh \neq 0$, as the integral E exists only if the regulation system is asymptotically stable, and in the region of stability $Kh < \tfrac{1}{2}\pi$, equation (7.2.69) has the form

$$2L(h) + 1 = 0.$$

Hence finally

$$M = \frac{\cos Kh}{2K(1 - \sin Kh)}. \qquad (7.2.85)$$

From this formula it can be readily seen that the ISE criterion (7.2.81) attains its minimum at $K_{opt} \in (0, \pi/2h)$ which is determined by the equation

$$K_{opt}h = \cos K_{opt}h \qquad (7.2.86)$$

whence $K_{opt}h \cong 0.74$.

7.3 Smith-type Regulation Systems

7.3.1 The Smith principle

In the sections which follow we shall discuss regulation systems in which the regulator includes a model of the controlled plant or process. The fact that the process model is an explicit part of the regulator creates a relatively easy design, based on the so-called Smith principle (Smith, 1959; Marshall, 1979). However, straightforward application of the Smith principle is restricted to processes with delayed input or output.

Consider the following two processes:

$$\begin{aligned}\dot{x}(t) &= Ax(t) + Bu(t-h) \quad t \geq 0 \\ x(0) &= x_0, \quad u(t) = v(t) \quad t \in [-h, 0) \\ y(t) &= Cx(t),\end{aligned} \qquad (7.3.1)$$

and

$$\begin{aligned}\dot{x}(t) &= Ax(t) + Bu(t) \quad t \geq 0 \\ x(t) &= \varphi(t) \quad t \in [-h, 0]. \\ y(t) &= Cx(t-h)\end{aligned} \qquad (7.3.2)$$

Here $x(t) \in R^n$, $u(t) \in R^m$, $y(t) \in R^p$, $h \geq 0$, A, B, C are matrices with compatible dimensions. As usual, u is the input (control) of the process, and $x(t)$ is its (instantaneous) state. The transfer functions of processes (7.3.1) and (7.3.2) are identical and given by

$$G_0(s) = C(sI - A)^{-1}Be^{-sh} = G_1(s)e^{-sh}. \qquad (7.3.3)$$

7.3 Smith-type Regulation Systems

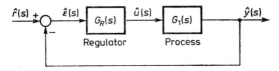

Figure 7.3.1 Reference regulation system S

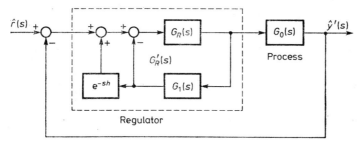

Figure 7.3.2 Smith-type regulation system with delay S_h

Let us compare the two regulation systems in Figures 7.3.1 and 7.3.2 which we shall denote by S and S_h, respectively. The first will be called the reference regulation system. The circumflexes over letters denote Laplace transforms. $G_R(s)$ is the transfer function of a linear regulator. Let $G(s)$ be the overall transfer function of the regulation system S, $G(s) = [I + G_1(s)G_R(s)]^{-1} G_1(s) G_R(s)$. A simple algebraic calculation shows that the overall transfer function $G'(s)$ of the system S_h depicted in Figure 7.3.2, is

$$G'(s) = G(s) e^{-sh}. \tag{7.3.4}$$

It is therefore evident that if $h = 0$, both systems S and S_h have identical transfer functions. Note that in the system S_h the regulator in broken contour consists of the regulator G_R of the system S and of two blocks, $G_1(s)$ and e^{-sh}, which together constitute a model of the controlled process. The transfer function of this regulator is

$$G'_R(s) = [I + (1 - e^{-sh}) G_R(s) G_1(s)]^{-1} G_R(s). \tag{7.3.5}$$

The Smith principle of design refers to controlled processes with positive delays $h > 0$. Suppose that we know how to design a linear regulator G_R for the reference system S, i.e. if the delay in the process does not occur. According to the Smith principle the regulator for the system S_h, that is for $h > 0$, is designed in such a way that all responses to reference inputs differ from those obtained in case $h = 0$ only by a positive shift of value h. More precisely, if $y: [0, \infty) \to R^1$ is a response to some input r in the system S, then the response of the system S_h with a positive delay $h > 0$ to the same input should be $y': [0, \infty) \to R^1$, $y'(t) = y(t-h)$, $t \geqslant h$. Another requirement is that if h tends to zero from above then the regulation system S_h becomes identical with the regulation system S. It should be stressed that here we consider responses only to inputs, tha tis, the initial conditions, possible disturbances, etc. are all assumed equal to zero. Notice that in the above example we have succeeded in constructing a regulation system satisfying the Smith principle by including a process model into the regulator.

As is apparent, the Smith principle is a rather attractive rule of design. It cannot be expected that the undesirable effects of delay might be reduced more than to a pure shift of the output trajectory, the value of which is equal to the delay. However, to make our analysis realistic we must take into account the effects of nonzero initial conditions, various disturbances and inaccuracy in modelling the process (so-called mismatch effects). All these factors may cause the performance of a regulation system designed according to the Smith principle to deteriorate considerably. Let us note also that the principle has been formulated for very simple processes (7.3.1) and (7.3.2).

We now go on to examine the effect of nonzero initial conditions. We wish to establish whether it is possible to design a linear regulation system satisfying the Smith principle if there are nonzero initial conditions in process (7.3.1) or (7.3.2), respectively. As before, we first consider the process without delay, $h = 0$. The corresponding regulation system is shown in Figure 7.3.1; it is immaterial whether we take process (7.3.1) or (7.3.2) as they are identical in the case $h = 0$, provided $\varphi(0) = x_0$. The output of the regulation system is

$$\hat{y} = (I+G_1 G_R)^{-1}[G_1 G_R \hat{r} + C(sI-A)^{-1}x_0]. \tag{7.3.6}$$

For the sake of brevity we omit the complex argument s. Now let us consider the regulation system for $h > 0$ with a new linear regulator G_R'' (Figure 7.3.3).

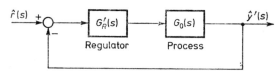

Figure 7.3.3 System with regulator G_R'

If we take the process (7.3.1), then

$$\hat{y}' = (I+G_0 G_R')^{-1}[G_0 G_R' \hat{r} + C(sI-A)^{-1}x_0 + G_0 \hat{v}] \tag{7.3.7}$$

where

$$\hat{v} = \int_{-h}^{0} e^{-st} v(t) \, dt.$$

7.3 Smith-type Regulation Systems

For the process (7.3.2)
$$\hat{y}' = (I+G_0 G'_R)^{-1}[G_0 G'_R \hat{r} + C(sI-A)^{-1}e^{-sh}x_0 + Ce^{-sh}\hat{\varphi}] \qquad (7.3.8)$$
where
$$\hat{\varphi} = \int_{-h}^{0} e^{-st}\varphi(t)\,dt.$$

Our task is to design a physically realizable linear regulator G'_R in such a way that the resulting regulation system satisfies the Smith principle. More precisely, for every initial condition and reference input we want

(i) $\hat{y}' = \hat{y}e^{-sh}$, or at least $y'(t) = y(t-h)$, $t \geqslant h$,

(ii) when $h \to 0$ the system S_h of Figure 7.3.3 becomes identical with the system S of Figure 7.3.1.

As long as we assume that the initial conditions are not measured or identified, it is evident from (7.3.7) and (7.3.8) that our problem has no exact solution unless x_0 and $\hat{v}$ or, respectively, $\hat{\varphi}$ are equal to zero. Thus the answer to our original question is, in principle no. However, this fact does not especially constrain the area of possible application of the Smith principle because a regulation system is as a rule asymptotically stable, so that the component of output generated by initial conditions vanishes with time.

If the initial state of the process is accessible in the sense that it can be either directly measured or reconstructed by some kind of observer, the Smith principle may be easily applied to the process with delay in the output (7.3.2) and pointwise initial condition $\varphi(0) = x_0$, $\hat{\varphi} = 0$. The reference system S with $h = 0$ is shown in Figure 7.3.4, and the corresponding Smith-type regu-

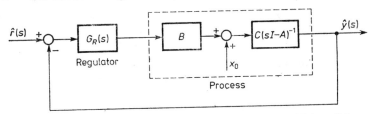

Figure 7.3.4 Reference regulation system S with pointwise initial condition x_0

lation system S_h with $h \geqslant 0$ is shown in Figure 7.3.5. It is evident from the figures that $\hat{y}'(s) = \hat{y}(s)e^{-sh}$. It should be stressed that in reality x_0 fed to the regulator may be an estimate obtained by operations which are in some sense inverse to $(r, x_0) \mapsto y'$. For greater clarity, the blocks realizing these operations are not shown in Figure 7.3.5.

We may conclude that the applicability of the Smith principle in the case of nonzero initial conditions is very limited if one wishes it to be satisfied

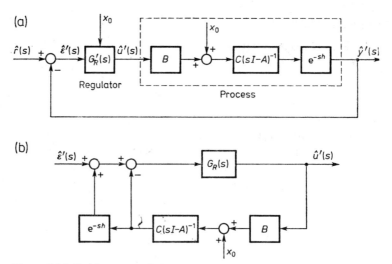

Figure 7.3.5 Smith-type regulation system S_h with pointwise initial condition x_0 (a); regulator G'_R (b)

precisely. The area of precise application can be much enlarged if we admit feedforward regulation; up to now only pure feedback has been used. There is no general agreement as to the definition of 'feedforward'; in the sequel a feedforward control is a control signal directly added to the output of the controlled process. This sum is the output signal of the control (or regulation) system. Needless to say, any feasible signal, also an output of some subsystem, may be used as the feedforward control. The construction of feedforward may seem easy theoretically, but in practice it may be difficult or even impossible due to the fact that it is not enough to deal with information signals like measurements, etc., but the actual output has to be dealt with in a direct way. We shall illustrate on a single example how much the use of feedforward increases the area of application of the Smith principle. Let us return once more to process (7.3.2) with arbitrary initial conditions. The reference regulation system is shown in Figure 7.3.4 and the corresponding Smith-type regulation system S_h is shown in Figure 7.3.6. The regulator G_R is the same as that presented in Figure 7.3.5b. It is evident from the figure that the Smith principle is satisfied.

Now we shall discuss the applicability of the Smith principle to controlled processes subject to disturbances. Following the well-established tradition we consider only additive disturbances, dividing them into three kinds: input, internal and output disturbances. Of course this division is in some sense a matter of convention. The regulation system with disturbances is shown in Figure 7.3.7.

7.3 Smith-type Regulation Systems 237

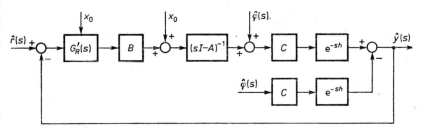

Figure 7.3.6 Smith-type regulation system S_h with functional initial condition

We assume that G_1 represents the easily invertible part of the transfer function of the process, e.g. $G_1(s)$ may be a ratio of two polynomials in s of the same degree. In that case if the internal disturbance z_2 can be measured or estimated, then it can be cancelled or reduced by a proper feedforward containing $G_1(s)^{-1}$. This feedforward, however, unlike the case shown in Figure 7.3.6, is fed to the input of the process. Provided the input disturbance z_1 can be measured, its elimination is still easier. Notice that even if we can measure z_3, it cannot be cancelled by adding an appropriate component to the input. Thus, if we assume zero initial conditions, and $z_1 = 0$, $z_3 = 0$, z_2 is measurable and $G_1(s)^{-1}$ is physically realizable, then the Smith-type regulation system takes the form presented in Figure 7.3.8. Regulator $G'_R(s)$ is shown in Figure 7.3.2.

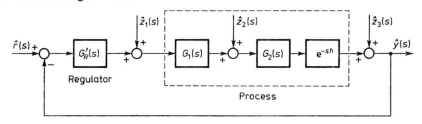

Figure 7.3.7 Smith-type regulation system S_h with disturbances

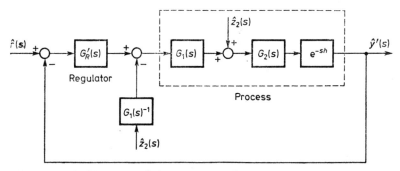

Figure 7.3.8 Smith-type regulation system S_h with disturbances and feedforward

In most cases disturbances cannot be measured and only statistical information about them is available. In such a situation it is impossible to implement the Smith principle precisely. In order to reduce the undesired or harmful effects of disturbances as well as those of initial conditions, the Smith-type regulation system is sometimes complemented by additional feedback loops and/or additional blocks in the loops, whose parameters are determined by parametric optimization. This approach, called the extended Smith method (Marshall, 1979), yields the regulation system presented in Figure 7.3.9.

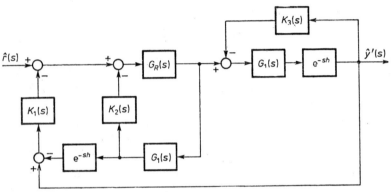

Figure 7.3.9 Extended Smith-type regulation system

For greater clarity the initial conditions and disturbances are omitted in the figure. In this way the regulation system may be stabilized where an unstable process is involved. Of course the extended Smith method is reduced to the previous one if $K_1 = 1$, $K_2 = 1$, and $K_3 = 0$.

7.3.2 Mismatch, stability and sensitivity

In this section we shall study the effects of inaccurate modelling in the Smith method. Let us consider the basic Smith-type regulation system of Figure 7.3.2 assuming zero initial conditions and zero disturbances. We replace the exact model of the controlled process in the feedback loop by an approximate one, that is, we put $\overline{G}_1$ instead of G_1, and $\overline{h}$ instead of h (Figure 7.3.10). Denote $H(s) = \exp(-sh)$, $\overline{H}(s) = \exp(-s\overline{h})$.

The reader is encouraged to verify that the transfer function of the regulation system of Figure 7.3.10 is

$$G' = (I + G_0 \overline{G}'_R)^{-1} G_0 \overline{G}'_R \qquad (7.3.9)$$

where $\overline{G}'_R$ is the transfer function of the Smith regulator with the inaccurate model,

$$\overline{G}'_R = [I + (1 - \overline{H}) G_R \overline{G}_1]^{-1} G_R. \qquad (7.3.10)$$

7.3 Smith-type Regulation Systems

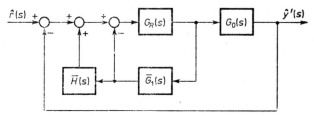

Figure 7.3.10 Smith-type regulation system with mismatch

After some algebra, formula (7.3.9) yields

$$G' = G_1 G_R H[I+(\bar{G}_1 - \bar{G}_1 \bar{H} + G_1 H)G_R]^{-1}$$
$$= G_1 G_R [I + G_1 G_R + (\bar{G}_1 - G_1 + G_1 H - \bar{G}_1 \bar{H})G_R]^{-1} H. \quad (7.3.11)$$

Thus, the transfer function of the regulation system with an inaccurate model differs from that required by the Smith principle by the term $(\bar{G}_1 - G_1 + G_1 H - \bar{G}_1 \bar{H})G_R$, which is called the mismatch term (Marshall, 1979).

An important question that has to be answered is whether mismatch, even a small one, may cause instability. It is obvious that the Smith method would be of little practical value if the regulation system became unstable due to even the slightest inaccuracy in modelling. We can analyse mismatch both in terms of transfer functions and in terms of differential equations. In the first case we fix some rational function of complex variable ΔG_1 and a real Δh, not equal to zero simultaneously, and set $\bar{G}_1 = G_1 + \varepsilon \Delta G_1$, $\bar{h} = h + \varepsilon \Delta h$. If we choose to use the differential description, we write the inaccurate model of the process in the form

$$\dot{\bar{x}}(t) = \bar{A}\bar{x}(t) + \bar{B}u(t) \quad t \geq 0$$
$$\bar{y}(t) = \bar{C}\bar{x}(t-\bar{h}) \quad \bar{x}(t) = \bar{\varphi}(t) \quad t \in [-\bar{h}, 0] \quad (7.3.12)$$

where $\bar{A} = A + \varepsilon \Delta A$, $\bar{B} = B + \varepsilon \Delta B$, $\bar{C} = C + \varepsilon \Delta C$, $\bar{h} = h + \varepsilon \Delta h$, ΔA, ΔB, ΔC are some fixed matrices of compatible dimensions, Δh is a real, ΔA, ΔB, ΔC and Δh are not simultaneously equal to zero. We constrain ourselves here only to process (7.3.2); (7.3.1) can be considered in an analogous way.

Let us assume, as is natural, that the reference system (without delay) is asymptotically stable. Then, in case of perfect matching the form of transfer function (7.3.4) suggests that the Smith-type regulation system is also asymptotically stable because the poles of (7.3.4) are identical to those of the reference transfer function. However, this is not true in general. The fact that the poles of (7.3.4) have negative real parts implies only that there is a linear subspace Σ of the state space, such that all output trajectories corresponding to zero input $r = 0$ and initial conditions from Σ, tend to zero. It is easy to prove that Σ is time invariant, that is, states generated by any input from any initial state in Σ always remain in Σ. In practical considerations we may define Σ as the space of all states that can be generated by the inputs from zero initial conditions. Here we have an instance of an analysis based on transfer function being inadequate: we shall discuss this later on in an example. It will be shown that if the process is unstable, the Smith-type regulation system also may be unstable even in the case of perfect matching and a stable reference system. However, the Smith-type system is asymptotically stable on Σ. Thus, the transfer function analysis gives important though fragmentary information on stability.

Now assume that the Smith-type regulation system is asymptotically stable in the case of perfect matching ($\varepsilon = 0$). The following question naturally arises. Given $\Delta A, \Delta B, \Delta C, \Delta h$, does there exist an $\varepsilon_0 > 0$ such that the mismatched system (with the model (7.3.12)) is asymptotically stable for every $\varepsilon \in (0, \varepsilon_0)$? To apply a regulation system in practice and in the presence of disturbances, we should have a positive answer for all types of mismatch determined by $\Delta A, \Delta B, \Delta C, \Delta h$, which are likely to occur. The same may be said in terms of transfer functions though we must remember that using this formulation we are likely to obtain fragmentary results. We shall show how transfer functions are used to study the effects of mismatch in the delay. For simplicity we consider the one-dimensional case, and assume that the undelayed part of the process is modelled exactly, $G_1 = \overline{G}_1$, but there is a mismatch in the model of the delay. Then

$$G' = \frac{G_1 G_R e^{-sh}}{1 + G_1 G_R + G_1 G_R e^{-sh}(1 - e^{-s\Delta h})}. \qquad (7.3.13)$$

G_1 and G_R are rational functions, $G_1 = M_1/N_1$, $G_R = M_R/N_R$ where the degrees of the polynomials M_1, M_R are less than those of the polynomials N_1, N_R, respectively. Then

$$G' = \frac{M_1 M_R e^{-sh}}{N_1 N_R + M_1 M_R + M_1 M_R e^{-sh}(1 - e^{-s\Delta h})}. \qquad (7.3.14)$$

It is evident that if the exactly matched system ($\Delta h = 0$) is asymptotically stable then there is a real $\varepsilon_0 > 0$ such that the mismatched system is asymptotically stable for every $|\Delta h| < \varepsilon_0$. Moreover, we may be sure that the additional poles which appear for $\Delta h \neq 0$, have real parts tending to $-\infty$ when $\Delta h \to 0$. Thus we see that as far as stability is concerned we can afford a certain inaccuracy in modelling the delay. This may be important if the model delay is realized by means of magnetic tapes.

7.3 Smith-type Regulation Systems

Perhaps the following simple example will be helpful in clarifying the stability problems of Smith-type regulation systems. The process is described by the equations

$$\dot{x}(t) = Ax(t)+u(t) \quad x(t) \in R \quad u(t) \in R \quad t \geq 0$$
$$x(t) = \varphi(t) \quad t \in [-1, 0] \quad (7.3.15)$$
$$y(t) = x(t-1).$$

We assume that the only possible mismatch in modelling is in the parameter A, so the model equations are the same as (7.3.15) with A replaced by $\bar{A}$. The regulator is of proportional action, $G_R(s) = K$. The whole regulation system is of the Smith type shown in Figure 7.3.10. If we denote by v the output of the delay-free part of the model G_1, we obtain for $t \geq 0$

$$y'(t) = x(t-1) \quad (7.3.16)$$
$$\dot{x}(t) = Ax(t)+Kr(t)-Kx(t-1)-Kv(t)+Kv(t-1) \quad (7.3.17)$$
$$\dot{v}(t) = \bar{A}v(t)+Kr(t)-Kx(t-1)-Kv(t)+Kv(t-1). \quad (7.3.18)$$

The transfer function of the system is

$$G' = \frac{K(s-\bar{A})}{(s-A)(s-\bar{A}+K)e^s+K(A-\bar{A})} \quad (7.3.19)$$

which yields in the case of perfect matching

$$G' = \frac{Ke^{-s}}{s-A+K}. \quad (7.3.20)$$

Let us first discuss stability when the matching is exact. From (7.3.20) we immediately obtain a necessary condition of asymptotic stability: $A-K < 0$. However, this condition is a sufficient one only for initial conditions taken from the subspace Σ of the state space. It is easy to see that the initial conditions in Σ satisfy the equality $x(t) = v(t)$, $t \leq 0$. Physically, this means that every input generates identical states of the process and of the model, provided there are zero initial conditions. Let us examine the solution of (7.3.16)–(7.3.18) for $\bar{A} = A$. Subtracting (7.3.17) from (7.3.18) and integrating we obtain

$$v(t)-x(t) = e^{At}[v(0)-x(0)]. \quad (7.3.21)$$

Substituting this into (7.3.17) after some calculations we get for $t \geq 1$ and $r = 0$,

$$x(t) = e^{At}[1-e^{-A}(1-e^{K(1-t)})][x(0)-v(0)]+e^{(A-K)t}v(0)+$$
$$+K\int_{-1}^{0} e^{(A-K)(t-s-1)}[v(s)-x(s)]ds.$$

It is evident that the system is asymptotically stable in the whole state space if and only if $A-K < 0$ and $A < 0$. Moreover, the output trajectory for $r = 0$ and arbitrary initial conditions tends to zero if and only if $A-K < 0$ and $A \leqslant 0$. However, we note that $A-K < 0$ is a necessary and sufficient condition of asymptotic stability if $x(t) = v(t)$, $t \leqslant 0$. Thus, we see that a complete discussion of stability on the basis of the transfer function is impossible. Still, it is not necessary to solve the system equations to obtain the conditions. The characteristic quasipolynomial of the system (7.3.17), (7.3.18) is

$$g(s) = \det\left[sI - \begin{bmatrix} A & -K \\ 0 & A-K \end{bmatrix} + e^{-s}\begin{bmatrix} -K & K \\ -K & K \end{bmatrix}\right] = (s-A)(s-A+K).$$

As was to be expected, for the matched Smith regulation system the characteristic quasipolynomial degenerates into a polynomial. Its roots are A and $A-K$, whence the above conditions of stability follow.

We come to the conclusion that if the controlled process is unstable we can hardly expect the Smith-type regulation system to work properly even provided there is perfect matching. This is because in real systems we cannot neglect the disturbances and we cannot assume that the initial states of the process and of the model are exactly equal. On the other hand, asymptotic stability of the process implies asymptotic stability (in the whole state space) of the regulation system.

The results for the mismatched system, $A \neq \bar{A}$, are somewhat similar. If the process is unstable, the mismatched system is also unstable, however small the difference $|A-\bar{A}|$. It is noteworthy that it is unstable also on Σ and so the transfer function (7.3.19) has at least one pole with a positive real part. For example, let $K = 2$, $A = -1$. If matching were perfect, the regulation system would be asymptotically stable on Σ and the only pole of G would be $A-K = -1$. It is evident that for every $\bar{A} \in (-1.1, -1)$, $\bar{G}'(7.3.19)$ has a real pole in the interval $(0.9, 1)$ tending to 1 when $\bar{A} \to -1$. Now let us assume that the process is asymptotically stable, $A < 0$. It then follows easily from the well-known properties of the characteristic roots (see Chapter 3) that there is a real $\varepsilon_0 > 0$ such that for every $\bar{A}$, $|A-\bar{A}| < \varepsilon_0$, the mismatched system is asymptotically stable.

The conclusion of this example can be easily generalized to multidimensional systems with arbitrary regulators. In particular, relation (7.3.21) remains valid for a very wide class of systems, hence the Smith-type regulation system may be globally asymptotically stable only if the controlled process is stable.

Now we proceed to the quantitative description of the effects of mismatch using sensitivity analysis. We shall investigate the rate of change of the Smith

7.3 Smith-type Regulation Systems

regulation system due to small changes in model parameters. The transfer function technique will be used. We assume that the delay-free part of the model depends on parameters $a_1, \ldots, a_k$, and the transfer function $\bar{G}_1$ is identical with the delay-free part of the process G_1 if $a_i = a_i^0$, $i = 1, \ldots, k$. Also the delay h will be treated as a parameter, h^0 being the delay of the process. Let $a = \text{col}(a_1, \ldots, a_k, h)$. In order to stress the dependence of system output on the parameters, we use the notation $\hat{y}'(s, a)$ and $y'(t, a)$ instead of $\hat{y}'(s)$ and $y'(t)$. In the sequel we assume that all the necessary derivatives exist and are continuous, moreover

$$\frac{\partial}{\partial a} \hat{y}'(\cdot, a) = L\left[\frac{\partial}{\partial a} y'(\cdot, a)\right]. \tag{7.3.22}$$

The expression

$$\hat{w}_i(s, a) = L\left[\frac{\partial}{\partial a_i} y'(\cdot, a)\right](s) \tag{7.3.23}$$

where y' corresponds to zero initial conditions, is called the sensitivity function with respect to the parameter a_i. One of the main tasks of sensitivity analysis is the examination of these functions or their originals in the time domain. To this end special schemes are designed, the outputs of which are originals of the sensitivity functions. In most cases these schemes are created as extensions of actual regulation systems.

Let us calculate the sensitivity function of the Smith-type regulation system (see Figure 7.3.10) with respect to a_i. In the standard approach that we are following here the output used in order to determine the sensitivity function is generated by the input from zero initial conditions. The disturbances are assumed to be equal to zero. Using the rules of matrix differentiation to formula (7.3.11) we readily obtain

$$\frac{\partial G'}{\partial a_i} = G'\left(\frac{\partial}{\partial a_i} \bar{G}_1\right) G_R(\bar{H}-1)[I+(\bar{G}_1-\bar{G}_1\bar{H}+G_1 H)G_R]^{-1}. \tag{7.3.24}$$

Unless a deliberate mismatch is assumed, we are interested in the derivatives calculated for perfect matching. In this case

$$\left.\frac{\partial G'}{\partial a_i}\right|_{a=a^0} = G'\left(\frac{\partial}{\partial a_i} \bar{G}_1\right)_{a=a^0} G_R(H-1)(I+G_1 G_R)^{-1}. \tag{7.3.25}$$

If G_1^{-1} exists (to this end the dimension of control must be equal to that of output, $m = p$) we may rewrite (7.3.25) in the form

$$\left.\frac{\partial G'}{\partial a_i}\right|_{a=a^0} = G'\left(\frac{\partial}{\partial a_i} \bar{G}_1\right)_{a=a^0} G_1^{-1}(G'-G) \tag{7.3.26}$$

where G is the transfer function of the reference regulation system,:
$$G = G_1 G_R (I + G_1 G_R)^{-1}.$$
The sensitivity function may be determined in a straightforward manner:
$$\hat{w}_i(s, a) = \frac{\partial G'(s, a)}{\partial a_i} \hat{r}(s). \tag{7.3.27}$$
Combining formulas (7.3.26) and (7.3.27) we obtain the following scheme to determine the sensitivity function (Figure 7.3.11). It may be helpful in the observation or simulation of the original of the sensitivity function in the time domain. Note that the physical realizability of this system depends on the realizability of $\dfrac{\partial \bar{G}_1}{\partial a_i} G_1^{-1}$.

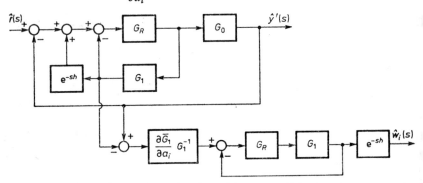

Figure 7.3.11 Scheme for determining a sensitivity function

Now we shall discuss the sensitivity of the Smith regulation system with respect to the mismatch in the delay. The respective sensitivity function is denoted by $\hat{w}_h$. As before we have
$$\hat{w}_h(s, a) = \frac{\partial G'(s, a)}{\partial h} \hat{r}(s) \tag{7.3.28}$$
where
$$\frac{\partial G'}{\partial h} = -sG'\bar{G}_1 G_R \bar{H}[I + (\bar{G}_1 - \bar{G}_1 \bar{H} + G_1 H)G_R]^{-1}. \tag{7.3.29}$$
If matching is perfect, we have
$$\left. \frac{\partial G'}{\partial h} \right|_{a=a^0} = -s(G')^2. \tag{7.3.30}$$
Obviously, here $G' = G_1 G_R (I + G_1 G_R)^{-1} H$. Since s represents a physically unrealizable operation (pure differentiation), it is impossible to draw a general physically realizable block-diagram such that $\hat{w}_h$ would be its output. However for most particular cases this may be done if advantage is taken of the properties of G. For example, sG may be physically realizable.

8

Controllability, Observability and Related Problems for Linear Systems

8.1 Definitions of Basic Notions

In order to introduce the concepts of controllability, observability and pole assignment, it is convenient to start from an abstract description of a dynamic system (Delfour, 1977b).

Let us assume an autonomous abstract evolution state equation in the Hilbert state space X is given:

$$\dot{x}(t) = Ax(t) + Bu(t) \quad t \geq 0$$
$$x(0) \in X \tag{8.1.1}$$

A is a linear closed operator in X, $\overline{D(A)} = X$, $B \in \mathscr{L}(U_0, X)$, U_0 a Hilbert space, $u: [0, \infty) \to U_0$, $u \in L^1([0, \infty); U_0) = U$ together with its mild solution

$$x(t) = F(t)x(0) + \int_0^t F(t-\tau)Bu(\tau)d\tau \quad t \geq 0 \tag{8.1.2}$$

where the family $\{F(t)\}_{t \geq 0}$ of linear continuous maps $F(t): X \to X$, $\forall t$, $t \geq 0$ fulfils

$$F(0) = I$$
$$F(t+s) = F(t)F(s) \quad t, s \geq 0$$
$$\frac{d}{dt}(F(t)x_0) = AF(t)x_0 \quad \forall t \geq 0 \quad \forall x_0 \in D(A) \tag{8.1.3}$$

the map $[0, \infty) \ni t \mapsto F(t)x_0$ is continuous for any $x_0 \in X$.

The operator A in (8.1.1) is called the infinitesimal generator of the family $\{F(t)\}_{t \geq 0}$. The family $\{F(t)\}_{t \geq 0}$ which fulfils (8.1.3) is called a strongly continuous semigroup generated by A.

There is a theory connecting the abstract homogeneous equation

$$\dot{x} = Ax \quad t \geq 0 \quad x(0) \in D(A)$$

with its generalized solution given by some strongly continuous semigroup, and conversely, connecting any strongly continuous semigroup $\{F(t)\}_{t \geq 0}$, with some operator A which is the generator of $\{F(t)\}_{t \geq 0}$.

Theorem 8.1.1 (Hille and Phillips, 1968) *Let a closed linear (not necessarily bounded) operator A be given, $A: D(A) \to X$, $\overline{D(A)} = X$, such that*

$$\|R(\lambda, A)^n\| \leq \frac{M}{(\lambda - \lambda_0)^n} \quad \forall \lambda > \lambda_0$$

for some $M > 0$, $\lambda_0 \in R^1$, where $R(\lambda, A) = (\lambda I - A)^{-1}$ is the resolvent of A. There then exists a family $\{F(t)\}_{t \geq 0}$ of linear bounded operators, $F(t): X \to X$, $t \geq 0$, such that

$$F(0) = I$$
$$F(t+s) = F(t)F(s) \quad t, s \geq 0$$

the map $[0, \infty) \ni t \mapsto F(t)x_0$ is continuous for any $x_0 \in X$

$$\frac{d}{dt}(F(t)x_0) = A(F(t)x_0) \quad t \geq 0 \quad \forall x_0 \in D(A)$$

$$\|F(t)\| \leq Me^{\lambda_0 t} \quad t \geq 0.$$

Theorem 8.1.2 (Hille and Phillips, 1968) *Let a family $\{F(t)\}_{t \geq 0}$ of linear bounded operators, $F(t): X \to X$, $t \geq 0$ fulfil the following:*

$$F(0) = I$$
$$F(t+s) = F(t)F(s) \quad t, s \geq 0$$

and the map $[0, \infty) \ni t \mapsto F(t)x_0$ is continuous for any $x_0 \in X$. Then there exist constants $M > 0$, $\lambda_0 \in R^1$ and a closed linear operator A, $\overline{D(A)} = X$ such that

$$\|F(t)\| \leq Me^{\lambda_0 t} \quad t \geq 0$$

$$\|R(\lambda, A)^n\| \leq \frac{M}{(\lambda - \lambda_0)^n} \quad \forall \lambda > \lambda_0$$

$$\frac{d}{dt}(F(t)x_0) = A(F(t)x_0) \quad \forall x_0 \in D(A)$$

$$Ax_0 = \lim_{t \to 0^+} \frac{1}{t}(F(t) - I)x_0 \quad \forall x_0 \in D(A).$$

8.1 Definitions of Basic Notions

Moreover,

$$R(\lambda, A)x = \int_0^\infty F(t)e^{-\lambda t}x\,dt \quad \forall x \in X \quad \forall \lambda > \lambda_0. \tag{8.1.4}$$

One can introduce the adjoint homogeneous system to (8.1.1), namely

$$\dot{x}(t) = A^*x(t) \quad x(0) \in X \quad t \geq 0 \tag{8.1.5}$$

where A^* is adjoint to A.
The solution of (8.1.5) is defined by

$$x(t) = F(t)^*x(0) \quad t \geq 0$$

where $\{F(t)^*\}_{t \geq 0}$ is the family of operators adjoint to operators of the family $\{F(t)\}_{t \geq 0}$. The family $\{F(t)^*\}_{t \geq 0}$ is also a strongly continuous semigroup. Its generator is A^*.

Let the so-called output equation for system (8.1.1)

$$y(t) = Cx(t) \quad t \geq 0 \tag{8.1.6}$$

be given, where $C: X \to Y$ is bounded linear, Y is the output Hilbert space.

Definition 8.1.1 *System* (8.1.1) *is called output null controllable at time* T *if and only if for any initial state* $x_0 \in X$ *there exists a control* $u \in U$, *such that* $y(T) = 0$.

Definition 8.1.2 *System* (8.1.1) *is called approximately output controllable at time* T *if and only if for any initial state* $x_0 \in X$, $\forall y_1 \in Y$, $\forall \varepsilon > 0$, *there exists a control* $u \in U$, *such that* $\|y_1 - y(T)\| < \varepsilon$.

If in the above definitions the output space $Y = R^p$, then we talk about R^p-controllability.
If in the above definitions the output relation (8.1.6) is of the form

$$Y = X$$
$$y(t) = x(t)$$

then we talk about null state controllability and approximate state controllability, respectively.

Definition 8.1.3 *System* (8.1.1) *is called observable at time* T, *if and only if for* $u = 0$, *the equality* $y(t) = 0$ *for* $t \in [0, T]$ *implies that the initial state* $x(0)$ *is equal to* 0.

Definition 8.1.4 *Let the operator A have a purely point spectrum, $\mathrm{Sp}(A) = \{\lambda_i\}_{i=1}^{\infty}$. System (8.1.1) is called spectrally controllable if and only if for any finite sequence of numbers $\{\bar{\lambda}_{r_i}\}_{i=1}^{n}$, $\{r_i\}_{i=1}^{n}$, r_i an integer, there exists a linear feedback operator $K: X \to U_0$ such that the spectrum of $(A+BK)$ is of the form*

$$\mathrm{Sp}(A+BK) = \{\lambda_1, \lambda_2, \ldots, \bar{\lambda}_{r_1}, \lambda_{r_1+1}, \ldots, \bar{\lambda}_{r_2}, \lambda_{r_2+1}, \ldots\}.$$

Definition 8.1.5 *Let the operator A have a purely point spectrum. System (8.1.1) is called exponentially stabilizable if and only if there is a feedback operator K such that the system*

$$\dot{x} = (A+BK)x$$

is exponentially stable. This means that

$$\|x(t)\| \leq Me^{-\lambda t}\|x(0)\| \quad \forall x(0)$$

for some $M, \lambda > 0$.

Now we shall specify the operator A and the family $\{F(t)\}_{t \geq 0}$ for the autonomous FDE system.

Let $X = R^n \times L^2(-h, 0; R^n)$. The element $x \in X$ will be denoted by the pair (x_0, x_1), $x_0 \in R^n$, $x_1 \in L^2(-h, 0; R^n)$. The scalar product $(x|y)$ of elements $x, y \in X$ is of the form

$$(x|y) = x_0^T y_0 + \int_{-h}^{0} x_1(s)^T y_1(s)\,ds. \tag{8.1.7}$$

Let us consider the FDE system

$$\dot{x}(t) = \int_{-h}^{0} dA(s)x(t+s) + Bu(t) \quad t \geq 0$$

$$x(0) = \hat{x}_0 \in R^n$$

$$x(s) = \hat{x}_1(s) \quad s \in [-h, 0)$$

$$(\hat{x}_0, \hat{x}_1) \in X \tag{8.1.8}$$

$$u(t) \in R^m = U_0 \quad t \geq 0$$

$$u \in L^1(0, \infty; R^m) = U$$

$A: [-h, 0] \to R^{n \times n}$ and is of bounded variation

B is an $(n \times m)$-matrix.

Under these conditions we have uniqueness and absolute continuity of the solution $x(t)$, $t \geq 0$ for any initial condition $(\hat{x}_0, \hat{x}_1) \in X$, and the continuous

8.1 Definitions of Basic Notions

dependence on the initial data (Hale, 1977). For system (8.1.8) we have the following solution formula:

$$x(t) = H(t)\hat{x}_0 + \int_0^t H(t-\tau)Bu(\tau)d\tau + W(t, \hat{x}_1) \qquad (8.1.9)$$

$$W(t, \hat{x}_1) = \begin{cases} \int_0^t H(t-\tau) \int_{-h}^{-\tau} dA(s)\hat{x}_1(\tau+s)d\tau & h \geqslant t \geqslant 0 \\ \int_0^h H(t-\tau) \int_{-h}^{-\tau} dA(s)\hat{x}_1(\tau+s)d\tau & t > h \\ \hat{x}_1(t) & t \in [-h, 0) \end{cases}$$

where H is the solution of the homogeneous system

$$\dot{H}(t) = \int_{-h}^0 dA(s)H(t+s)$$

$$H(0) = I \in R^{n \times n}$$

$$H(s) = 0 \quad s \in [-h, 0].$$

Defining the state trajectory of (8.1.8) in the space X (see Delfour, 1977b) as follows:

$$\begin{bmatrix} x_0(t) \\ (x_1(t))(s) \end{bmatrix} = \begin{bmatrix} H(t)\hat{x}_0 + W(t, \hat{x}_1) \\ H(t+s)\hat{x}_0 + W(t+s, \hat{x}_1) \end{bmatrix} +$$

$$+ \int_0^t \begin{bmatrix} H(t-\tau) \\ 0 \end{bmatrix} u(\tau)d\tau \qquad t \geqslant 0, s \in [-h, 0] \qquad (8.1.10)$$

one can introduce the strongly continuous semigroup $\{F(t)\}_{t \geqslant 0}$ of linear bounded operators acting in X

$$F(t)\begin{bmatrix} \hat{x}_0 \\ \hat{x}_1 \end{bmatrix} = \begin{bmatrix} H(t)\hat{x}_0 + W(t, \hat{x}_1) \\ H(t+(\cdot))\hat{x}_0 + W(t+(\cdot), \hat{x}_1) \end{bmatrix} \in X \qquad t \geqslant 0.$$

together with the operator A of this semigroup defined by

$$A\begin{bmatrix} x_0 \\ x_1 \end{bmatrix} = \begin{bmatrix} \int_{-h}^0 dA(s)x_1(s)ds \\ \dfrac{d}{ds}x_1(\cdot) \end{bmatrix} \qquad (x_0, x_1) \in D(A) \qquad (8.1.11)$$

where $D(A)$ is the set of pairs (x_0, x_1) such that

$$\frac{d}{ds}x_1 \in L^2(-h, 0; R^n) \qquad (8.1.12)$$

$$x_0 = x_1(0).$$

Now the suitable mild solution formula for (8.1.8) is

$$\begin{bmatrix} x_0(t) \\ x_1(t) \end{bmatrix} = F(t)\begin{bmatrix} x_0(0) \\ x_1(0) \end{bmatrix} + \int_0^t F(t-\tau)\left(\sum_{i=1}^m \begin{bmatrix} b_i \\ 0 \end{bmatrix} u_i(\tau)\right) d\tau \quad (8.1.13)$$

where b_i denotes the i-th column of B, u_i denotes the i-th component of vector u and $(b_i, 0) \in X$, $i = 1, \ldots, m$.

The appropriate evolution equation is

$$\begin{bmatrix} \dot{x}_0(t) \\ \dot{x}_1(t) \end{bmatrix} = \begin{bmatrix} \int_{-h}^0 dA(s) x_1(s+t) \\ \dfrac{d}{d(\cdot)} x_1(t+(\cdot)) \end{bmatrix} + \sum_{i=1}^m \begin{bmatrix} b_i \\ 0 \end{bmatrix} u_i(t) \quad (x_0, x_1) \in D(A) \quad t \geq 0. \quad (8.1.14)$$

From formula (8.1.9) it follows that the family $\{F(t)\}_{t \geq h}$ consists of compact operators because the state $x(t) \in X$, $t \geq h$ is defined by an integral operator.

The resolvent of the operator A has the form

$$(\lambda I - A)^{-1} \begin{bmatrix} x_0 \\ x_1 \end{bmatrix}$$

$$= \begin{bmatrix} \Delta(\lambda)\left[x_0 + \int_{-h}^0 dA(s)\left(\int_{-s}^0 e^{\lambda(s-\tau)} x_1(\tau) d\tau\right)\right] \\ e^{(\cdot)\lambda}\Delta(\lambda)\left[x_0 + \int_{-h}^0 dA(s)\left(\int_{-s}^0 e^{\lambda(s-\tau)} x_1(\tau) d\tau + \int_{(\cdot)}^0 e^{\lambda((\cdot)-\tau)} x_1(\tau) d\tau\right)\right] \end{bmatrix}$$

$$\forall (x_0, x_1) \in X \quad (8.1.15)$$

where

$$\Delta(\lambda) = \left[\lambda I - \int_{-h}^0 e^{\lambda s} dA(s)\right]^{-1} \quad \lambda \notin \mathrm{Sp}(A). \quad (8.1.16)$$

For $\begin{bmatrix} x_0 \\ x_1 \end{bmatrix} = \begin{bmatrix} b_i \\ 0 \end{bmatrix}$ $(i = 1, \ldots, m)$ we have

$$(\lambda I - A)^{-1} \begin{bmatrix} b_i \\ 0 \end{bmatrix} = \begin{bmatrix} \Delta(\lambda) b_i \\ e^{(\cdot)\lambda} \cdot \Delta(\lambda) b_i \end{bmatrix}. \quad (8.1.17)$$

It is easy to see that the spectrum of A coincides with the set of those λ for which $\Delta(\lambda)$ does not exist. Obviously the function $\lambda \mapsto \Delta^{-1}(\lambda)$ is an entire function of a complex variable, has no poles and has zeros of finite multiplicity. Thus, $\Delta(\lambda)$ has poles of finite multiplicity. Moreover, for any $r \in R^1$, the set

$$\{\lambda : \lambda \in \mathrm{Sp}(A), \mathrm{Re}(\lambda) > r\} \quad (8.1.18)$$

is finite (Hale, 1977).

8.2 R^p-controllability of Linear FDE Systems

In this section all the notation of Section 8.1 is valid. Let there be given an FDE system described by its mild solution (8.1.2), (8.1.13), together with an output relation

$$y(t) = \begin{bmatrix} (c_1|x(t)) \\ \vdots \\ (c_p|x(t)) \end{bmatrix} = cx(t) \quad c_i \in X \quad y(t) \in R^p \quad (8.2.1)$$

$c: X \to R^p$.

We have

$$y(t) = cF(t)x_0 + \int_0^t cF(t-\tau)\begin{bmatrix} B \\ 0 \end{bmatrix}u(\tau)\mathrm{d}\tau t \geq 0 \quad u \in U. \quad (8.2.2)$$

One can observe that the term $cF(t-\tau)\begin{bmatrix} B \\ 0 \end{bmatrix}$ is a $p \times m$ real matrix and $cF(t)x_0$ is a vector from R^p.

It is easy to see that system (8.1.2), (8.2.1) is R^p-controllable at time $T > 0$, if and only if the range of the map

$$S_T: U \ni u \mapsto \int_0^T cF(T-\tau)\begin{bmatrix} B \\ 0 \end{bmatrix}u(\tau)\mathrm{d}\tau = \int_0^T cF(t)\begin{bmatrix} B \\ 0 \end{bmatrix}u(T-\tau)\mathrm{d}\tau \in R^p \quad (8.2.3)$$

is the whole space R^p.

The following is valid:

$$R(S_{T_1}) \subset R(S_{T_2}) \quad T_2 \geq T_1 \quad (8.2.4)$$

where $R(S_T)$ denotes the range of S_T. This fact follows easily from (8.2.3). In order to derive R^n-controllability conditions we shall present a general tool allowing us to answer questions concerning the range of an operator P

$$P: U \mapsto R^n, \quad (8.2.5)$$

where U is some linear space of controls. Obviously the map P can be written in the form

$$Pu = (\langle f_1, u \rangle, \ldots, \langle f_n, u \rangle)^\mathrm{T} \in R^n \quad (8.2.6)$$

where $f_1, \ldots, f_n$ are linear functionals defined on U, $(\cdot)^\mathrm{T}$ denotes transposition and $\langle f_i, u \rangle$ denotes the value of the linear functional f_i on element $u \in U$.

Let V be a linear space and V^D denote its algebraic adjoint. The set $G \subset V^\mathrm{D}$ is called total on V if and only if

$$(\langle g, u \rangle = 0, \forall g \in G) \Rightarrow u = 0 \in V.$$

If G is a linear subspace of V^D, then it will be called a total space.

By $\text{lin}(E)$, where $E \subset V$, V a linear space, we mean the set of all finite linear combinations of elements of E.

Lemma 8.2.1 (Rolewicz and Przeworska-Rolewicz, 1970) *Let V be a linear space, V_1 a linear subspace of V and $x_0 \notin V_1$, $x_0 \in V$. There exists $f \in V^D$ such that*

$$\langle f, x \rangle = 0 \quad \forall x \in V_1$$
$$\langle f, x_0 \rangle = 1.$$

Theorem 8.2.1 *We assume the following*:
A.1 V *is a linear space*
A.2 $\{x_i\}_{i=1}^n \subset V$
A.3 $U \subset V^D$ *is an arbitrary subspace total on V*
A.4 *There is given an operator P defined on V^D by*

$$P: V^D \to R^n$$
$$Pg = (\langle x_1, g \rangle, \ldots, \langle x_n, g \rangle)^T, \quad g \in V^D.$$

Then

$$P(U) = P(V^D) \subset R^n. \tag{8.2.7}$$

Moreover

$$P(U) = R^n$$

if and only if the elements x_i, $i = 1, \ldots, n$ are linearly independent.

Proof Let us assume $P(U) \neq P(V^D)$. Obviously, $P(U) \subset P(V^D)$. Let $x_0 \in P(V^D)$, $x_0 \notin P(U)$. By Lemma 8.2.1 there exists $g \in R^n$, such that

$$g^T P u = 0 \quad \forall u \in U$$
$$g^T P x_0 = 1. \tag{8.2.8}$$

The space U is total on V, hence

$$\left(g^T P u = 0 = \left\langle \left(\sum_{i=1}^n g_i x_i \right), u \right\rangle, \forall u \in U \right) \Rightarrow \sum_{i=1}^n g_i x_i = 0 \tag{8.2.9}$$

which gives

$$g^T P x_0 = 0$$

and leads to contradiction with (8.2.8). Thus we have obtained the first point of thesis (8.2.7). From implication (8.2.9) we see that if $P(U) \neq R^n$, then the elements x_i, $i = 1, \ldots, n$ are linearly dependent. If $P(U) = R^n$ and x_i, $i = 1, \ldots, n$ are linearly dependent, then one can choose $g \in R^n$, $g \neq 0$ such that $\sum_{i=0}^n g_i x_i = 0$, and $g^T P u = 0$, $\forall u \in U$. But this contradicts $P(U) = R^n$ and completes the proof. □

8.2 R^p-controllability of Linear FDE Systems

From Theorem 8.2.1 it follows that the images of any total space by the map P are the same. In some cases one can construct the total space directly by means of the elements $\{x_i\}_{i=1}^n$ defining the map P.

Theorem 8.2.2 We assume that V is a linear space, $\{x_i\}_{i=1}^n \subset V$ and $U \subset V^D$ is a subspace total on V. Let there be given a linear map $W: V \to V^D$ such that

$$(\langle x, Wx \rangle = 0) \Rightarrow x = 0 \in V.$$

Given also a map

$$P: V^D \mapsto R^n$$

$$Pu = (\langle x_1, u \rangle \ldots \langle x_n, u \rangle)^T \quad u \in U.$$

Then

$$P(U) = \operatorname{lin}(\{AWx_i\}_{i=1}^n) = \operatorname{lin}\left(\begin{bmatrix} \langle x_1, Wx_1 \rangle \\ \vdots \\ \langle x_n, Wx_1 \rangle \end{bmatrix}, \ldots, \begin{bmatrix} \langle x_1, Wx_n \rangle \\ \vdots \\ \langle x_n, Wx_n \rangle \end{bmatrix}\right).$$

Proof Let us define the space

$$X = \operatorname{lin}(\{x_i\}_{i=1}^n) \subset V.$$

We will prove that the set $\{Wx_i\}_{i=1}^n \subset V^D$ is total on X. Let $x \in X$, $x = \sum a_i x_i$, $\in R^n$ and

$$\langle x, Wx_i \rangle = 0 \quad (i = 1, \ldots, n).$$

Obviously

$$a_i \langle x, Wx_i \rangle = 0 \quad (i = 1, \ldots, n)$$

and

$$\sum_{i=1}^n a_i \langle x, Wx_i \rangle = \left\langle x, W\left(\sum_{i=1}^n a_i x_i\right)\right\rangle = \langle x, Wx \rangle = 0.$$

But this implies $x = 0$, and shows that the set $\{Wx_i\}_{i=1}^n$ is total on X. By Theorem 8.2.1 we see that

$$P(U) = \operatorname{lin}(\{PWx_i\}_{i=1}^n)$$

for any total space $\hat{U}$ on X. U is of course total on V, and so it is also total on $X = \operatorname{lin}(\{x_i\}_{i=1}^n) \subset V$, which ends the proof. $\square$

It may be observed that when V is a Hilbert space we can define

$$\langle h, Wh \rangle = (h|h)$$

where $(h|h)$ denotes the scalar product in V, and

$$P(U) = \mathrm{lin}\left(\left\{\begin{bmatrix}(h_1|h_i)\\ \vdots \\ (h_n|h_i)\end{bmatrix}\right\}_{i=1}^n\right) \qquad (8.2.10)$$

for any total space U on V. The matrix composed of columns occurring in (8.2.10) is the well-known Gram matrix of elements $\{h_i\}_{i=1}^n \subset V$.

We write the following sequence of inclusions:

$$C^{AN}(D) \subset C^{\infty}(D) \subset \ldots \subset C^m(D) \subset \ldots \subset C^0(D) \subset L^{\infty}(D)$$
$$\ldots \subset L^p(D) \subset L^1(D) \qquad (8.2.11)$$

where D is a compact domain in R^k and

$C^{AN}(D)$ is the set of all analytic functions defined by a convergent power series in some open domain D', $D \subset D' \subset R^k$,

$C^{\infty}(D)$ is the set of all infinitely continuously differentiable functions on the open domain D', $D \subset D' \subset R^k$,

$C^m(D)$ is the class of m-times continuously differentiable functions,

$L^p(D)$, $1 \leqslant p < \infty$, is the set of all Lebesgue integrable functions in power p. For $p = \infty$, $L^{\infty}(D)$ is the set of all essentially bounded functions with respect to the Lebesgue measure.

It may be easily noted that if U is a total family on some space from (8.2.11), then it is also total on the previous spaces in relation (8.2.11).

Let us define some families of total sets on spaces from (8.2.11).

For $a \in R^k$, a_i are nonnegative integers, $i = 1, \ldots, k$, and for fixed $z_0 \in D \subset R^k$ we write the set $\{u_{a,z_0}^{AN}\}$ where

$$\langle u_{a,z_0}^{AN}, f\rangle = \frac{\partial^{|a|}}{\partial z_1^{a_1}, \ldots, \partial z_k^{a_k}}(f(z))|_{z=z_0} = \frac{\partial^{|a|}}{\partial z^a}(f(z))|_{z=z_0} \qquad (8.2.12)$$

which is total on $C^{AN}(D)$. Indeed, if for some $f \in C^{AN}(D)$ the sides in (8.2.12) are equal to zero, then all coefficients in the power expansion of f are equal to zero, hence $f = 0$.

For $a \in R^k$, $|a| \leqslant m$, a_i, $i = 1, \ldots, k$, are nonnegative integers, and for fixed $z_0 \in D$ we define the total set $\{u_a^m\}_{|a| \leqslant m}$ of functionals on $C^m(D)$ which consists of functionals of the form

$$\langle u_a^m, f\rangle = \frac{\partial^{|a|}}{\partial z^a}f(z)|_{z=z_0} \quad \text{for} \quad a: |a| < m \quad f \in C^m(D)$$
$$(8.2.13)$$
$$\langle u_a^m, f\rangle = \frac{\partial^{|a|}}{\partial z^a}f(z) \quad \text{for} \quad |a| = m \quad z \in D.$$

8.2 R^p-controllability of Linear FDE Systems

For fixed $a_0 \in R^k$ and for $a \in R^k$ such that $||s-a_0|| \leqslant \varepsilon, \varepsilon > 0$, we define the total set of functionals $\{z \mapsto e^{a^T z}\}_{||a-a_0|| \leqslant \varepsilon}$ on the space $L^1(D)$ as follows:

$$f \in \{z \mapsto e^{a^T z}\}_{||a-a_0|| \leqslant \varepsilon} \Leftrightarrow \langle f, x \rangle = \int_D e^{a^T z} x(z) \mathrm{d}z. \tag{8.2.14}$$

This set is total, because the values of the functionals can be treated as values of the finite Laplace transform of $x \in L^1(D)$ taken at points a, $||a-a_0|| \leqslant \varepsilon$, $\varepsilon > 0$.

As the simplest example of the application of the theory presented above, let us consider the ordinary differential system

$$\dot{x} = Ax+bu \quad x(t) \in R^n \quad A \in R^{n \times n} \quad b \in R^n \quad u(t) \in R^1 \quad u \in L^1(0, \infty). \tag{8.2.15}$$

The solution of (8.2.15) is of the form

$$x(t) = e^{At} x_0 + \int_0^t e^{A(t-\tau)} bu(\tau) \mathrm{d}\tau.$$

Given T, $T > 0$, the question is what values of

$$\int_0^T e^{A(T-\tau)} bu(\tau) \mathrm{d}\tau \in R^n \tag{8.2.16}$$

can be obtained using all u from the space $L^1(0, \infty)$. We rewrite formula (8.2.16) in the form

$$R^n \ni \int_0^T e^{A(T-\tau)} bu(\tau) \mathrm{d}\tau = \int_0^T \begin{bmatrix} g_1(\tau) \\ \vdots \\ g_n(\tau) \end{bmatrix} u(\tau) \mathrm{d}\tau = Pu$$

where g_i, $i = 1, \ldots, n$ are analytic scalar functions. We define the space $V = C^{AN}([0, T])$, $g_i \in V$. Obviously, the control space $L^1(0, \infty)$ can be treated as a total space on V. Using the total set (8.2.12) with $D = [0, T]$, $z_0 = T$, and applying the thesis of Theorem 8.2.1 we have

$$P(L^1(0, \infty)) = \mathrm{lin}\left(\begin{bmatrix} g_1(T) \\ \vdots \\ g_n(T) \end{bmatrix}, \begin{bmatrix} g_1'(T) \\ \vdots \\ g_n'(T) \end{bmatrix}, \ldots, \begin{bmatrix} g_1^{(k)}(T) \\ \vdots \\ g_n^{(k)}(T) \end{bmatrix}\right)$$

$$= \mathrm{lin}(b, -Ab, A^2 b, -A^3 b, \ldots).$$

Defining $V = C^m([0, T])$, $g_i \in V$, $i = 1, \ldots, n$, and using the total set (8.2.13) with $z_0 = T$ we have

$$P(L^1(0, \infty)) = \mathrm{lin}\left(b, -Ab, \ldots, (-1)^m A^m b, \begin{bmatrix} g_1^{(m)}(t) \\ \vdots \\ g_n^{(m)}(t) \end{bmatrix}_{t \in [0, T]}\right).$$

Taking $V = L^1(0, T)$, $g_i \in V$, $i = 1, \ldots, n$ and the total set (8.2.14) with $a_0 \notin \mathrm{Sp}(A)$, and sufficiently small ε, we obtain

$$P(L^1(0, \infty)) = \mathrm{lin}\left(\left[\int_0^T e^{A(T-\tau)} b e^{a\tau} d\tau\right]_{||a-a_0||<\varepsilon}\right)$$

$$= \mathrm{lin}\left([e^{AT}(aI-A)^{-1}(e^{(aI-A)T}-I)b]_{||a-a_0||<\varepsilon}\right)$$

$$= \mathrm{lin}\left([e^{aT}(aI-A)^{-1}b - e^{AT}b]_{|a-a_0|<\varepsilon}\right) = \mathrm{lin}\left([[aI-A]^{-1}b]_{|a-a_0|<\varepsilon}\right).$$

Taking $V = L^2(0, T)$ and using Theorem 8.2.1 we have

$$P(L^1(0, \infty)) = \mathrm{lin}\left(\left[\int_0^T \begin{bmatrix} g_1(\tau) \\ \vdots \\ g_n(\tau) \end{bmatrix} g_i(\tau) d\tau\right]_{i=1, \ldots, n}\right)$$

$$\dim P(L^1(0, \infty)) = \mathrm{rank}\left[\int_0^T e^{A\tau} BB^T e^{A^T\tau} d\tau\right].$$

Now we turn back to FDE systems and to maps of the same type as (8.2.3). Let the nonautonomous system

$$\dot{x}(t) = \int_{-h_1}^0 d_s A(t, s) x(t+s) + \int_{-h_2}^0 d_s B(t, s) u(t+s) \quad t \geq t_0 \quad (8.2.17)$$

and the output relation

$$y(t) = C_0(t) x(t) \quad (t \geq t_0) \tag{8.2.18}$$

be given where

$x(t) \in R^n$ $\quad u(t) \in R^m$ $\quad y(t) \in R^p$ $\quad t \in [t_0, \infty)$ $\quad u \in U = L^2((-h_2, \infty); R^m)$
$$h_1, h_2 \geq 0.$$

Matrices A, B, C_0 are of appropriate dimensions, matrices A, B are of bounded variation with respect to their second arguments for all t, matrix C_0 is continuous, matrix B has the representation

$$\int_{-h_2}^0 d_s B(t, s) u(t+s) = \sum_{i=1}^\infty B_i(t) u(t - h_i(t)) + \int_{-h_2}^0 B_0(t, s) u(t+s) ds$$

where $B_i \in L^2[0, \infty)$, $h_i(t) \in [-h_2, 0]$, $t \in [t_0, \infty)$, and the map $[t_0, \infty) \ni t \mapsto ||B(t, (\cdot))||_{L^2(-h_2, 0)}$ is continuous. We do not specify all the topological properties of the components of formula (8.2.17). We only assume that for any initial data $(x_0, x_1) \in R^n \times L^2(-h_1, 0; R^n) = X$, and for any $u \in L^2(-h_2, \infty)$ there exists a unique absolutely continuous solution $x(\cdot)$ of equation (8.2.17). Moreover, this solution depends continuously on the initial data from X and control from U.

8.2 R^p-controllability of Linear FDE Systems

Introducing the $n \times n$ absolutely continuous matrix $H(t, \tau)$ $t \geqslant \tau \geqslant t_0$, which is the solution to

$$\frac{d}{dt} H(t, \tau) = \int_{-h_1}^{0} d_s A(t, s) H(t+s, \tau) \tag{8.2.19}$$

with conditions

$$H(\tau, \tau) = I \quad \tau \geqslant t_0$$
$$H(t, \tau) = 0 \quad t < \tau$$

one can write the solution of (8.2.17) with initial data equal to zero in the following form:

$$x(t) = \int_{t_0}^{t} H(t, \tau) \int_{-h_2}^{0} d_s B(\tau, s) u(\tau+s) d\tau$$

$$= \int_{t_0}^{t} \left(\int_{-h_2}^{0} H(t, \tau-s) d_s B(\tau-s, s) \right) u(\tau) d\tau \tag{8.2.20}$$

for $u(s) = 0$ $s \leqslant t_0$. Denote

$$f(t, \tau) = \int_{-h_2}^{0} H(t, \tau-s) d_s B(\tau-s, s). \tag{8.2.21}$$

The output relation (8.2.18) can be written as follows:

$$y(t) = \int_{t_0}^{t} C_0(t) f(t, \tau) u(\tau) d\tau. \tag{8.2.22}$$

By Theorem 8.2.2 we have the fact that system (8.2.17) with $u(s) = 0$, $s < t_0$ is R^p-controllable at time T, $T > t_0$, if and only if

$$\operatorname{rank}\left[\int_{t_0}^{T} f(T, \tau)^T C_0(T)^T C_0(T) f(T, \tau) d\tau \right] = p. \tag{8.2.23}$$

Using the family of functionals $\{u_a^0\}$ from (8.2.13) we have that system (8.2.17) is R^p-controllable if and only if

$$\dim \operatorname{lin}\left(\{C_0(T) f(T, \tau)\}_{\tau \in [t_0, T]} \right) = p \tag{8.2.24}$$

on condition that $f(T, \cdot)$ is continuous. The term in braces is a $(p \times n)$-matrix, and the sign 'lin' should be understood as a linear combination of the matrix columns $C_0(T) f(T, \tau)$, $\tau \in [t_0, T]$.

These conditions of R^p-controllability are relatively complicated. They are not immediately computable because they require a knowledge of the fundamental matrix $H(t, \tau)$. However, we cannot say any more until we restrict our considerations to systems which are less general than (8.2.17).

The most interesting are systems with a piecewise analytic or piecewise of class C^m fundamental matrix of solutions. In these cases one can apply families of functionals $\{u_{a,z_0}^{AN}\}$ or $\{u_a^m\}$, (8.2.12), (8.2.13) in order to obtain computable conditions.

Let the following system be given (Manitius, 1972):

$$\dot{x}(t) = \int_{-h}^{0} dA(s)x(t+s) + \int_{-h}^{0} dB(s)u(t+s) \qquad (8.2.25)$$

$$x(t) \in R^n \quad u(t) \in R^m \quad t \geq 0$$

$$y(t) = x(t) \quad t \geq 0.$$

We assume that

(i) $A(s)$, $B(s)$ are $(n \times n)$- and $(n \times m)$-matrix-valued functions of bounded variation, respectively, normalized so that

$$A(s) = 0, \quad B(s) = 0 \quad \text{for} \quad s \geq 0$$

$$A(s) = A(-h), \quad B(s) = B(-h) \quad \text{for} \quad s \leq -h,$$

(ii) A, B are continuous from the left in $(-h, 0)$,

(iii) A, B are the sum of absolutely continuous functions and jump functions, with discontinuities placed at points $s = -h_i^1$, $i = 0, 1, \ldots, p_1$ where $0 = h_0^1 < h_1^1 < , \ldots, < h_{p_1}^1 \leq h$,

(iv) the absolutely continuous part of A, B is q-times (for A), $(q+1)$-times (for B) continuously differentiable on $(-h-\varepsilon, \varepsilon)$, $\varepsilon > 0$, except at a finite number of points $s = -h_i^{j+1}$, $i = 0, 1, \ldots, p_{j+1}$, where the index j is connected to the j-th derivative. In addition, let there exist left and right limits of all derivatives at these points.

We use the notation

$$\Delta f^{(j)}(t) = f^{(j)}(t^+) - f^{(j)}(t^-).$$

The output map is of the form

$$y(t) = \int_0^t f(t, \tau)u(\tau)d\tau \quad u(s) = 0 \quad \text{for} \quad s \in [-h, 0]$$

where

$$f(t, \tau) = \int_{-h}^{0} H(t-\tau-s)dB(s).$$

Let us define the matrices

$$A_i^{j+1} = (-1)^j \Delta A^{(j)}(-h_i^{j+1}) \quad (j = 0, 1, \ldots, q)$$

$$B_i^{j+1} = (-1)^j \Delta B^{(j)}(-h_i^{j+1}) \quad (j = 0, 1, \ldots, q+1)$$

2 R^p-controllability of Linear FDE Systems

and the recurrently obtainable $(n \times m)$-matrices $Q_k(s)$

$$Q_{k+1}(s) = \sum_{j=0}^{k+1} \sum_{i=0}^{p+1} [A_i^{j+1} Q_{k-j}(s-h_i^{j+1}) + B_i^{j+1} U_{k+1-j}(s-h_i^{j+1})] \quad (8.2.26)$$

real, $k = -1, 0, 1, \ldots, q$

$$U_0(s) = \begin{cases} I & \text{for } s = 0 \\ 0 & \text{for } s \neq 0 \end{cases} \quad ((m \times n) \text{ matrix})$$

$$U_k(s) = 0 \text{ for } k \neq 0$$

$$Q_{-1}(s) = 0.$$

The matrices $Q_k(s)$ are directly connected to derivatives of $f(t, \cdot)$. We have the following sufficient condition of controllability (Manitius, 1972).
System (8.2.25) is R^n-controllable on the interval $[0, T]$ if

$$\text{rank}[\{Q_k(s)\}_{k,s}] = n \quad s \in [0, T] \quad k = 0, 1, \ldots, q+1. \quad (8.2.27)$$

This condition is also necessary in the case where A, B are piecewise constant with a finite number of jumps. In this case condition (8.2.27) for system 8.2.25) with $B(s) = B = \text{const}$ can be written in the form (Gabasov and Kirillova, 1971)

$$\text{rank}[\{Q_k(s)\}_{k,s}] = n \quad k = 0, 1, 2 \ldots \quad s \in [0, (n-1)h_{p_1}] \quad (8.2.28)$$

$$Q_k(s) = \sum_{i=0}^{p_1} A_i^1 Q_{k-1}(s-h_i^1) \quad (8.2.29)$$

$$Q_0(s) = \begin{cases} B & s = 0 \\ 0 & s \neq 0 \end{cases}$$

and A_i^1, $i = 0, 1, \ldots, p_1$ are as above. Equations of the type (8.2.26), (8.2.29) are called 'determining equations'. Now we will present another approach to R^p-controllability. Let us recall the mild solution (8.1.2) to an abstract evolution equation (8.1.1):

$$x(t) = F(t)x_0 + \int_0^t F(t-\tau) Bu(\tau) d\tau \quad t \geq 0 \quad u \in L^1(0, \infty; R^m) = U$$

together with the output relation

$$y(t) = Cx(t) \quad C \colon X \mapsto R^p.$$

Let us denote

$$S_T = \left\{ y \in R^p \colon y = \int_0^T CF(T-s) Bu(s) ds, \, u \in L^1(0, \infty; R^m) \right\}.$$

The $n \times m$ matrix function $CF(T-(\cdot))B$ is continuous by strong continuity of the semigroup $\{F(t)\}_{t \geq 0}$. The set of controls
$$\{u \colon u(s) = e^{-\lambda(T-s)}v, v \in R^m, \lambda \in R^1\}$$
is total (see (8.2.14)) on the set of continuous functions and moreover, it dense in $L^2(0, \infty; R^m)$. Hence
$$S_T = \left\{\int_0^T e^{-\lambda(T-s)}CF(T-s)Bv\,ds = \int_0^T e^{-\lambda s}CF(s)Bv\,ds, v \in R^m, \lambda \in R^1\right\}.$$
This formula remains valid when we take $\lambda \geq \lambda_0$ where λ_0 is as in Theorem 8.1.1, 8.1.2 (see also (8.2.14)). Thus we have
$$S_{T_2} \supset S_{T_1} \quad \text{for} \quad T_2 \geq T_1.$$
Denoting
$$R^p \supset S_\infty = \left\{\int_0^\infty e^{-\lambda s}CF(s)Bv\,ds, v \in R^m, \lambda \geq \lambda_0\right\}$$
and applying formula (8.1.4) we have
$$S_\infty = \mathrm{lin}(\{CR(\lambda, A)B\}_{\lambda \geq \lambda_0}) \qquad (8.2.30)$$
and
$$\mathrm{rank}\,[\{CR(\lambda, A)B\}_{\lambda \geq \lambda_0}] = \dim S_\infty. \qquad (8.2.31)$$
Let us consider system (8.1.8) with the output relation
$$y(t) = \begin{bmatrix} (c_1|x(t)) \\ \vdots \\ (c_k|x(t)) \end{bmatrix} \quad \begin{array}{l} c_i \in X = R^n \times L^2(-h, 0; R^n) \\ c_i = (c_i^0, c_i^1) \quad c_i^0 \in R^n \quad c_i^1 \in L^2(-h, 0; R^n) \end{array}$$
$$i = 1, \ldots, k \qquad (8.2.32)$$
Taking (8.1.17) into account, we find system (8.1.8) to be R^p-controllable if and only if
$$\mathrm{rank}\left[\left\{\bar{c}_0^T \varDelta(\lambda) B + \int_{-h}^0 e^{\lambda s}\bar{c}_1^T(s)\varDelta(\lambda)B\,ds\right\}_{\lambda \geq \lambda_0}\right] = p \qquad (8.2.33)$$
where
$$\bar{c}_0 = [c_1^0, \ldots, c_p^0] \quad \bar{c}_1 = [c_1^1, \ldots, c_p^1]$$
$$\varDelta(\lambda) = \left[\lambda I - \int_{-h}^0 e^{\lambda s}dA(s)\right]^{-1}.$$
It may be easily observed that the functions appearing in (8.2.33) are analytic with respect to λ, $\lambda > \lambda_0$, hence condition (8.2.33) may be replaced by
$$\mathrm{rank}\left[\left\{\frac{d^i}{d\lambda^i}\left(\bar{c}^T \varDelta(\lambda)B + \int_{-h}^0 \bar{c}_1(s)e^{\lambda s}\varDelta(\lambda)B\,ds\right)\Big|_{\lambda = \lambda_1}\right\}_{i=0,\ldots}\right] = p \qquad (8.2.34)$$

where λ_1 is arbitrary, $\lambda_1 > \lambda_0$. One can also make use of the equality

$$\frac{d^i}{d\lambda^i} R(\lambda, A) = (-1)^i R(\lambda, A)^{i+1} \cdot i!$$

in relation (8.2.31) and obtain

$$\dim S_\infty = \text{rank}[\{CR(\lambda_1, A)^i B\}_i = 0, \ldots] \quad \text{for some } \lambda_1 > \lambda_0. \quad (8.2.35)$$

Let us consider the system

$$\dot{x} = \int_{-h}^{0} dA(s)x(t+s) + \int_{-h}^{0} dB(s)u(t+s)$$

$$x(t) \in R^n \quad u(t) \in R^m \quad t \geqslant -h$$

and A, B are of bounded variation and of dimensions $n \times n$, $n \times m$, respectively.

Let the output relation be described by (8.2.32). By use of controls $u(s) = e^{-\lambda(T-s)}v$, $v \in R^m$, $\lambda \geqslant \lambda_0$, we have the following necessary and sufficient condition of R^p-controllability which is analogous to (8.2.33):

$$\text{rank}\left[\left\{\bar{c}_0^T \varDelta(\lambda) B(\lambda) + \int_{-h}^{0} e^{\lambda s} \bar{c}_1^T(s) \varDelta(\lambda) B(\lambda) ds\right\}_{\lambda \geqslant \lambda_0}\right] = p$$

where $B(\lambda) = \int_{-h}^{0} dB(s) e^{\lambda s}$.

The appropriate versions of (8.2.34), (8.2.35) are of course valid.

8.3 The Null-state Controllability of FDE Systems

The null-state controllability problem is much more complicated than the R^p-controllability one. For general FDE systems there are no exact conditions except for the simplest cases with one lumped delay or so-called 'full' control.

Interconnection between R^p and null-state controllability is rather exceptional, although there are systems for which these notions coincide.

Let the following system be given:

$$\dot{x}(t) = \int_{-h}^{0} dA(s)x(t+s) + Bu(t)$$

$$x(t) \in R^n \quad u(t) \in R^m \quad t \geqslant 0 \quad u \in L^2(0, \infty; R^m) \quad (8.3.1)$$

$$x(0) = x_0 \in R^n$$

$$x(s) = x_1(s) \quad s \in [-h, 0] \quad x_1 \in L^2(-h, 0; R)^n$$

A is of bounded variation and appropriate dimensions. As the state space X we take the set $R^n \times L^2(-h, 0; R^n)$ which consists of pairs (x_0, x_1), $x_0 \in R^n$ $x_1 \in L^2(-h, 0; R^n)$. This is a Hilbert space with scalar product

$$(x|y) = x_0^T y_0 + \int_{-h}^{0} x_1(s)^T y_1(s) \, ds \quad x, y \in X. \tag{8.3.2}$$

The state trajectory $\hat{x}(t) = (\hat{x}_0(t), \hat{x}_1(t))$ is defined by

$$\hat{x}_0(t) = x(t) \quad (\hat{x}_1(t))(s) = x(t+s) \quad \hat{x}(t) \in X \quad \forall t \in [0, \infty).$$

System (8.3.1) is called null-state controllable at time T, $T > 0$ if and only if for any initial state $\hat{x}(0) \in X$ there exists a control u such that the state $\hat{x}(T)$ is equal to zero.

System (8.3.1) is called IP-null-state controllable at time T, $T > 0$ if and only if for any initial state $\hat{x}(0)$ of the form $\hat{x}(0) = (x_0, 0)$ there exists a control u such that the state $\hat{x}(T)$ is equal to zero.

System (8.3.1) is controllable (null-state or IP-null-state) if and only if there is $T > 0$ such that it is controllable in the respective sense at time T.

It may be easily observed that if the system is controllable (in both these meanings) at time T, then it is controllable at any T_1, $T_1 \geq T$.

Lemma 8.3.1 *Null-state controllability coincides with IP-null-state controllability.*

Proof Null-state controllability obviously implies IP-null-state controllability. We focus on the converse relation. Let $\hat{x}^i(0) = (e_i, 0)$, where e_i, $i = 1, \ldots, n$ is a standard basis in R^n. Let u^i, $i = 1, \ldots, n$, be the appropriate control which give $\hat{x}^i(T) = 0$. Denote

$$v_i(t) = Bu^i(t) \quad t \in [0, T]$$
$$v_i(t) = 0 \quad \text{for} \quad t \geq T \quad \text{and for} \quad t \leq 0 \tag{8.3.3}$$
$$V(t) = [v_1(t) \ldots v_n(t)]_{n \times n}.$$

We have

$$0 = H(T+s) + \int_{0}^{T+s} H(T+s-\tau) V(\tau) \, d\tau \quad \text{for} \quad s > -h \tag{8.3.4}$$

where H is the fundamental matrix of solutions of (8.3.1). The solution of (8.3.1) with arbitrary initial condition (x_0, x_1) is equal to

$$x(t) = H(t) x_0 + \int_{0}^{t} H(t-\tau) Bu(\tau) \, d\tau + \int_{0}^{h} H(t-\tau) m(\tau) \, d\tau \quad t \geq h \tag{8.3.5}$$

8.3 The Null-state Controllability of FDE Systems

where

$$R^n \ni m(\tau) = \int_{-h}^{-\tau} dA(s) x_1(\tau+s) \quad \tau \in [0, h], m \in L^2(0, h; R^n). \tag{8.3.6}$$

Define

$$\overline{V}(t) = V(t)x_0 + \int_0^h V(t-\tau)m(\tau)d\tau \quad t \geq 0. \tag{8.3.7}$$

It obviously follows that $\overline{V}(t) = B\bar{u}(t)$ with appropriately chosen $\bar{u}$, because $V(\cdot)$ is a linear combination of u^i. Applying $\overline{V}$ in (8.3.5) we have

$$x(t) = H(t)x_0 + \int_0^t H(t-\tau)V(\tau)d\tau x_0 +$$

$$+ \int_0^t H(t-\tau)\int_0^h V(\tau-z)m(z)dz d\tau + \int_0^h H(t-\tau)m(\tau)d\tau \quad \text{for} \quad t \geq h. \tag{8.3.8}$$

By (8.3.4) we have

$$x(t) = \int_0^t H(t-\tau)\int_0^h V(\tau-z)m(z)dz d\tau + \int_0^h H(t-\tau)m(\tau)d\tau \quad \text{for} \ t \geq T-h.$$

Changing the order and the variable of integration we get

$$x(t) = \int_0^h \int_0^t H(t-\tau)V(\tau-z)d\tau m(z)dz + \int_0^h H(t-\tau)m(\tau)d\tau$$

$$= \int_0^h \int_{-z}^{t-z} H(t-r-z)V(r)dr m(z)dz + \int_0^h H(t-\tau)m(\tau)d\tau$$

$$= \int_0^h \left[\int_0^{t-z} H(t-r-z)V(r)dr + H(t-z)\right]m(z)dz \quad t \geq T-h. \tag{8.3.9}$$

But by (8.3.4) the term in brackets vanishes for $t \geq T$. So we see that if system (8.3.1) is IP-controllable at time T, then it is null-state controllable at time $T+h$. This ends the proof. □

This lemma shows that in order to obtain null-state controllability it is sufficient to check controllability for a few initial states of the form

$$\hat{x}^i(0) = (e^i, 0).$$

Now we shall give a number of simple statements.

If system (8.3.1) has B of full range, rank $B = n$, then it is null-state controllable at any time $T > h$.

Conversely, if system (8.3.1) is null-state controllable at any time $T > h$ then rank $B = n$.

If the system is described by

$$x^{(n)}(t) = \int_{-h}^{0} \left(\sum_{i=0}^{k} dA_i(s) x^{(i)}(t+s) \right) + Bu(t) \quad k < n$$

$$\text{rank } B = p$$

$$x(t) \in R^p \quad t \geq 0,$$

then it is null-state controllable.

If the system is described by

$$\dot{x}(t) = Ax(t-h) + Bu(t) \quad x(t) \in R^p \tag{8.3.10}$$

and it is R^p-controllable, then it is null-state controllable (Gabasov and Kirillova, 1971). From the determining equation (see Section 8.2) we know that (8.3.10) is null-state controllable if

$$\text{rank } [B, AB, \ldots, A^{n-1}B] = n.$$

By Lemma 8.3.1, system (8.3.1) is null-state controllable if and only if for any $x_0 \in R^n$ there exists control u of bounded support (i.e. $u(t) = 0$ for $t \geq T$) such that the corresponding trajectory $x(t)$, $x(0) = x_0$, $x(t) = 0$ for $t < 0$, is also of bounded support.

Taking the Laplace transform of (8.3.1) we have

$$\hat{x}(s) = [sI - \hat{A}(s)]^{-1} [x_0 + B\hat{u}(s)],$$

where $\hat{x}$, $\hat{u}$ are the Laplace transforms of x, u respectively, and

$$\hat{A}(s) = \int_{-h}^{0} dA(z) e^{sz}.$$

From the Paley–Wiener theorem (Doetsch, 1956) we know that $\hat{u}(s)$ is the transform of a function of bounded support from $L^2(0, \infty; R^m)$ if and only if it is an entire function of complex variable (has no poles) and

$$\|\hat{u}(s)\| \leq a e^{b|s|} \quad \text{for some} \quad a, b > 0 \quad \text{for all } s \in C \tag{8.3.12}$$

$$\int_{-\infty}^{\infty} \|\hat{u}(it)\|^2 dt < \infty. \tag{8.3.13}$$

In some particular cases of system (8.3.1) it is possible to obtain controllability results constructing $\hat{u}(s)$ in the form of an entire function so that $\hat{x}(s)$ in (8.3.11) has no poles. More precisely system (8.3.1) is null-state controllable

8.3 The Null-state Controllability of FDE Systems

if and only if there exists an entire $(n \times m)$-matrix function $M(s)$ fulfilling (8.3.12), (8.3.13), such that the matrix

$$[sI - \hat{A}(s)]^{-1}[I + B\hat{M}(s)] \tag{8.3.14}$$

has all poles cancelled and fulfils (8.3.12), (8.3.13). However, there are no algorithms available based on this approach except in the simplest cases.

Now we will develop null-state controllability conditions for systems with commensurate time delays

$$\dot{x}(t) = \sum_{i=0}^{r} A_i x(t - ih) + Bu(t) \quad t \geq 0 \quad h > 0 \tag{8.3.15}$$

$$x(t) \in R^n \quad u(t) \in R^m$$
$$x(t) = f(t) \quad t \in [-rh, 0) \quad x(0) = f_0$$

where A_i, B are matrices of suitable dimensions. In order to derive the necessary and sufficient conditions of null-state controllability we follow Olbrot (1973b) and reduce system (8.3.15) to an appropriate system of ordinary differential equations. We introduce the notation

$$x_{j+1}(t) = x(t + jh) \quad j = -r, \ldots, -1, 0, 1, \ldots, k-1 \quad t \in [0, h]$$
$$u_{j+1}(t) = u(t + jh) \quad j = 0, 1, \ldots, k-1 \quad t \in [0, h] \tag{8.3.16}$$

$$\bar{x} = \begin{bmatrix} x_1 \\ \vdots \\ x_k \end{bmatrix} \quad \bar{u} = \begin{bmatrix} u_1 \\ \vdots \\ u_k \end{bmatrix} \quad \bar{f} = \begin{bmatrix} x_0 \\ \vdots \\ x_{1-r} \end{bmatrix} \quad \bar{f_0} = \begin{bmatrix} f_0 \\ 0 \\ \vdots \\ 0 \end{bmatrix} \in R^{kn} \tag{8.3.17}$$

$$\bar{A} = \begin{bmatrix} A_0 & 0 & & \\ A_1 & A_0 & \ddots & \\ \vdots & A_1 & \ddots & \\ A_r & \vdots & & \\ 0 & A_r & \ddots & \end{bmatrix}_{nk \times nk} \quad \bar{B} = \begin{bmatrix} B & & 0 \\ & B & \\ & & \ddots \\ 0 & & B \end{bmatrix}_{nk \times mk} \tag{8.3.18}$$

$$P = \begin{bmatrix} A_1 & \cdots & A_r \\ A_2 & \cdots & \\ \vdots & \ddots & \\ A_r & & \\ 0 & & \end{bmatrix}_{nk \times nr} \quad \bar{H} = \begin{bmatrix} 0 & & & 0 \\ I & 0 & & \\ & I & \ddots & \\ & & \ddots & \\ 0 & & I & 0 \end{bmatrix}_{nk \times nk}.$$

Using the method of steps described in Section 3.2 we transform system (8.3.15) into the equation

$$\dot{\bar{x}}(t) = \bar{A}\bar{x}(t) + \bar{B}\bar{u}(t) + P\bar{f}(t) \quad t \in [0, h] \tag{8.3.19}$$

$$\bar{x}(t) \in R^{kn} \quad \bar{u}(t) \in R^{km} \quad \bar{f}(t) \in R^{rn}$$

with split boundary conditions

$$\bar{x}(0) = \bar{f}_0 + \bar{H}\bar{x}(h). \tag{8.3.20}$$

By virtue of relations (8.3.16), (8.3.17) the solution of system (8.3.15) on interval $[0, kh]$ is equivalent to the solution of system (8.3.19), (8.3.20) on interval $[0, h]$.

Let us assume that the interval of control action is $[0, kh]$, $k \geq r$. System (8.3.15) is null-state controllable at time kh, if and only if for any $\bar{f}, \bar{f}_0$ there exists a control $\bar{u}$ such that the solution $\bar{x}$ of (8.3.19), (8.3.20) fulfils

$$x_k(t) = x_{k-1}(t) = \ldots = x_{k-r+1}(t) = 0 \quad \text{for} \quad t \in [0, h].$$

By Lemma 8.3.1 it is sufficient to consider the case $\bar{f} = 0$, f_0 arbitrary. Let us denote

$$Y_0 = \left\{ y \in R^{kn} : y = \begin{bmatrix} f_0 \\ 0 \\ \vdots \\ 0 \end{bmatrix}, \quad f_0 \in R^n \right\}$$

$$Y = \left\{ y \in R^{kn} : y = \begin{bmatrix} x_1 \\ x_2 \\ \vdots \\ x_{k-r} \\ 0 \\ \vdots \\ 0 \end{bmatrix}, \quad x_1, \ldots, x_{k-r} \in R^n \right\}.$$

The following lemma is obvious.

Lemma 8.3.2 *System* (8.3.15) *is null-state controllable at time* kh, $k \geq r$ *if and only if for every* $\bar{f}_0 \in Y_0$ *there exists a control* $\bar{u}$ *such that the corresponding trajectory* $\bar{x}(t)$ *of system* (8.3.19), (8.3.20) *lies in* Y *for every* $t \in [0, h]$.

For an ordinary differential system

$$\dot{z}(t) = Az(t) + Bu(t) \quad z(t) \in R^p \tag{8.3.21}$$

and for some subspace $W \subset R^p$ we introduce two notions: W-maintainability and W-reachability.

The state $z_0 \in W$ is said to be W-maintainable if there exists a trajectory of (8.3.25) starting from z_0 and lying in W for all $t \geq 0$.

The state z_1 is said to be W-reachable from $0 \in R^n$ at time t_1 iff there exists a trajectory $z(t)$ of (8.3.21) such that $z(0) = 0$, $z(t_1) = z_1$, and $z(t)$ lies in W.

8.3 The Null-state Controllability of FDE Systems

We have the following criteria of W-maintainability and W-reachability (Olbrot, 1973b).

The set MAIN (A, B, W) of all W-maintainable states of system (8.3.25) is equal to the subspace W_l, where

$$W_i = W_{i-1} \cap A^{-1}(R(B) + W_{i-1}) \quad i = 1, \ldots, l, \quad l = \dim W$$
$$W_0 = W$$

and $A^{-1}(R(B) + W_{i-1})$ denotes the counterimage of the set $(R(B) + W_{i-1})$, $R(B)$ is the range of B.

The set MAIN(A, B, W) can also be obtained in another way:

$$\mathrm{MAIN}(A, B, W) = W_l^{\perp}$$
$$W_0 = W, \quad W_i = W^{\perp} + A^{\mathrm{T}}(W_{i-1} \cap R(B)) \quad i = 1, \ldots, l \quad l = \dim W,$$

where $W^{\perp}$ denotes the orthogonal complement subspace to W. Of course, if some state is W-maintainable, then it is MAIN(A, B, W)-maintainable, and vice versa. Moreover, MAIN$(A, B, \mathrm{MAIN}(A, B, W)) = \mathrm{MAIN}(A, B, W)$, which is an immediate consequence of the definition.

The set REACH (A, B, W) of all states W-reachable from the origin is a subspace of W of the form

$$\mathrm{REACH}(A, B, W) = W_q \quad q = \dim \mathrm{MAIN}(A, B, W)$$
$$W_0 = 0 \quad W_i = \mathrm{MAJN}(A, B, W) \cap (AW_{i-1} + R(B)) \quad (i = 1, \ldots, q).$$

Let us turn back to the situation described in Lemma 8.3.2. The set of all values of trajectories lying in Y and starting from $\bar{x}(0) \in Y$ is equal to MAIN$(\bar{A}, \bar{B}, Y)$. The values $\bar{x}(h)$ can be expressed by

$$\bar{x}(h) = e^{\bar{A}h}\bar{x}(0) + \int_0^h e^{\bar{A}(h-\tau)}\bar{B}\bar{u}(\tau)\,\mathrm{d}\tau. \quad (8.3.22)$$

Of course, $\int_0^h e^{\bar{A}(h-\tau)}\bar{B}\bar{u}(\tau)\,\mathrm{d}\tau \in \mathrm{REACH}(\bar{A}, \bar{B}, Y)$, as the final point of a trajectory lying in Y.

From (8.3.20) and (8.3.22) we have

$$(I - \bar{H}e^{\bar{A}h})\bar{x}(0) = \bar{f}_0 + \bar{H}\int_0^h e^{\bar{A}(h-\tau)}\bar{B}\bar{u}(\tau)\,\mathrm{d}\tau \quad \bar{x}(0) \in \mathrm{MAIN}(\bar{A}, \bar{B}, Y), \quad \bar{f}_0 \in Y_0.$$

$$(8.3.23)$$

Now we can state the main result.

Theorem 8.3.1 *System* (8.3.15) *is null-state controllable at time* kh, $k \geq r$, *if and only if for any* $\bar{f}_0 \in Y_0$ *there are* $y_1 \in \text{MAIN}(\bar{A}, \bar{B}, Y)$, $y_2 \in \text{REACH}(\bar{A}, \bar{B}, Y)$ *such that*

$$(I - \bar{H}e^{\bar{A}h})y_1 - \bar{H}y_2 = \bar{f}_0. \tag{8.3.24}$$

In other words, system (8.3.15) *is null-state controllable at time* $t = kh$ *if and only if*

$$Y_0 \subset (I - \bar{H}e^{\bar{A}h})\text{MAIN}(\bar{A}, \bar{B}, Y) + \bar{H} \cdot \text{REACH}(\bar{A}, \bar{B}, Y). \tag{8.3.25}$$

Olbrot (1973b) gave (8.3.25) as a necessary and sufficient condition of IP-controllability for a system with one delay. This result is one of the strongest in the field of controllability. It is of great importance that this condition is valid for a relatively wide class of systems and is checkable in a finite number of steps. Moreover it allows a zeroing control to be constructed by means of a finite moments approach.

8.4 Controllability, Observability and Pole Assignment

We now consider the system

$$\dot{x}(t) = \int_{-h}^{0} dA(s)x(t+s) + Bu(t) \quad t \geq 0 \tag{8.4.1}$$

$$x(t) \in R^n \quad u(t) \in R^m$$

where the matrix $A_{n \times n}$ is of bounded variation on the interval $[-h, 0]$. We take $X = R^n \times L^2(-h, 0; R^n)$ as the state space for system (8.4.1) and for $x \in X$ we write $x = (x_0, x_1)$, $x_1 \in L^2(-h, 0; R^n)$, $x_0 \in R^n$. X is a Hilbert space with scalar product $(x|y)_X = x_0^T y_0 + \int_{-h}^{0} x_1^T(\tau) y_1(\tau) d\tau$. The state trajectory in X will be denoted by $\bar{x}(t)$, $t \geq 0$. Generally speaking, the notion of controllability is related to properties of the set of all states $\bar{x}(t_1) \in X$, $t_1 \geq 0$ which can be obtained by use of all integrable controls u. This attainable set is never equal to the whole state space X, which follows from the continuity of the trajectory $x(t) \in R^n$, $t \geq 0$, while X contains also states with discontinuous functions.

Thus the approximate controllability is the proper notion of controllability for system (8.4.1) in the state space X. Moreover, this notion of controllability allows stabilizability conditions to be obtained, which is important from the practical point of view. One might consider alternative notions of controllability connected with the density of attainable states in spaces other than X, but we shall restrict ourselves to controllability with respect to X.

8.4 Controllability, Observability and Pole Assignment

In what follows we will use the notions and notation introduced in Section 8.1.

Let us denote the set of all states attainable from $\bar{x}(0) = 0$ in system (8.4.1) at time $T > 0$ by $Z(T)$,

$$Z(T) = \left\{ x \in X: x = \int_0^T F(T-\tau)\bar{B}u(\tau)d\tau, u \in (L^1(0,T))^m \right\} \quad (8.4.2)$$

where $\{F(t)\}_{t \geq 0}$, $\bar{B}$ are defined in Section 8.1. This means that $Z(T)$ consists of pairs $x = (x_0, x_1) \in X$, $x_1 \in L^2(-h, 0; R^n)$, $x_0 \in R^n$, of the form

$$x_0 = \int_0^T H(T-\tau)Bu(\tau)d\tau$$

$$x_1(s) = \int_0^{T+s} H(T+s-\tau)Bu(\tau)d\tau \quad u \in (L^1(0,T))^m \quad s \in [-h, 0] \quad (8.4.3)$$

and $H(t)$ is the fundamental matrix of solutions to (8.4.1)

$$\dot{H}(t) = \int_{-h}^{0} dA(s)H(t+s) \quad H(0) = I \quad H(s) = 0 \quad -h \leq s < 0.$$

(8.4.4)

Obviously, the attainable set can also be expressed as follows:

$$Z(T) = \left\{ x \in X: x = \int_0^T F(t)\bar{B}z(t)dt, z \in (L^1(0,T))^m \right\}. \quad (8.4.5)$$

Hence we have

$$Z(T_2) \supset Z(T_1) \quad \text{if} \quad T_2 \geq T_1. \quad (8.4.6)$$

The set $Z(T)$ is a linear subspace in X. From the definition of approximate controllability we have the following lemma.

Lemma 8.4.1 *System* (8.4.1) *is approximately controllable if and only if*

$$\overline{Z(T)} = X$$

where the bar denotes the closure in X-topology.

$\overline{Z(T)}$ is a closed subspace of X and is different from X if and only if there exists $\bar{y} \in X$, $\bar{y} \neq 0$ such that

$$(\bar{y}|x) = 0 \quad \forall x \in \overline{Z(T)}$$

or equivalently (by the density of $Z(T)$ in $\overline{Z(T)}$)
$$(\bar{y}|x) = 0 \quad \forall x \in Z(T).$$
Thus $\bar{y}$ is orthogonal to $Z(T)$.
This leads us to the following lemma.

Lemma 8.4.2 *System* (8.4.1) *is approximately controllable if and only if for any* $\bar{y} \in X, \bar{y} \neq 0, \exists x \in Z(T)$ *such that* $(\bar{y}|x) \neq 0$.

From (8.4.2) we have the fact that system is not approximately controllable at time T iff there exists $\bar{y} \in X$, $\bar{y} \neq 0$, such that

$$0 = \int_0^T (\bar{y}|F(t)\bar{B})u(t)\,dt \quad \forall u \in (L^1(0, T))^m \tag{8.4.7}$$

or, by the arbitrariness of u, if and only if the R^m-valued transposed vector function $t \mapsto (\bar{y}|F(t)\bar{B})$ vanishes,

$$(\bar{y}|F(t)\bar{B}) = 0 \quad \forall t \in [0, T]. \tag{8.4.8}$$

Lemma 8.4.3 *The system is approximately controllable if and only if for any* $y \in X, y \neq 0$, *the function* $[0, T] \ni t \mapsto (\bar{y}|F(t)\bar{B}) \in R^m$ *is not identically equal to 0.*

Lemma 8.4.4 *System* (8.4.1) *is not approximately controllable at any* $T \in [0, h]$.

We have the following lemma concerning the relation between R^1-and approximate controllability.

Lemma 8.4.5 *System* (8.4.1) *is approximately controllable at time T, if and only if for any $c \in X$ system* (8.4.1) *together with the additional scalar output relation*

$$y(t) = (c|\bar{x}(t)) \quad t \in [0, T]$$

is R^1-controllable, with respect to the output $y(t) \in R^1$.

Another characterization of controllability may be given by the properties of the adjoint equation. We assume $m = 1$. Let us rewrite the left-hand side of (8.4.8) in the form

$$(\bar{y}|F(t)\bar{B}) = (\bar{B}|F^*(t)\bar{y}).$$

According to the observability concept given in Section 8.1 we have the following

Lemma 8.4.6 *System* (8.4.1) *is approximately controllable on* $[0, T]$ *if and only if the adjoint system*

$$\dot{z}(t) = A^* z(t)$$

with the output

$$y(t) = (\bar{B}|z(t))$$

observable on the interval $[0, T]$.

Taking (8.4.8) into account together with the fact that

$$F(t)\bar{B} = (\bar{x}_0(t), \bar{x}_1(t)) \in X \quad t \in [0, T]$$

where

$$\bar{x}_0(t) = H(t)B$$
$$(\bar{x}_1(t))(s) = H(t+s)B \quad s \in [-h, 0],$$

and with the definition of scalar product in X we have

$$(\bar{y}|F(t)\bar{B}) = \bar{y}_0^T H(t)B + \int_{-h}^{0} \bar{y}_1^T(s) H(t+s) B \, ds = B^T H(t)^T \bar{y}_0 +$$

$$+ \int_{-h}^{0} B^T H(t+s)^T \bar{y}_1(s) \, ds \quad t \in [0, T] \quad \bar{y} \in X. \qquad (8.4.9)$$

Consider the R^n-valued function f

$$f(t) = H(t)^T \bar{y}_0 + \int_{-h}^{0} H(t+s)^T \bar{y}_1(s) \, ds = H(t)^T \bar{y}_0 +$$

$$+ \int_{0}^{h} H(t-\tau)^T \bar{y}_1(-\tau) \, d\tau \quad t \in [0, T].$$

By the definition of $H(t)$ we have

$$\dot{H}(t) = \int_{-h}^{0} dA(s) H(t+s) = \int_{-h}^{0} H(t+s) \, dA(s) \quad t \geq 0 \quad H(0) = I$$

Hence

$$\dot{H}(t)^T = \int_{-h}^{0} dA(s)^T H(t+s)^T \quad t \geq 0 \quad H(0)^T = I.$$

Thus we see that $f(t)$ fulfils the following equation, dual to (8.4.1):

$$\dot{f}(t) = \int_{-h}^{0} dA(s)^T f(t+s) + w(t) \qquad (8.4.10)$$

where $w(t) = \bar{y}_1(-t)$, $t \in [0, h]$, $w(t) = 0$, $t > h$, and the initial condition
$$f(0) = \bar{y}_0$$
$$f(s) = 0 \quad s \in [-h, 0].$$

Lemma 8.4.7 *System (8.4.1) is approximately controllable at time T, if and only if for any $f_0 \in R^n$, $w \in L^2(0, h; R^n)$, $(w, f_0) \neq 0$, the output*
$$y(t) = B^T f(t)$$
of the system
$$\dot{f}(t) = \int_{-h}^{0} \mathrm{d}A(s)^T f(t+s) + w(t) \quad f(0) = f_0 \quad f(s) = 0 \quad \text{for} \quad s < 0$$
(8.4.1)
is not identical to 0 on the interval $[0, T]$.

We now give a spectral characterization of approximate controllability. Let us denote by $\mathrm{FLT}(-h, 0)$ the set of all Laplace transforms of functions with finite support

$$\mathrm{FLT}(-h, 0) = \left\{ \hat{x} : \hat{x}(s) = \int_{-h}^{0} e^{ts} x(t) \mathrm{d}s, \; x \in L^2(-h, 0; R^n) \right\}.$$

This set consists of entire functions of a complex variable fulfilling properties given by the Paley–Wiener theorem (see Section 8.3). Taking the Laplace transform of equation (8.4.9) and putting the left-hand side of (8.4.9) equal to 0, we have the following result.

Lemma 8.4.8 *System (8.4.1) is approximately controllable at some sufficiently large T if and only if for $y_0 \in R^n$, $\hat{y} \in \mathrm{FLT}(-h, 0)$ the condition*
$$(y_0^T + \hat{y}(s)^T) \Delta(s)^{-1} B = 0 \tag{8.4.1}$$
implies
$$y_0 + \hat{y}(s) = 0$$
where $\Delta(s) = sI - \int_{-h}^{0} \mathrm{d}A(t) e^{st}$.

This means that if there are $y_0 \in R^n$, $\hat{y} \in \mathrm{FLT}(-h, 0)$ such that (8.4.1) is fulfilled, then the system is not approximately controllable.

Making use of Lemma 8.4.5 and condition (8.2.36) one obtains the controllability condition for systems with delays in control.

4 Controllability, Observability and Pole Assignment 273

Lemma 8.4.9 *The system*

$$\dot{x}(t) = \int_{-h}^{0} dA(s)x(t+s) + \int_{-h}^{0} dB(s)u(t+s)$$

approximately controllable at some sufficiently large T if and only if for $y_0 \in R^n$, $\hat{y} \in \text{FLT}(-h, 0)$ the condition

$$(y_0^T + \hat{y}(s)^T)\hat{A}(s)^{-1}\hat{B}(s) = 0$$

implies

$$y_0 + \hat{y}(s) = 0$$

where $\hat{A}(s) = sI - \int_{-h}^{0} dA(t)e^{st}$ and $\hat{B}(s) = \int_{-h}^{0} dB(t)e^{st}$.

Lemma 8.4.10 *System (8.4.1) is approximately controllable at time T if the symmetric operator $V: X \to X$ is positively definite*

$$(Vx|x) = \int_0^T \left(B\left(H(t)x_0 + \int_{-h}^{0} H(t+s)x_1(s)\,ds\right) \right)^T \times$$

$$\times B\left(H(t)x_0 + \int_{-h}^{0} H(t+s)x_1(s)\,ds \right) dt > 0 \quad \text{for} \quad x \neq 0.$$

For the proof, let us observe that the positive definiteness of V is a necessary and sufficient condition for the nonexistence of $x_0 \neq 0$, $x_1 \neq 0$ such that

$$BH(t)x_0 + B\int_{-h}^{0} H(t+s)x_1(s) = 0 \quad 0 \leq t \leq T.$$

This lemma finishes our presentation of general controllability results. We will not discuss the observability problem separately, because it is dual to that of controllability (see Lemmas 8.4.6 and 8.4.7).

The remaining problem for general FDE systems of the form (8.4.1) is to find checkable criteria for controllability which are expressed explicitly in terms of $A(s)$, $s \in [-h, 0]$. Even in the case of the simplest FDE system

$$\dot{x} = A_0 x(t) + A_1 x(t-h) + Bu(t) \tag{8.4.13}$$

there are no such criteria. Almost all that are known are based on Lemma 8.4.8. However, this approach requires a knowledge of properties of the zero of $\hat{A}(s)$, which is not easily checkable in general (Delfour and Manitius, 1980; Manitius, 1974a; Manitius and Triggiani, 1978).

In our opinion, a relatively simple, and at the same time strong way of obtaining checkable criteria is presented by Olbrot (1973b, 1976b) and we will follow in the steps of that author.

His method is based on Lemma 8.4.7. For the system

$$\dot{x}(t) = \sum_{i=0}^{r} A_i x(t-ih) + Bu(t) \quad x(t) \in R^n \quad u(t) \in R^m \quad (8.4.1)$$

where A_i, B are matrices of appropriate dimensions, we form the dual system (8.4.11)

$$\dot{x}(t) = \sum_{i=0}^{r} A_i^T x(t-ih) + w(t) \quad t \in [0, T] \quad T = kh \quad k \geqslant r \quad (8.4.1)$$

$$x_0 \in R^n \quad x(t) = 0 \quad \text{for} \quad t < 0$$

$$w|_{[0, rh]} \in L^2(0, rh; R^n) \quad w(t) = 0 \quad \text{for} \quad t \geqslant rh \quad y(t) = B^T x(t).$$

In order to obtain controllability conditions for system (8.4.14) let us extend the results of Lemma 8.4.7.

It may easily be observed that the arbitrariness of x_0 and $w \in L^2(0, rh; R^n)$ enables us to obtain any continuous function g, such that $\dot{g} \in L^2(0, rh; R^n)$ as a solution to (8.4.15) on the interval $[0, rh]$.

In fact, for any g of the class mentioned above, putting $g(t) = 0$ for $t \leqslant 0$ we can define

$$w(t) = \dot{g}(t) - \sum_{i=0}^{r} A_i^T g(t-ih).$$

Hence g is a solution of (8.4.15) with $w \in L^2(0, rh; R^n)$. Thus we have:

Lemma 8.4.11 *There are $x_0 \in R^n$, $w \in L^2(0, rh; R^n)$, not both equal to zero such that $y(t) = 0$ for $t \geqslant 0$ in system (8.4.15) if and only if for system*

$$\dot{x}(t) = \sum_{i=0}^{r} A_i^T x(t-ih)$$

$$y(t) = B^T x(t) \quad t \geqslant 0 \quad (8.4.1)$$

there is a continuous initial condition g, $g \neq 0$, $\dot{g} \in L^2(-rh, 0; R^n)$, $g(t) \in \text{Ker } B$ such that

$$y(t) = 0 \quad \text{for} \quad t \geqslant 0.$$

It is easy to observe that if there is an initial state g in (8.4.16) giving $y(t) = 0$ for $t \geqslant 0$, then there are infinitely many states with the same property. Indeed, denoting the trajectory corresponding to g by $x_g(t)$, $-rh \leqslant t$, we have that the function $f(s)$,

$$f(s) = \int_0^\infty x_g(t+s)\,dm(t) \quad s \in [-rh, 0] \quad (8.4.1)$$

8.4 Controllability, Observability and Pole Assignment

where m is an arbitrary scalar function of bounded variation and compact support, is an initial function for (8.4.16) which also guarantees $y(t) = 0$ for $s \geq 0$. Of course, the same formula (8.4.17) gives the solution of the dual system (8.4.16) for $s > 0$.

Lemma 8.4.12 *If system* (8.4.14) *is controllable, then*

$$\mathrm{rank}[A_r, B] = n.$$

Proof If above does not hold, then there exists $f \in R^n$, $f \neq 0$ such that $f^T A_r = 0$, $f^T B = 0$ or, equivalently, such that $A_r^T f = B^T f = 0$. Taking the initial condition for system (8.4.16) in the form

$$g(t) = a(t)f \quad -rh \leq t \leq 0$$

where the scalar continuous function a has support in the interval $[-rh, (-r+1)h]$, we have the solution $x(t) = 0$, $t \geq 0$ for system (8.4.16), and non-controllability of system (8.4.14). □

Lemma 8.4.13 *The condition*

$$\mathrm{rank}[A_r, B] = n$$

is necessary and sufficient for $(x(t) = 0, t \geq 0) \Leftrightarrow g = 0$ *in system* (8.4.16).

Proof Necessity is obvious by the proof of Lemma 8.4.12. Let us assume that $g \neq 0$, $g(t) \in \mathrm{Ker}\, B^T$ and $x(t) = 0$ for $t \geq 0$. Denoting $g_i(t) = g(t+ih) \in \mathrm{Ker}\, B^T$, $t \in [0, h]$, $i = 0, \ldots, r-1$, $g_{r-1}(h) = 0$ we have

$$\dot{x}(t) = 0 = \sum_{i=1}^{r} A_i^T x(t - ih) \quad t \in [0, rh]$$

which can be rewritten in the form

$$\begin{bmatrix} A_r^T & \cdot & & & 0 \\ A_{r-1}^T & \cdot & \cdot & & \\ \vdots & & A_{r-1}^T & \cdot & \cdot \\ A_1^T & & & \cdot & A_r^T \end{bmatrix} \begin{bmatrix} g_{r-1}(t) \\ \vdots \\ g_0(t) \end{bmatrix} = 0 \quad t \in [0, h].$$

But the condition $\mathrm{rank}[A_r, B] = n$ implies $g_{r-1} = 0$, and in turn, $g_{r-2} = 0, \ldots, g_0 = 0$. □

Lemma 8.4.14 *If for system* (8.4.15) $\mathrm{rank}[A_r, B] = n$ *and* $x(t) = 0$ *for* $t \geq 0$ *is fulfilled, then* $x_0 = 0$ *and* $w = 0$.

Now we may start to derive the necessary and sufficient conditions for approximate controllability for system (8.4.1) by use of Lemma 8.4.7 and the dual system (8.4.15).

This system is transformed to an equivalent non-delayed system by use of the technique presented in Section 8.3, formulas (8.3.16)–(8.3.18). We have

$$\dot{\bar{x}}(t) = \bar{A}\bar{x}(t) + \bar{D}\bar{w}(t) \quad \bar{x}(t) \in R^{nk} \quad \bar{w}(t) \in R^{nr}$$
$$\bar{y}(t) = \bar{B}\bar{x}(t) \quad \bar{y}(t) \in R^{mk} \quad t \in [0, h] \tag{8.4.18}$$

where

$$\bar{A} = \begin{bmatrix} A_0^T & & & & 0 \\ A_1^T & A_0^T & & & \\ \vdots & A_1^T & \ddots & & \\ A_r^T & \vdots & & \ddots & \\ 0 & A_r^T & & & \ddots \end{bmatrix}_{nk \times nk} \quad \bar{D} = \begin{bmatrix} I_{nr \times nr} \\ 0 \end{bmatrix}_{nk \times nr}$$

$$\bar{w}(t) = \begin{bmatrix} w_1(t) \\ \vdots \\ w_r(t) \end{bmatrix}_{nr \times 1} \quad \bar{B} = \begin{bmatrix} B^T & & & 0 \\ & \ddots & & \\ & & B^T & \\ 0 & & & \ddots & B^T \end{bmatrix}_{mk \times nk}$$

$$w_i(t) = w(t + (i-1)h) \quad i = 1, \ldots, r$$

and the initial–final condition's relation is

$$\bar{x}(0) = \begin{bmatrix} x_0 \\ 0 \\ \vdots \\ 0 \end{bmatrix}_{nk \times 1} + \bar{H}\bar{x}(h) \quad \bar{H} = \begin{bmatrix} 0 & & & & 0 \\ I & 0 & & & \\ & I & 0 & & \\ & & \ddots & \ddots & \\ 0 & & & I & 0 \end{bmatrix}_{nk \times nk} \tag{8.4.19}$$

A problem arises when there exist $\bar{x}(0)$, $\bar{w}$ not both equal to zero such that $\bar{y}(t) = 0$. The solution of this problem is easy by use of the notions of maintainability and reachability introduced in Section 8.3. In that section algorithms producing sets of maintainable and reachable states are also given. In our case (Ker $\bar{B}$)-maintainability and reachability will play a crucial role.

Theorem 8.4.1 *System* (8.4.1) *fulfilling* rank $[A_r, B] = n$ *is uncontrollable at time* $T = kh$, $k \geq r$ *if and only if there exists a nonzero trajectory of system* (8.4.18) *fulfilling* (8.4.19) *with* $x_0 \in \text{Ker } B^T$ *and lying completely in* Ker $\bar{B}$.

Proof Necessity is obvious. Sufficiency follows from Lemma 8.4.14. □

Theorem 8.4.2 *System* (8.4.1) *is controllable at time* $T = kh$, $k \geq r$, *if and only if*
 (a) rank$[A_r, B] = n$,

(b) *the unique solution* x_0, $x_0 \in \text{Ker } B^T$, *for any* $y_1, y_2, y_1 \in \text{MAIN}(\bar{A}, \bar{D}, \text{Ker }\bar{B})$, $y_2 \in \text{REACH}(\bar{A}, \bar{D}, \text{Ker }\bar{B})$ *of the equation*

$$[I - \bar{H}e^{\bar{A}h}]y_1 - \bar{H}y_2 = \begin{bmatrix} x_0 \\ \vdots \end{bmatrix}_{nk \times 1}$$

is equal to zero, $x_0 = 0$,

(c) $\text{MAIN}(\bar{A}, \bar{D}, \text{Ker }\bar{B}) \cap R(\bar{D}) = 0$.

Proof We only outline the argument. Condition (c) guarantees that $\bar{w} \in R(\bar{D})$ which gives a trajectory lying in $\text{Ker }\bar{B}$ must be equal to zero. Condition (b) says that the initial condition $x_0 = 0$, and by (c) only the zero trajectory of system (8.4.16) fulfils $B^T x(t) = 0$ for $t \geq 0$. Condition (a) guarantees controllability. Necessity can be proved similarly. □

Now we will present some results concerning pole assignability and stabilizability. This part is based on Pandolfi (1975, 1976), Manitius and Olbrot (1979), Hale (1977), and Delfour and Manitius (1980a, 1980b). First we recall the fundamental spectral properties of the system

$$\dot{x}(t) = \int_{-h}^{0} dA(s) x(t+s) \quad t \geq 0 \quad x(t) \in R^n \quad (8.4.20)$$

$$x(s) = f(s) \quad s \in [-h, 0] \quad f \in X.$$

By $\hat{A}$ we denote the infinitesimal generator (see Section 8.1) of the semigroup of linear operators. corresponding to equation (8.4.20),

$$\hat{A}: X \supset D(A) \to X \quad X = L^2(-h, 0; R^n) \times R^n. \quad (8.4.21)$$

X is a Hilbert space with scalar product denoted by $(x|y)$ for $x, y \in X$ (see 8.1.7)). Denoting $x = (x_1, x_0) \in X$, $x_1 \in L^2(-h, 0; R^n)$, $x_0 \in R^n$ we have

$$D(A) = \{x \in X: \dot{x}_1 \in L^2(-h, 0: R^n), \ x_0 = x_1(0)\}$$

and $\hat{A}$ is defined by

$$\hat{A}x = y \quad y = (\dot{x}_1, \int_{-h}^{0} dA(z) x_1(z)) \quad \text{for} \quad x \in D(\hat{A})$$

we use the same notation X for the complexified state space X). The resolvent $R(s, \hat{A})$, s a complex variable, is equal to

$$R(s, \hat{A}): \quad X \to X$$

$$R(s, \hat{A})x = (sI - A)^{-1}x = y = (y_1, y_0) \quad x, y \in X$$

$$y_0 = \Delta(s)^{-1}x_0 + \Delta(s)^{-1} \int_{-h}^{0} \mathrm{d}A(z) \int_{z}^{0} e^{s(z-t)}x_1(t)\mathrm{d}t \qquad (8.4.22)$$

$$y_1(z) = e_p^{sz}y_0 + \int_{z}^{0} e^{s(z-t)}x_1(t)\mathrm{d}t \qquad z \in [h, -0]$$

where

$$\Delta(s) = sI - \int_{-h}^{0} \mathrm{d}A(z)e^{sz}. \qquad (8.4.23)$$

We see that $\det\Delta$ is a complex variable function and thus has no poles and has zeros only of finite multiplicity. Hence the singularities of the resolvent $R(s, \hat{A})$ are connected only with zeros of $\Delta(s)$, while zeros of $\det\Delta$ are poles of the resolvent. Moreover, from the form of $\hat{A}$ it can be observed that for any s_0, $\det\Delta(s_0) = 0$, there is a nontrivial solution x of the equation

$$(s_0 I - \hat{A})x = 0.$$

Finally, we have the spectrum of $\hat{A}$ being a purely point spectrum consisting of eigenvalues

$$\mathrm{Sp}(\hat{A}) = \{s: \det\Delta(s) = 0\}.$$

Moreover, the set $\{s \in \mathrm{Sp}(\hat{A}): \mathrm{Re}(s) > r, r \in R^1\}$ has a finite number of points for any $r \in R^1$.

The multiplicity of a pole of $R(s, \hat{A})$ is equal to the maximal multiplicity of this pole in the components of the matrix $\Delta(s)^{-1}$. Hence the multiplicity of poles of the resolvent is never greater than the multiplicity of the pole $s_i \in \mathrm{Sp}(\hat{A})$ of the resolvent. The eigensubspace E_{s_i} of the operator $\hat{A}$ connected with s_i is of finite dimension and equals (Delfour and Manitius, 1980)

$$E_{s_i} = \mathrm{Ker}(s_i I - \hat{A})^{k_i} \qquad (8.4.24)$$

and consists of functions $f = (f_1, f_0) \in X$ of the form

$$f_1(z) = \sum_{j=0}^{k_i-1} \frac{z^j}{j!} e^{s_i z} a_j \qquad z \in [-h, 0] \qquad a_j \in R^n \qquad j = 0, \ldots, k_i-1$$

$$f_0 = f_1(0) \qquad (8.4.25)$$

where the vector $a = \mathrm{col}(a_0, a_1, \ldots, a_{k_i-1}) \in R^{nk_i}$ fulfils

$$A_{k_i}(s_i)a = 0$$

8.4 Controllability, Observability and Pole Assignment

with

$$A_{k_i}(s) = \begin{bmatrix} \Delta(s) & \dfrac{\Delta'(s)}{1!} & \cdots & \dfrac{\Delta^{(k_i-1)}(s)}{(k_i-1)!} \\ & \ddots & \ddots & \vdots \\ & & \Delta(s) & \dfrac{\Delta^{(k_i-2)}(s)}{(k_i-2)!} \\ 0 & & \ddots & \Delta(s) \end{bmatrix}_{nk_i \times nk_i}.$$

It can be proved (Delfour and Manitius, 1980) that

$$\dim E_{s_i} = (\text{multiplicity of zeros of the } \Delta(s) \text{ at } s_i)$$
$$= \dim \operatorname{Ker} A_{k_i}(s_i). \tag{8.4.26}$$

A continuous projection operator P_{s_i} on the eigensubspace E_{s_i} is given by the formula (Hille and Phillips, 1968)

$$P_{s_i}: X \to X \qquad P_{s_i} x = \frac{1}{2\pi j} \int_C R(s, \hat{A}) x \, ds \tag{8.4.27}$$

where j denotes $\sqrt{-1}$ and the curve C is a rectifiable closed curve containing the point $s_i \in \operatorname{Sp}(\hat{A})$ inside and other points of the spectrum outside. The operator P_{s_i} has the properties

$$R(P_{s_i}) = E_{s_i}$$
$$\hat{A} P_{s_i} x = P_{s_i} \hat{A} x \quad \text{for } x \in D(A)$$
$$P_{s_i} P_{s_i} = P_{s_i} \tag{8.4.28}$$
$$P_{s_i} P_{s_l} = 0 \quad \text{for } i \neq l, s_i, s_l \in \operatorname{Sp}(\hat{A})$$

which means that $E_{s_i} \subset \operatorname{Ker} P_{s_l}$ for $i = l$. Moreover the operator $(I - P_{s_i})$ is also a projection on $\operatorname{Ker} P_{s_i}$ and

$$\operatorname{Ker}(I - P_{s_i}) = R(P_{s_i})$$
$$\operatorname{Ker} P_{s_i} = R(I - P_{s_i}). \tag{8.4.29}$$

The operator P_{s_i} can be immediately obtained in particular cases from (8.4.27) by use of the calculus of residues:

$$P_{s_i}(x) = \lim_{s \to s_i} \frac{1}{(k_i - 1)!} \frac{d^{k_i - 1}}{ds^{k_i - 1}} [(s - s_i)^{k_i} R(s, \hat{A}) x] \tag{8.4.30}$$

since we have an explicit form (8.4.22) of the resolvent. There is also another description of P_{s_i}. Since the operator is finite-dimensional, it must have the form

$$P_{s_i}(x) = \sum_{j=1}^{\bar{k}_i} f_{s_i}^j (g_{s_i}^j | x) \qquad \bar{k}_i = \dim E_{s_i} \tag{8.4.31}$$

where the elements
$$\{f_{s_l}^j\}_{j=1}^{\bar{k}_l} \subset E_{s_l}$$
form a basis in E_{s_l} and are described by (8.4.25), and the elements
$$\{g_{s_l}^j\}_{j=1}^{\bar{k}_l} \subset X$$
are orthogonal to $\mathrm{Ker}\, P_{s_l}$ and $\mathrm{Ker}\, P_{s_l} = \bigcap_j \mathrm{Ker}\, g_{s_l}^j$, $\mathrm{Ker}\, g_{s_l}^j = \{x\colon (g_{s_l}^j|x) = 0\}$. The elements $\{g_{s_l}^j\}_{i=1}^{\bar{k}_l}$ can be chosen (Hale, 1977; Delfour and Manitius, 1980; Pandolfi, 1975) as the basis of the set $\bar{E}_{s_l}$
$$\bar{E}_{s_l} = \mathrm{Ker}(s_l I - \hat{A}^\mathrm{T}) \quad \dim \bar{E}_{s_l} = \bar{k}_l$$
where $\hat{A}^\mathrm{T}$ is such that
$$(y|\hat{A}x) = (\hat{A}^\mathrm{T} y|x) \quad \text{for} \quad x, y \in D(\hat{A}),$$
and $\bar{E}_{s_l}$ consists of functions of the form
$$\sum_{j=1}^{k_l} \frac{(-z)^{k_l-j}}{(k_l-j)!} e^{-s_l z} \cdot c_i \tag{8.4.32}$$
where the vector $c = \mathrm{col}(c_1, \ldots, c_{k_l})$ fulfils $A_{k_l}^\mathrm{T}(s_i)c = 0$ with the matrix $A_{k_l}(s_i)$ as before. Moreover, from the elements $\{g_{s_l}^j\}_{j=1}^{\bar{k}_l}$ by linear combinations one can form a basis $\{\bar{g}_{s_l}^j\}_{j=1}^{k_l}$ such that
$$(f_{s_l}^j|\bar{g}_{s_l}^l) = \begin{cases} 0 & \text{for } j \neq l \\ 1 & \text{for } j = l. \end{cases} \tag{8.4.33}$$
The operator P_{s_l} decomposes the state space X into a direct sum of the closed subspaces
$$X = E_{s_l} \oplus \mathrm{Ker}\, P_{s_l} \tag{8.4.34}$$
which are invariant with respect to the operator A (see (8.4.28)). For the finite set S, $S \subset \mathrm{Sp}(\hat{A})$, we can define an operator
$$P_S = \sum_{s_i \in S} P_{s_i}. \tag{8.4.35}$$
We may observe that P_S is a projection on the subspace
$$E_S = \bigoplus \sum_{s_i \in S} E_{s_i}$$
and decomposes X in the following way:
$$X = E_S \oplus \bigcap_{s_i \in S} \mathrm{Ker}\, P_{s_i}.$$

8.4 Controllability, Observability and Pole Assignment

We now state the pole assignability problem for the system

$$\dot{x}(t) = \int_{-h}^{0} dA(z)x(t+z) + Bu(t) \quad x(t) \in R^n \quad u(t) \subset R^m \quad (8.4.36)$$

$$\mathrm{Sp}(\hat{A}) = \{s_1, s_2, \ldots\}.$$

System (8.4.36) is called pole assignable (or spectrally controllable) if there exists a feedback

$$u(t) = \int_{-h}^{0} dH(z)x(t+z), \quad (8.4.37)$$

where H is an $(m \times n)$-matrix function of bounded variation, such that for any finite set of complex numbers $S = \{\bar{s}_i\}_{i=1}^{l}$ the closed-loop system

$$\dot{x}(t) = \int_{-h}^{0} d\big(A(z) + BH(z)\big)x(t+z) \quad (8.4.38)$$

has the spectrum

$$\{\bar{s}_1, \bar{s}_2, \ldots, s_{l+1}, \ldots\}.$$

Theorem 8.4.3 *System (8.4.36) is pole assignable if and only if*

$$\mathrm{rank}\,[\Delta(s), B] = n \quad \text{for any complex } s. \quad (8.4.39)$$

Proof Let us denote

$$\int_{-h}^{0} dH(z)x(z) = \langle \bar{H}, x \rangle \in R^m \quad x \in D(\hat{A}) \quad (8.4.40)$$

where $\bar{H}$ can be interpreted as an m-vector of (not necessarily continuous) linear functionals on X. The characteristic equation for the closed-loop system is

$$(sI - \hat{A})x = (\bar{0}, B\langle H, x \rangle) \quad x = (x_1, x_0) \in D(\hat{A})$$

Hence

$$\Delta(s)x_0 = B\langle H, x \rangle \quad x_1(z) = e^{sz}x_0 \quad (8.4.41)$$

and s_i is the eigenvalue of the closed-loop system if and only if

$$\Delta(s_i)x_0 = B\langle \bar{H}, x \rangle \quad (8.4.42)$$

for some x and $\bar{H}$ (see (8.4.21)).

Observe that condition (8.4.39) holds for any $s \notin \mathrm{Sp}(\hat{A})$.

Proof of necessity Let us assume that (8.4.39) does not hold for s_i, $s_i \in \text{Sp}(\hat{A})$. Formula (8.4.42) can be written as

$$\Delta(s_i)x_0 = BMx_0, \quad M \text{ is an } (m \times n) \text{ matrix.} \quad (8.4.43)$$

But if rank $[\Delta(s_i), B] < n$, then (8.4.43) always has a nontrivial solution with respect to x_0, s_i remains the eigenvalue which contradicts pole assignability of the closed-loop system.

Proof of sufficiency We shall construct an $\bar{H}$ which replaces the point $s_i \in \text{Sp}(\hat{A})$ by $\bar{s}_i$ without moving any other points of the spectrum $\text{Sp}(\hat{A})$. Let us consider the following feedback:

$$\langle \bar{H}, x \rangle = (w | P_{s_i} x)$$

where w is arbitrary m-vector of elements of X. The characteristic equation is

$$(sI - \hat{A})x = (0, B(w|P_{s_i}x)). \quad (8.4.44)$$

It may be observed that if $s = s_j$, $s_j \neq s_i$, $s_j \in \text{Sp}(A)$, then the eigenvectors from E_{s_j} remain unchanged since the right-hand side of (8.4.44) equals zero for $x \in E_{s_j}$. Thus the proposed feedback does not move any eigenvalue s_j different from s_i, and does not move any eigenspace E_{s_j}. All possible changes of the spectral properties of the closed-loop system occur in a finite-dimensional subspace of dimension $\bar{k}_i$, complementary to $\text{Ker} P_{s_i}$. Of course

$$(w|P_{s_i}x) = (P_{s_i}^* w|x) = (\bar{g}|x)$$

where $\bar{g}$ is an m-vector of elements from E_{s_i}.

Let us decompose the evolution state equation

$$\dot{\hat{x}} = \hat{A}\hat{x} + \hat{B}(w|P_{s_i}\hat{x})$$

where $\hat{x}$ is the state trajectory of the original system and $\hat{B}(w|P_{s_i}x) = (0, B(w|P_{s_i}x)) \in X$. Denoting

$$\hat{x}(t) = P_{s_i}\hat{x}(t) + (I - P_{s_i})\hat{x}(t) = \hat{y}(t) + \hat{v}(t)$$
$$\hat{y}(t) = P_{s_i}\hat{x}(t)$$
$$\hat{v}(t) = (I - P_{s_i})\hat{x}(t),$$

we have

$$\dot{\hat{y}} = \hat{A}\hat{y} + P_{s_i}\hat{B}(w|\hat{y}) \quad (8.4.45)$$

$$\dot{\hat{v}}(t) = \hat{A}\hat{v} + (I - P_{s_i})\hat{B}(w|\hat{y}). \quad (8.4.46)$$

Formula (8.4.46) describes the behaviour of the projection of the trajectory $\hat{x}$ on $\text{Ker} P_{s_i}$ and as may be seen, the eigenvalues of this component of the system are $\text{Sp}(\hat{A}) \setminus \{s_i\}$. Formula (8.4.45) gives the finite-dimensional part of the system.

As can be shown (Pandolfi, 1975), this finite-dimensional system of ordinary differential equations of dimension k_i is controllable if and only if rank $[\varDelta(s_i), B] = n$. By the fact that the finite-dimensional system is controllable if and only if it is spectrally controllable (Hautus, 1969) the spectrum of (8.4.45) may be located at any $\bar{k}_i$ point of the complex plane, in particular at point $\bar{s}_i$. Thus if rank$[\varDelta(s_i), B] = n$, then s_i can be replaced by $\bar{s}_i$. Continuing this procedure of replacement for $\bar{s}_i$, $i = 1, \ldots, r$ one obtains the new eigenvalues $\bar{s}_i$, $i = 1, \ldots, r$ and the spectrum of the infinitesimal generator of the closed-loop system becomes $\{\bar{s}_1, \ldots, \bar{s}_r, s_{r+1}, \ldots\}$. □

The condition rank $[\varDelta(s), B] = n$ can be formulated in a slightly different form. Denoting

$$G(s) = \int_{-h}^{0} \mathrm{d}A(z) \mathrm{e}^{sz}$$

and taking into account the fact that

$(\text{rank}\,[\lambda I - H, B] = n \quad \forall \lambda) \Leftrightarrow \text{rank}\,[B, HB, \ldots, H^{n-1}B] = n$

for any matrices H, B, one has

$$\text{rank}\,[\varDelta(s), B] = \text{rank}[B, G(s)B, \ldots, G^{n-1}(s)B].$$

Lemma 8.4.15 *If system (8.4.36) is approximately controllable, then it is spectrally controllable.*

Proof Let us suppose the system is not spectrally controllable. There then exists $s_i \in \text{Sp}(\hat{A})$ which cannot be changed. This means that system (8.4.36) with output y in $R^{\bar{k}_i}$

$$y(t) = (w | P_{s_i} \hat{x}(t))$$

where w is a $\bar{k}_i$-vector built from elements of E_{s_i}, is not controllable with respect to the output y, and hence is not controllable with respect to the one-dimensional output $y_1 = h^T y$ with appropriately chosen h, $h \neq 0$, $h \in R^n$. Consequently, by Lemma 8.4.5 it is not approximately controllable, which completes the proof. □

Lemma 8.4.16 *If the linear hull of the set of eigenvectors of system (8.4.36) is dense in the state space X (in short, if the set of eigenvectors is complete) and the system is spectrally controllable, then it is approximately controllable.*

Proof If the system is spectrally controllable, then it is controllable with respect to any finite-dimensional eigensubspace. If additionally the system of eigensubspaces covers the whole state space X, then for any one-dimensional output $y_1 = (c|\hat{x})$, $c \in X$, $c \neq 0$ there is at least one eigenvalue s_i and vector $h \in E_{s_i}$ such that $(c|h) \neq 0$. Finally, system (8.4.36) is controllable with respect to any one-dimensional output and by Lemma 8.4.5, approximately controllable. □

The condition of spectral controllability $\text{rank}[\Delta(s), B] = n$ has been generalized on more complex systems (Pandolfi, 1976). If for the system

$$\dot{x}(t) = \int_{-h}^{0} dA(z)x(t+z) + \int_{-h}^{0} dM(z)\dot{x}(t+z) + \int_{-h}^{0} dB(z)u(t+z)$$

the following is fulfilled:

$$\text{rank}\left[sI - \int_{-h}^{0} (dA(z) + s\,dM(z))e^{sz}, \int_{-h}^{0} dB(z)e^{sz}\right] = n$$

then the system is spectrally controllable.

System (8.4.36) is called stabilizable if and only if there exists a feedback, such that closed-loop system has eigenvalues with real parts less than zero.

From the proof of Theorem 8.4.3 we obtain that system (8.4.36) is stabilizable if and only if

$$\text{rank}[\Delta(s), B] = n \quad \text{for } s, \quad \text{Re}(s) \geq 0.$$

Optimal Control for Continuous-time Systems

9.1 Formulation of Optimal Control Problem

Let a general FDE system be given:

$$\dot{x}(t) = f(x_t, u(t), t) \quad t \in [t_0, t_1]$$
$$x(t) \in R^n \quad t \in [-h+t_0, t_1] \quad (9.1.1)$$
$$u(t) \in M \subset R^m \quad t \in [t_0, t_1]$$

where

$$x_t: [-h, 0] \ni s \mapsto x(t+s) \in R^n \quad x_t \in X \quad t \in [t_0, t_1]$$
$$f: X \times R^m \times [t_0, t_1] \mapsto R^n.$$

The set X in (9.1.1) is the function state space for the system. We shall assume that X may be one of the following Banach spaces:

$$L^2(-h, 0; R^n) \times R^n$$

with norm of the element x equal to

$$\|x\|_{L^2} = \left(\sum_{i=1}^{n} \int_{-h}^{0} x_i^2(s)\,ds + \sum_{i=1}^{n} x_i^2(0) \right)^{\frac{1}{2}}$$

$$C([-h, 0], R^n)$$

with norm of the element x equal to

$$\|x\|_C = \max_{i=1,\ldots,n} \sup_{s \in [-h,0]} |x_i(s)|$$

or
$$W_1^2(-h, 0; R^n)$$
with norm
$$\|x\|_W = \left(\|x(0)\|_{R^n}^2 + \|\dot{x}\|_{L^2}^2\right)^{\frac{1}{2}}.$$

By x_i we denote the i-th component of the vector function x.

Let the functional F be given

$$F: X \times [t_0, \infty) \mapsto R^1. \tag{9.1.2}$$

Let the convex set $K \subset X$ be given.

Now we state the optimal control problem. Find the triple (u^*, x^*, t^*) connected with system (9.1.1) fulfilling the constraints

$$x_{t_0}^* \in K \quad u^*(t) \in M \subset R^m \quad t^* \in [t_0, \infty) \tag{9.1.3}$$

and minimizing the functional F. Elements of the triple (u^*, x^*, t^*) will be called optimal control, optimal trajectory and optimal time, respectively.

We do not make any topological assumptions on f, G, F for the moment. We shall describe them in detail when we derive the necessary or sufficient conditions for optimality.

9.2 Necessary Conditions of Optimality

The presentation of results will be divided into several parts. In Section 9.2.1 we obtain general necessary optimality conditions for systems defined as a family of curves. In Section 9.2.2 the properties of solutions of the FDE system (9.1.1) which will ensure the applicability of results from Section 9.2.1 are checked. In Section 9.2.3 we formulate the final result of Section 9.2.1 in terms of the properties of an FDE system. In the next section we present other necessary optimality conditions for more particular cases (Olbrot, 1976a; Manitius, 1974b). In Section 9.2.3 we formulate the necessary optimality condition for a linear FDE.

9.2.1 Necessary conditions. General approach

Let X, Y be Banach spaces.

Assumption 9.2.1 The space X is densely contained in Y, $X \subset Y$, and the topology of X is stronger than the topology of Y. This means that there exists $r > 0$ such that $\|x\|_Y \leqslant r\|x\|_X$ for any $x \in X$.

For example, $X = C([-h, 0], R^n)$, $Y = L^2(-h, 0; R^n)$.

9.2 Necessary Conditions of Optimality

Let a nonempty set K, $K \subset X$ be given.
Let there be given an arbitrary family Z of continuous functions defined on interval $[t_0, \infty)$ with values in X

$$Z \subset C([t_0, \infty), X).$$

By x_t we denote the value of $x \in Z$ at point t, $t \in [t_0, \infty)$.

Assumption 9.2.2 *For any $y \in K \subset X$, there exists $x \in Z$ such that $x_{t_0} = y \in K$.*

For the set Z we define the family $\{C_t\}_{t \geqslant t_0}$ of cross-sections

$$C_t = \{x_t \colon x \in Z\} \subset X.$$

Assumption 9.2.3 *For any $x \in Z$, there exists a nonempty subset $S_x \subset [t_0, \infty)$ such that for any $t \in S_x$ there are a number $\varepsilon(t)$, $\varepsilon(t) > 0$, a nonvoid set $E(t)$, and a family $\{y_x^{t,\,a}\}_{a \in E(t)}$ of continuous functions $y_x^{t,\,a} \colon [0, \varepsilon(t)] \ni w \mapsto y_x^{t,\,a}(w) \in C_t \subset X$ such that*

$$y_x^{t,\,a}(0) = x_t. \tag{9.2.1}$$

There exists a right derivative in the space Y

$$\frac{\mathrm{d}}{\mathrm{d}w} y_x^{t,\,a}(w)\big|_{w=0^+} = \mathrm{d}_x^{t,\,a} \in Y \quad \forall a \in E(t) \quad \forall t \in S_x. \tag{9.2.2}$$

Assumption 9.2.4 *For any $t, s \in [t_0, \infty)$, $s \geqslant t$ and for any $x \in Z$ there exists a continuous map $R(x, s, t) \colon X \ni y \mapsto R(x, s, t)(y) \in X$ such that*

$$\begin{aligned} R(x, s, t)(C_t) &\subset C_s \text{ for } s \geqslant t \geqslant t_0 \\ R(x, s, t)(x_t) &= x_s. \end{aligned} \tag{9.2.3}$$

For any fixed $y \in X$ the map

$$R(x, \cdot, \cdot)(y) \colon (s, t) \mapsto R(x, s, t)(y) \in X \quad s \geqslant t \geqslant t_0$$

is continuous. We assume that the map $R(x, t, s)$ is Fréchet differentiable with derivative denoted by $P(x, t, s)$, and $P(x, t, s) \colon X \to X$ can be extended by continuity to the bounded operator $\bar{P}(x, t, s) \colon Y \to Y$. Such an extension is unique by density of X in the space Y. Moreover, let $\bar{P}(x, \cdot, \cdot)y \colon (s, t) \to \bar{P}(x, s, t)y$ be continuous for $s \geqslant t \geqslant t_0$.

Obviously, under the above assumptions the map

$$(s, t, y) \mapsto \bar{P}(x, s, t)y \in Y$$

is continuous.

Assumption 9.2.5 Let F be a continuous and continuously differentiable function $F: Y \times [t_0, \infty) \to R^1$. By F_x and F_t we denote Fréchet derivatives after the first and the second argument, respectively.

Now we start to formulate a sequence of optimization problems of increasing complication. The appropriate necessary optimality condition expressed by the notions and notation introduced above will be assigned to each of these problems.

Problem 9.2.1 Let $K = \{x_0\}$, $x \in X$, t_1 be fixed, $t_1 > t_0$. Find $x^* \in Z$ such that

$$F(x_{t_1}, t_1) \geqslant F(x^*_{t_1}, t_1) \quad \forall x \in Z. \tag{9.2.4}$$

Lemma 9.2.1 If $x^* \in Z$ is optimal for Problem 9.2.1 then

$$F_x \bar{P}(x^*, t_1, t) d_{x^*}^{t,a} \geqslant 0 \quad \forall t \in S_{x^*} \quad \forall a \in E(t) \tag{9.2.5}$$

where F_x and $\bar{P}$ are calculated at points $x^*_{t_1}$ and x^*_t respectively.

Proof From (9.2.4) and Assumption 9.2.4 we have

$$F\big(R(x^*, t_1, t)(y), t_1\big) - F(x^*_{t_1}, t_1) \geqslant 0 \quad \forall t \in [t_0, \infty) \quad \forall y \in C_t. \tag{9.2.6}$$

Taking $t \in S_{x^*}$, making use of the differentiability of F, R and defining $y^{t,a}_{x^*}$ as in Assumption 9.2.3, we have

$$F\big(R(x^*, t_1, t)\big(y^{t,a}_{x^*}{}_x(w)\big), t\big) - F(x^*_{t_1}, t_1) \geqslant 0$$
$$\forall w \in [0, \varepsilon(t)] \quad \forall a \in E(t) \quad \forall t \in S_{x^*}. \tag{9.2.7}$$

From the continuity of $y^{t,a}_{x^*}$ at point $w = 0$ it follows that for sufficiently small w and fixed t, a

$$F_x \bar{P}(x^*, t_1, t)\big(y^{t,a}_{x^*}(w) - x^*_t\big) + r\big(y^{t,a}_{x^*}(w) - x^*_t\big) \geqslant 0 \tag{9.2.8}$$

where r is such that

$$\lim_{\|x\|\to 0} \frac{\|r(x)\|_y}{\|x\|_y} = 0.$$

Dividing both sides of (9.2.8) by w, $w > 0$ and passing to the zero limit with w, we finally obtain (9.2.5). □

Problem 9.2.2. Let K be a convex set and let t_1, $t_1 > t_0$ be fixed. Find $x^* \in Z$ such that

$$F(x_{t_1}, t_1) \geqslant F(x^*_{t_1}, t_1) \quad \forall x \in Z \quad x^*_{t_0} \in K.$$

Lemma 9.2.2 Let $x^* \in Z$ be optimal for Problem 9.2.2. Then

$$F_x \bar{P}(x^*, t_1, t) d_{x^*}^{t,a} \geqslant 0 \quad \forall t \in S_{x^*} \quad \forall a \in E(t)$$
$$F_x \bar{P}(x^*, t_1, t_0)(y - x^*_{t_0}) \geqslant 0 \quad \forall y \in K \tag{9.2.9}$$

9.2 Necessary Conditions of Optimality

where derivatives F_x and $\bar{P}$ are calculated at points $x_{t_1}^*$ and x_t^* respectively.

Proof Arguing as in the proof of Lemma 9.2.1, we obtain the first inequality. Putting $y_{x^*}^{t_0, a}(w) = x_{t_1}^* + w(a - x_{t_1}^*)$, $a \in K$, we have the second inequality. □

Problem 9.2.3 Let $K = \{x_0\}$, $x_0 \in X$, be fixed. Find a pair (x^*, t^*), $x^* \in Z$, $t^* \in [t_0, \infty)$ such that

$$F(x_t, t) \geq F(x_{t^*}^*, t^*) \quad \forall t \geq t_0 \quad \forall x \in Z. \qquad (9.2.10)$$

Lemma 9.2.3 Let (x^*, t^*) be the optimal pair for Problem 9.2.3. Then

$$F_x \bar{P}(x^*, t^*, t) \mathrm{d}_{x^*}^{t, a} \geq 0 \quad \forall t \in S_{x^*} \quad \forall a \in E(t) \qquad (9.2.11)$$

and, if a limit

$$\lim_{\Delta \to 0^+} \frac{1}{\Delta}(x_{t^* + \Delta} - x_{t^*}^*) = y \in Y \qquad (9.2.12)$$

exists for some $x \in Z$ which fulfils $x_{t^*} = x_{t^*}^*$, then

$$F_x y + F_t \geq 0 \qquad (9.2.13)$$

where the derivatives F_x, F_t and $\bar{P}$ are taken at points $(x_{t^*}^*, t^*)$ and x_t^* respectively.

Proof Inequality (9.2.11) follows from the fact that for fixed t^* one can apply Lemma 9.2.1. Relation (9.2.13) follows immediately from the definition of the optimal pair and from the properties of partial derivatives. □

Problem 9.2.4 Let $K \subset X$ be a convex set. Find the pair (x^*, t^*), $x^* \in Z$, $t^* \in [t_0, \infty)$ such that

$$x_{t_0}^* \in K \quad F(x, t) \geq F(x^*, t^*) \quad \forall t \in [t_0, \infty) \quad \forall x \in Z.$$

Lemma 9.2.4 If the pair (x^*, t^*) is optimal for Problem 9.2.4, then

$$\begin{aligned} F_x \bar{P}(x^*, t^*, t) \mathrm{d}_{x^*}^{t, a} &\geq 0 \quad \forall t \in S_{x^*} \quad \forall a \in E(t) \\ F_x \bar{P}(x^*, t^*, t_0)(y - x_{t_0}^*) &\geq 0 \quad \forall y \in K, \end{aligned} \qquad (9.2.14)$$

and if there exists a limit

$$\lim_{\Delta \to 0^+} \frac{1}{\Delta}(x_{t^* + \Delta} - x_{t^*}^*) = y \in Y$$

for some $x \in Z$ which fulfils $x_{t^*} = x_{t^*}^*$, then

$$F_x y + F_t \geq 0. \qquad (9.2.15)$$

Proof of the Lemma 9.2.4 involves a simple combination of arguments used in the proofs of Lemmas 9.2.2 and 9.2.3, and so will be omitted.

We now formulate a more general optimization problem.

Problem 9.2.5. Find the pair (x^*, t^*), $x^* \in Z$, $t^* \in [t_0, \infty)$ such that

$$F(x_t, t) \geq F(x_{t^*}^*, t^*) \quad \forall t \in [0, \infty) \quad \forall x \in Z$$

under the conditions

$$x(t_0) \in K \quad K \subset X$$
$$x(t^*) \in K_1 \quad K_1 \subset X \subset Y$$

where K, K_1 are convex sets and K_1 is a closed set in Y.

Let us denote by $G(y)$ the set of supporting functionals from Y' (dual to Y) to the set K_1 at point $y \in K_1 \subset Y$. This means that

$$G(y) = \{g \in Y': \langle g, z \rangle \geq \langle g, y \rangle, \forall z \in K_1\}$$

where $\langle g, z \rangle$ denotes the value of the functional g at point z. For $x \in Z$, $t \in S_x$, $s \geq t$, we define the sets

$$D(x, s, t_0) = \{y \in K: \langle g, \bar{P}(x, s, t_0)(y - x_{t_0}) \rangle \geq 0, \forall g \in G(x_s)\}$$
$$D(x, s, t) = \{d_x^{t;a}: \langle g, \bar{P}(x, s, t) d_x^{t;a} \rangle \geq 0, \forall g \in G(x_s)\} \subset Y$$

where $d_x^{t;a}$ is as in Assumption 9.2.3.

Lemma 9.2.5 Let (x^*, t^*) be the optimal pair for Problem 9.2.5. Then

$$F_x \bar{P}(x^*, t^*, t) d \geq 0 \quad \forall d \in D(x^*, t^*, t) \quad t \in S_{x^*}$$
$$F_x \bar{P}(x^*, t^*, t_0)(y - x_{t_0}^*) \geq 0 \quad y \in D(x^*, t^*, t_0) \quad (9.2.16)$$

and if there exists a limit

$$\lim_{\Delta \to 0^+} \frac{1}{\Delta}(x_{t^*+\Delta} - x_t^*) = y^0 \in Y \quad (9.2.17)$$

$$\langle g, y^0 \rangle \geq 0 \quad \forall g \in G(x_{t^*}^*)$$

for some $x \in Z$ which fulfils $x_{t^*} = x_{t^*}^*$, then

$$F_x y^0 + F_t \geq 0$$

where $F_x, F_t, \bar{P}$ are derivatives taken at points $x_{t^*}^*, t^*, x_t^*$ respectively.

Proof It can be observed that these conditions differ from conditions (9.2.14), (9.2.15) only in the form of $d_x^{t;a}$. As used here, $d_x^{t;a}$ ensures that $R(x^*, t^*, s) y \in K_1$ for y sufficiently near to x_t^*, which follows from the properties of the supporting functionals of the set K_1 at point $x_{t^*}^*$. □

9.2 Necessary Conditions of Optimality

The problems presented are not very general. They do not include, for instance, constraints on the behaviour of x on some interval of time. However, the approach we have described is one of the simplest and, simultaneously, relatively strong. It should be noted that the relationship

$$F_x P(x, t, s)\,\mathrm{d}_x^{t;a} \geq 0$$

is equivalent in many cases to Pontryagin's maximum principle. The term $P(x, t, s)\,\mathrm{d}_x^{t;a}$ is closely related to solutions of the so-called adjoint equation, as we shall show further on.

9.2.2 Fundamental properties of FDE systems

Let $X = C([-h, 0], R^n)$, $h > 0$. For b, $b \in C([t_0-h, t_1], R^n)$ we define $b_t \in X$, $t_1 \geq t \geq t_0$,

$$b_t(s) = b(t+s) \quad s \in [-h, 0]. \tag{9.2.18}$$

Let us observe that the function $[t_0, t_1] \ni t \mapsto b_t \ni X$ is continuous.

If we take $X = L^2(-h, 0; R^n) \times R^n$ and function b defined on interval $[t_0-h, t_0)$ as from $L^2(t_0-h, t_0; R^n)$ and on interval $[t_0, t_1]$ as from $C([t_0, t_1], R^n)$, then the function $b_t: [t_0, t_1] \ni t \mapsto b_t \in X = L^2(-h, 0; R^n) \times R^n$, where b_t is described by (9.2.18), is continuous.

Theorem 9.2.1 (Hale, 1977; Myshkis, 1972) *Let $X = C([-h, 0], R^n)$. Let the function f,*

$$f: C([-h, 0], R^n) \times R^m \times [t_0, \infty) \to R^1$$

for any fixed $u \in L^\infty((t_0, \infty), R^m)$ fulfil

$$\|f(y_1, u(t), t) - f(y_2, u(t), t)\|_{R^n} \leq \|y_1 - y_2\|_X m(t) \quad \forall y_1, y_2 \in X \tag{9.2.19}$$

for almost all t, $t \in [t_0, \infty)$, where m is locally integrable.

Then, for any $z \in X$, $u \in L^\infty((t_0, \infty), R^m)$, $t_1 \geq 0$ there exists a unique solution of the equation

$$\dot{x}(t) = f(x_t, u(t), t) \quad t_1 \geq t \geq t_0 \tag{9.2.20}$$

with condition $x_{t_0} = z$.

The function

$$[t_0, t_1] \ni t \mapsto x_t \in X \tag{9.2.21}$$

is continuous.

The function

$$X \ni z \mapsto x_t \in X \tag{9.2.22}$$

is continuous for any fixed t.

The function
$$X \ni z \mapsto x \in C([t_0-h, t_1], R^n) \qquad (9.2.23)$$
is continuous.

Theorem 9.2.2 (Hale, 1977; Myshkis, 1972; Hale and Cruz, 1970) *Let the FDE system fulfil the conditions of Theorem 9.2.1. Moreover, let the Fréchet partial derivative f_x exist, and for fixed $u \in L^\infty([t_0, \infty), R^n)$ let the following hold*:
$$\|f_x(y_1, u(t), t) - f_x(y_2, u(t), t)\|_{X'} \leq \|y_1 - y_2\|_X H(t) \qquad (9.2.24)$$
for almost all $t, t \in [t_0, \infty)$, $\forall y_1, y_2 \in X$, *where X' is the space of functions of bounded variation, adjoint (dual) to X and $H(\cdot)$ is locally integrable. Then for fixed u the mappings defined in (9.2.22), (9.2.23) are continuous and continuously Fréchet differentiable. Moreover, solutions x^1, x^2 of (9.2.20) corresponding to the initial conditions z_1, z_2 respectively, fulfil the following*:

$$\dot{\Delta}(t) = f_x(x_t^1, u(t), t) \cdot \Delta_t \qquad (9.2.25)$$

$$\Delta_{t_0} = z_1 - z_2$$

$$\Delta_t = x_t^1 - x_t^2 + r(\Delta_{t_0}, t)$$

$$\lim_{\|\Delta_{t_0}\| \to 0} \frac{\|r(\Delta_{t_0}, t)\|_{R^n}}{\|\Delta_{t_0}\|} = 0.$$

Now we recall some properties of linear FDE.

Theorem 9.2.3 (Hale, 1977; Myshkis, 1972; Banks, 1969) *Let us consider a linear FDE system*

$$\dot{x}(t) = \int_{-h}^{0} d_s A(t, s) x(t+s) + g(t) \qquad t \geq t_0 \qquad (9.2.26)$$

$$x_{t_0} = \varphi \in C([-h, 0], R^n)$$

where $A(t, s)$ is a matrix of functions which are locally integrable with respect to the first argument for any $s \in [-h, 0]$, or are of bounded variation with respect to the second argument for almost all $t \geq t_0$, and moreover, there exists a function m_1, locally integrable, such that

$$\operatorname*{Var}_{[-h, 0]} (A(t, \cdot)) < m_1(t) \quad \text{for a.a.} \quad t \geq t_0 \qquad (9.2.27)$$

where $\operatorname*{Var}_{[-h, 0]} (f(\cdot))$ *denotes the variation of the function f on interval* $[-h, 0]$.

9.2 Necessary Conditions of Optimality

Then, for any $\varphi \in C([-h, 0), R^n)$ there exists a unique solution $x(t)$, $\infty > t \geqslant t_0$ of equation (9.2.26) which is absolutely continuous with respect to the pair (t_0, t), fulfils $x_t = \varphi$ and is continuously dependent on initial data and control g. Moreover, there exists a unique, absolutely continuous matrix function $H(t, t_1)$, $t \geqslant t_1 \geqslant t_0$ which is the solution of the equation

$$\frac{\mathrm{d}}{\mathrm{d}t} H(t, t_1) = \int_{-h}^{0} \mathrm{d}_s A(t, s) H(t+s, t_1) \quad t \geqslant t_1 \geqslant t_0 \quad (9.2.28)$$

$$H(t_1, t_1) = I \quad t_1 \geqslant t_0$$
$$H(t, t_1) = 0 \quad t < t_1.$$

The solution of (9.2.26) can be expressed as

$$x(t) = H(t, t_0)\varphi(t_0) + \int_{t_0}^{t} H(t, z) \left(\int_{-h}^{-(z-t_0)} \mathrm{d}_s A(z, s)\varphi(z-t_0+s) \right) \mathrm{d}z +$$

$$+ \int_{t_0}^{t} H(t, z) g(z) \mathrm{d}z \quad \text{for } t \geqslant t_0 \quad (9.2.29)$$

where

$$\int_{-h}^{-(z-t_0)} \mathrm{d}_s A(z, s)\varphi(z-t_0+s) \stackrel{\mathrm{df}}{=} 0 \quad \text{for } z > t_0+h. \quad (9.2.30)$$

It is essential for the application of optimization theory explained in Section 9.2.1 to know in what cases the linear equation (9.2.26) has a solution for any $\varphi \in L_0^2(-h, 0; R^n)$.

Theorem 9.2.4 Let the linear FDE be as in Theorem 9.2.3. If in addition there exist $\varepsilon > 0$, $M > 0$ such that

$$\|g^t\|_{L^2(0, h; R^n)} \leqslant M\Delta \quad \Delta \in [0, \varepsilon] \quad t \geqslant t_0$$

where

$$(g^t)(s) = A(t+s, -s) - A(t+s, -s-\Delta) \quad s \in [0, h]$$
$$A(t, r) \stackrel{\mathrm{df}}{=} 0 \quad r \notin [-h, 0]$$

then for any $\varphi \in L_0^2(-h, 0; R^n)$ there exists a unique solution of equation (9.2.26) which is continuous as a function of time and continuously dependent on initial data from $L_0^2(-h, 0; R^n)$.

The proof of this fact follows in a simple way from consideration of solution (9.2.29) for the initial data as a characteristic function of some subinterval from $[-h, 0]$. Passing to the limit with the sum of characteristic functions approximating the initial data, one obtains the thesis. We restrict ourselves to these remarks, and refer the reader to Banks (1969) and Hale and Cruz (1970) where with a few changes to the argument one can reach this result.

It can be observed that the FDE system

$$\dot{x}(t) = \sum_{i=0}^{n} A_i(t) x(t-h_i) \qquad t \geqslant t_0 \qquad (9.2.31)$$

$$A_i \in L^2(t_0, \infty; R^n \times R^n) \qquad h_i \geqslant 0 \qquad (i = 0, \ldots, n)$$

fulfils the assumptions of Theorem 9.2.4.

In the sequel the so-called adjoint equation to equation (9.2.28) will be used (Hale, 1977; Myshkis, 1972; Banks, 1969). The adjoint equation may be written in a differential or integral form.

The integral form of the adjoint equation to equation (9.2.28) is

$$W(t, t_1) = I - \int_t^{t_1} d_a A^T(a, t-a) W(a, t_1) \qquad t \in [t_0, t_1] \qquad (9.2.32)$$

$$W(t_1, t_1) = I$$

$$W(t, t_1) = 0 \qquad t > t_1.$$

The differential form of the adjoint equation to equation (9.2.28) is

$$\frac{d}{dt} W(t, t_1) = - \int_{-h}^{0} d_a A^T(t-a, a) W(t-a, t_1) \qquad t \in [t_0, t_1] \qquad (9.2.33)$$

$$W(t_1, t_1) = I$$

$$W(t, t_1) = 0 \qquad t > t_1.$$

There is a simple relation between the solution H of (9.2.28) and the solution W of (9.2.32) or (9.2.33):

$$H(t_1, t) = W^T(t, t_1) \qquad t \in [t_0, t_1] \qquad t_1 > t_0. \qquad (9.2.34)$$

9.2.3 Necessary optimality conditions for FDE system

Let $X = C([-h, 0], R^n)$, $Y = L_0^2(-h, 0; R^n)$.

Let us consider the FDE system

$$\begin{aligned} \dot{x}(t) &= f(x_t, u(t), t) & t_0 \leqslant t \\ x_{t_0}(s) &= z(s) & s \in [-h, 0] & z \in X \\ u(t) &\in M \subset R^m \end{aligned} \qquad (9.2.35)$$

which fulfils the assumptions of Theorems 9.2.1, 9.2.2.

9.2 Necessary Conditions of Optimality

Let the linearized FDE system

$$\dot{\Delta}(t) = f_x(x_t, u(t), t)\Delta_t \stackrel{df}{=} \int_{-h}^{0} d_s A(t, s)\Delta(t+s) \qquad (9.2.36)$$

satisfy the conditions of Theorem 9.2.4 for any suitable x.

Let there be given a continuously differentiable functional F

$$F: X \times [t_0, \infty) \to R^1$$
$$F(x, t) = \hat{F}(x(0), t) \quad x \in X \quad t \in [t_0, \infty) \qquad (9.2.37)$$
$$\hat{F}: R^n \times [t_0, \infty) \to R^1.$$

Let $K \subset R^n$ be a convex set. By $\hat{K}$ we denote the set

$$X \supset \hat{K} = \{x: x(0) \in K \quad x(s) = 0 \quad \text{for } s \in [-h, 0)\}.$$

We now state the following optimal control problem.

Find the optimal triple (x^*, u^*, t^*) which fulfils relations (9.2.35) and

$$\hat{F}(x_{t*}^*(0), t^*) \leqslant \hat{F}(x_t(0), t) \quad x_{t_0}^* \in \hat{K} \qquad (9.2.38)$$

for any $t \in [t_0, \infty)$, and for any trajectory x of the FDE system fulfilling $x_{t_0} \in \hat{K}$.

In order to derive the necessary conditions of optimality we define the notions required by Lemma 9.2.4 in terms of FDE systems.

The family Z is obviously the set of all solutions of the FDE (9.2.35), taken in the functional state space X.

From now on, by x we shall mean the functional state space trajectory

$$x: [t_0, \infty) \ni t \mapsto x_t \in X.$$

The family of cross-sections $\{C_t\}_{t \geqslant t_0}$ is immediately defined by the set Z.

We define the family of mappings R (see (9.2.3)) as follows. For any trajectory x of the FDE system (9.2.35) there exists a corresponding control u. There may be many controls giving the same trajectory but one can choose one arbitrarily. For the control u, for any $y \in X$ and for any $t, s \in [t_0, \infty)$, $t \geqslant s$, we define the element $R(x, t, s)(y)$ as the value of the space trajectory of FDE (9.2.35) at the point t associated with control u and initial condition y at time s. This definition of R guarantees all the topological properties of R needed for the validity of Lemma 9.2.4, if the FDE and the linearized FDE fulfil the assumptions of Theorems 9.2.1–9.2.4. In particular we have the existence of the Fréchet derivative $\bar{P}$ in the space Y (see Assumption (9.2.4)). The element $\bar{P}(x, t, s)y$, $y \in Y$ can be treated as the value of the solution of the linearized FDE (9.2.36) at time t, starting from the initial state $y \in Y$ at time s, $t \geqslant s$.

Now we describe the construction of the family $\{y_x^{t,a}\}$ introduced in

Assumption 9.2.3. Let x be some functional state space trajectory of the FDE (9.2.35) and u the corresponding control. Let S_x denote the set of all $t \in [t_0, \infty)$ for which there exists a derivative of the function $z\colon [t_0, \infty) \ni t \mapsto x_t(0) \in R^n$. The set S_x has the property $\eta(I \cap S_x) = \eta(I)$, where η is the Lebesgue measure and I is any finite subinterval of $[t_0, \infty)$, since the solution of FDE (9.2.35) taken in R^n is absolutely continuous. For $t_1 \in S_x$, $0 < \varepsilon < h$, we define the set $E(t_1)$ of integrable functions

$$E(t_1) = \{a\colon [0, \infty) \ni p \mapsto a(p) \in M \subset R^m, a(0^+) = a(0)\} \quad (9.2.39)$$

and the family of controls $\{u_{w,a}\}$, $w \in [0, \varepsilon]$, $a \in E(t_1)$

$$u_{w,a}(t) = \begin{cases} u(t) & \text{for } t \in [t_0, t_1 - w) \\ a(t - (t_1 - w)) & \text{for } t \in [t_1 - w, \infty) \end{cases} \quad (9.2.40)$$

With any member $u_{w,a}$ of the family $\{u_{w,a}\}$ we associate the trajectory $x^{w,a}$ of the FDE (9.2.35) with control $u_{w,a}$. Defining

$$y_x^{t_1, a}(w) = x_{t_1}^{w, a} \in X$$

we see that the function $y^{t_1, a}(\cdot)$ has the properties required in Assumption 9.2.3. It is easy to observe that for the original trajectory x corresponding to control u and for $x^{w,a}$ we have

$$x_{t_1}(s) = (y_x^{t_1, a}(w))(s) \quad s \in [-h, -w].$$

Hence

$$\frac{d}{dw} y_x^{t_1, a}(w)|_{w=0}(s) = 0 \quad s \in [-h, 0)$$

and

$$\frac{d}{dw} y_x^{t_1, a}(w)|_{w=0}(s) = f(x_{t_1}, a(0), t_1) - f(x_{t_1}, u(t_1), t_1)$$

$$s = 0 \quad t_1 \in S_x \quad a \in E(t_1).$$

Finally,

$$d_x^{t_1, a}(s) = \begin{cases} f(x_{t_1}, a(0), t_1) - f(x_{t_1}, u(t_1), t_1) & \text{for } s = 0 \\ 0 & \text{for } s \in [-h, 0). \end{cases}$$

For our optimal control problem (9.2.38) we can now apply the optimality condition given in Lemma 9.2.4. Namely, if the triple $\{x^*, u^*, t^*\}$ is optimal then, taking into account the form of $d_x^{t, a}$ and fact that $\dfrac{\partial \hat{F}}{\partial x(0)}$ is a row vector in R^n, we have

$$\left.\frac{\partial \hat{F}}{\partial x(0)}\right|_{(x_{t*}^*(0), t^*)} H(t^*, t)[f(x_t^*, a, t) - f(x_t^*, u^*(t), t)] \geqslant 0 \quad (9.2.41)$$

$$\forall t \in [t_0, t^*] \quad \forall a \in E(t)$$

9.3 Some Particular Optimization Problems

where $H(t^*, t)$ (see (9.2.28)) is the fundamental solution of the linearized FDE (9.2.36) and the linearization is taken along the trajectory x^*.
Moreover,

$$\left.\frac{\partial \hat{F}}{\partial x}\right|_{(x_t^*(0),\,t^*)} H(t^*, t_0)\left(y - x_{t_0}^*(0)\right) \geq 0 \quad \forall y \in K \subset R^n \quad (9.2.42)$$

$$\left.\frac{\partial \hat{F}}{\partial x(0)}\right|_{(x_{t*}^*(0),\,t^*)} [f(x_{t*}^*, a, t^*) - f(x_{t*}^*, u^*(t^*), t^*)] + \left.\frac{\partial \hat{F}}{\partial t}\right|(x_{t*}^*(0), t^*) \geq 0$$
$$\forall a \in E(t^*).$$

From the computational point of view the way one obtains the matrices $H(t^*, t)$, $t \in [t_0, t^*]$ is important. Obviously, the immediate application of definition (9.2.28) of H requires a great number of calculations of solutions of the FDE (9.2.36). However, by introducing the adjoint equation (9.2.32), (9.2.33) one can make these calculations relatively simple. Taking into account (see (9.2.34)) the fact

$$H^T(t_1, t) = W(t, t_1) \quad t \in [t_0, t_1] \quad t_1 > t_0$$

where W is the solution of adjoint equation, and defining

$$\psi(t) = W(t, t^*) \left[\left.\frac{\partial \hat{F}}{\partial x}\right|_{(x_{t*}^*(0),\,t^*)}\right]^T \quad t \in [t_0, t^*]$$

one can write condition (9.2.41), (9.2.42) as follows:

$$\psi^T(t)[f(x_t^*, a, t) - f(x_t^*, u^*(t), t)] \geq 0 \quad \forall t \in [t_0, t^*] \quad \forall a \in E(t) \quad (9.2.43)$$

$$\psi^T(t_0)\left(y - x_{t_0}^*(0)\right) \geq 0 \quad \forall y \in K \subset R^n$$

$$\psi^T(t^*)[f(x_{t*}^*, a, t^*) - f(x_{t*}^*, u^*(t^*), t^*)] + \left.\frac{\partial F}{\partial t}\right|_{(x_{t*}^*(0),\,t^*)} \geq 0 \quad \forall a \in E(t^*).$$

We restrict ourselves to problem (9.2.38), although Lemma 9.2.5 enables us to obtain more general results in relation to state constraints.

9.3 Some Particular Optimization Problems

This part is based on Olbrot (1976a) and Manitius (1974b). We give the final results; the line of argument will only be sketched in roughly here.

Let the following optimal control problem be given (Olbrot, 1976a). Minimize the functional

$$\int_0^T f_0(x(t), u(t), t)\,dt \quad (9.3.1)$$

with constraints

$$\dot{x}(t) = f(x(t), x(t-h_1), \ldots, x(t-h_s), u(t), u(t-h_1), \ldots, u(t-h_s), t)$$
$$\text{a.e. on } [0, T] \quad (9.3.2)$$
$$x(t) = \varphi(t) \quad u(t) = w(t) \quad \text{for } t \in [-h_s, 0]$$

where φ, w are continuous,

$$\|x_T - \eta\| = \sup_{s \in [-h_s, 0]} \|x(T+s) - \eta(s)\|_{R^n} \leqslant \varepsilon \quad \varepsilon > 0 \quad (9.3.3)$$

$\|\cdot\|_{R^n}$ is euclidean, η is of class C^2 on $[-h_s, 0]$

$$u \in L^\infty((0, T); U) \quad U \subset R^r \quad U \neq \phi. \quad (9.3.4)$$

The following additional assumptions are made.

Assumption 9.3.1 *Let $T, h_1, \ldots, h_s$, $T > h_s > , \ldots, h_1 > 0$ be commensurable, i.e. let there exist $h > 0, k_1, \ldots, k_s, k \in N$ such that $T = kh, h_i = hk_i$.*

Assumption 9.3.2 *The functions f_0, f are continuous on their domains*

$$f_0: G \times U \times [0, T] \to R^1$$
$$f: G^{s+1} \times U^{s+1} \times [0, T] \to R^n$$

where the set $G \subset R^n$ is nonempty. Moreover, $f_0(x, u, t)$ is continuously differentiable with respect to x and $f(x, y_1, \ldots, y_s, u, w_1, \ldots, w_s, t)$ is continuously differentiable with respect to $(x, y_1, \ldots, y_s)$.

Theorem 9.3.1 (Olbrot, 1976a) *Let x^*, u^* be the optimal trajectory and optimal control for problem (9.3.1)–(9.3.4). Then there exists a nonzero triple $(p_0, p(\cdot), m(\cdot))$*

$$p_0 \leqslant 0$$
$$p: [0, T+h_s] \to R^n$$
$$m: [0, T+h_s] \to R^n$$

which satisfies the conditions

(i) $m(t) = m_{k-k_s+1}(t) + \ldots + m_k(t)$. *For each $i = k-k_s+1, \ldots, k$ the function m_i is nonincreasing on $[ih-h, ih]$ and is equal to zero on $[0, ih-h] \cup [ih, T]$. On subintervals of $[ih-h, ih]$ on which $\|x_T^* - \eta\| < \varepsilon$, the function m_i is constant. Moreover m is right continuous.*

(ii) *The function p is absolutely continuous on each of the intervals $[0, T-h_s]$, $[T-ih, T-ih+h]$, $i = 1, \ldots, k_s$. At points $T-ih$, $i = 1, \ldots, k_s$, the function p has the jumps*

$$p(T-ih-0) = p(T-ih+0) + 2m(T-ih)(x(T-ih) - \eta(T-ih)).$$

9.3 Some Particular Optimization Problems

Almost everywhere on $[0, T]$ the function p satisfies the difference-differential equation of advanced type

$$\dot{p}^T(t) = -\frac{\partial}{\partial x(t)} \hat{H}(t)$$

with boundary condition

$$p(t) = 0 \quad t \in [T, T+h_s]$$

where

$$\hat{H}(t) = H(t) + H(t+h_1) \ldots + H(t+h_s)$$

$$H(t) = H(p(t), x^*(t), x^*(t-h_1), \ldots, x^*(t-h_s), u^*(t), \ldots, u^*(t-h_s), t) \quad (9.3.5)$$

and

$$H(p, x, y_1, \ldots, y_s, u, w_1, \ldots, w_s, t)$$
$$= p_0 f_0(x, u, t) + (p - 2m(t)x - \eta(t))^T (\dot{\eta}(t) + f(x, y_1, \ldots, y_s, u, w_1, \ldots, w_s, t)).$$

(iii) *For almost all* $t \in [0, T]$

$$\hat{H}(t) = \min_{u(t) \in U} \hat{H}(u(t), t)$$

where $\hat{H}(u(t), t)$ is described by (9.3.5) but with u instead of u^*.

In the proof of Theorem 9.3.1 the commensurability of $T, h_1, \ldots, h_s$ is essential. This assumption enables us to formulate the optimization problem for FDE as a problem for ordinary differential equation with state constraints. Namely, defining

$$z_i(t) = x(t+(i-1)h) \quad v_i = u(t+(i-1)h)$$
$$t \in [0, h] \quad i = 1, k,$$

one can rewrite equation (9.3.2) in the form

$$\dot{z}_i(t) = f(z_i(t), z_{i-k_1}(t), z_{i-k_2}(t), \ldots, v(t), v_{i-k_1}(t), \ldots, v_{i-k_s}(t), t) \quad i = 1, k$$
$$t \in [0, h] \quad (9.3.6)$$

with the additional mixed initial-final state conditions

$$z_1 = \varphi(0)$$
$$z_i(h) = z_{i+1}(0) \quad (i = 1, \ldots, k-1). \quad (9.3.7)$$

The performance index (9.3.1) and constraint (9.3.3) may be changed in a similar way. The maximum principle for ordinary differential equations in the form given by Makowski and Neustadt (1974) may be applied to this reformulated problem which leads to the optimality conditions as described above.

Let a system with delay depending on control and time be given as follows (Manitius, 1974b):

$$\dot{x}(t) = f(x(t), x(t-h(\cdot)), u(t), t) \quad x(t) \in R^n \quad t \in [t_0, t_1]$$
$$h(\cdot) = h(x(t), u(t), t) \geq 0 \quad \forall u(t) \in U \quad \forall x(t) \in R^n \quad \forall t \in [t_0, t_1]. \quad (9.3.8)$$

Let the initial condition for the FDE (9.3.8) be defined on the interval $[h_0, t_0]$ where

$$h_0 = \inf(t - h(x, u, t)) \quad (x, u, t) \in R^n \times U \times [t_0, t_1]. \quad (9.3.9)$$

The following control problem will be considered. Find u^*, $u^*(t) \in U$, $t^* \in [t_0, t_1]$ such that

$$\int_{t_0}^{t^*} f_0(x(t), u(t), t) dt \quad (9.3.10)$$

is minimized under the additional constraint

$$(x^*(t^*), t^*) \in V \subset R^{n+1} \quad (9.3.11)$$

where V is some manifold of class C^1 and dimension less than $n+1$.

The following assumptions are made.

Assumption 9.3.3 Functions $f(x, y, u, t)$, $f_x(x, y, u, t)$, $f_y(x, y, u, t)$ and $h(x, u, t)$, $h_x(x, u, t)$, $f_0(x, u, t)$, $f_{0x}(x, u, t)$ are defined and continuous on $R^n \times R^n \times U \times [t_0, t_1]$ or $R^n \times U \times [t_0, t_1]$ respectively, where f_x, h_x, f_y, f_{0x} denote the partial derivatives. Moreover, there exist constants L_1, L_2, L_3, such that for every $x, \bar{x}, y, \bar{y} \in R^n$, $u \in U$, $t \in [t_0, t_1]$

$$\|f(\bar{x}, \bar{y}, u, t) - f(x, y, u, t)\|_{R^n} \leq L_1(\|\bar{x} - x\|_{R^n} + \|\bar{y} - y\|_{R^n})$$
$$|h(\bar{x}, u, t) - h(x, u, t)| < L_2 \|\bar{x} - x\| \quad (9.3.12)$$
$$\|f_0(\bar{x}, u, t) - f_0(x, u, t)\|_{R^n} \leq L_3 \|\bar{x} - x\|.$$

Assumption 9.3.4 The set U is a compact subset of R^m, and the controls u, $u(t) \in U \subset R^m$, $t \in [t_0, t_1]$ are Lebesgue measurable.

Assumption 9.3.5 The initial condition $\varphi(t)$, $t \in [h_0, t_0]$, is Lipschitz continuous, i.e. it exists L such that

$$\|\varphi(t_2) - \varphi(t_3)\|_{R^n} \leq L|t_2 - t_3| \quad t_2, t_3 \in [t_0, t_1].$$

Assumption 9.3.6 For any fixed $t \in [t_0, t_1]$, $\varphi \in C([t-(t_0-h_0), t], R^n)$, let the set

$$\{y \in R^n : y = f(\varphi(t), \varphi(t-h(\varphi(t), u, t)), u, t), u \in U\} \subset R^n \quad (9.3.13)$$

be compact and convex.

9.3 Some Particular Optimization Problems

Theorem 9.3.2 (Manitius, 1974b). *If Assumptions 9.3.3–9.3.6 are fulfilled and moreover if there exist $t_2 \in [t_0, t_1]$ and u, $u(t) \in U$, $t \in [t_0, t_2]$, such that $(x(t_2), t_2) \in V$, then there is a triple (x^*, u^*, t^*) minimizing (9.3.10) under the constraint (9.3.11).*

The proof follows from the compactness of the set of all trajectories of (9.3.8) in the space $C([t_0, t_1], R^n)$, which is guaranteed by Assumptions 9.3.3–9.3.4; see Jacobs (1968) and Manitius (1974b). Now we shall cite some of the necessary conditions of optimality for the problem under consideration. Let us denote

$$T(t) = t - h(x(t), u(t), t) \quad t \in [t_0, t_1] \tag{9.3.14}$$

$$E_t = \{s: T(s) \geq t\} \cap [t_0, t_1]. \tag{9.3.15}$$

By $K_E(t, s)$ we denote the characteristic function of the set E_t and by $T^{-1}(t)$ the inverse image of t under transformation $T: [t_0, t_1] \to R^1$.

Theorem 9.3.3 (Manitius, 1974b) *Let $h(x, u, t) = h(u, t)$, and let the FDE system (9.2.8) fulfil Assumptions 9.3.3–9.3.6. Let the optimal triple (x^*, u^*, t^*) be such that the Fréchet derivative of function F*

$$F: \text{LIP}[t^* - (t_0 - h_0), t^*] \ni x \mapsto f(x(t^*), x(t^* - h(u^*(t^*), t^*), t^*) \tag{9.3.16}$$

exists at point x^*: $[t^* - (t_0 - h_0), t^*] \to R^n$. Here $\text{LIP}[t^* - (t_0 - h_0), t^*]$ is the Banach space of functions satisfying a Lipschitz condition (Dunford and Schwartz, 1967).

There exists a nontrivial $(n+1)$-vector function $\hat{\psi}(t) = (\psi_0(t), \psi(t))$ of bounded variation on $[t_0, t^*]$, continuous at t^*, which satisfies

$$\psi_0(t) = \text{const} \leq 0 \quad \hat{\psi}(t^*) \neq 0$$

$$\psi(t) = \int_t^{t^*} \psi(s)[f_x^*(s) + K_E(t, s)f_y^*(s)]ds + \tag{9.3.17}$$

$$+ \psi_0 \int_t^{t^*} f_{0x}^*(s)ds + \psi(t^*) \quad t \leq t^* \quad \psi(t) = 0 \quad t \geq t^*$$

and

$$\psi_0 f_0(x^*(t), u^*(t), t) + \psi(t)f(x^*(t), x^*(t - h(u^*(t), t)), u^*(t), t)$$
$$= \min_{u \in U}[\psi_0 f_0(x^*(t), u, t) + \psi(t)f(x^*(t), x^*(t - h(u, t)), u, t)]$$

for almost all $t \in [t_0, t^*]$

and the $(n+1)$-vector $(-\psi_0 f_0^*(t^*) - f^*(t^*), \psi(t^*))$ is orthogonal to V at $(t^*, x^*(t^*))$. Moreover, ψ is continuous at points t satisfying the condition

$m(T^{*-1}(t)) = 0$, where m is the Lebesgue measure. The jumps of ψ occur at such that $m(T^{*-1}(t)) > 0$, and are equal to

$$\psi(t) - \psi(t^+) = \int_{T^{*-1}(t)} \psi(s) f_y^*(s) \, ds.$$

In the terms above we use for brevity the notation $f^*(s), f_x^*(s), f_y^*(s), f_0^*(s), f_{0x}^*(s)$ for values of the functions f, f_0 and their partial derivatives taken along the optimal trajectory x^* and optimal control u^*.

10

Linear-quadratic Problem of Optimal Control

10.1 Problem Formulation

The linear-quadratic problem is a central one in the theory and applications of optimal control. Its role in the synthesis and analysis of linear control systems is well established; for general background and results we refer the reader to the vast literature on the subject, a good representative of which is the book by Kwakernaak and Sivan (1972). Nearly all the basic results that can be obtained for control systems described by ordinary differential equations have their counterparts in the theory of control systems described by functional-differential equations. However, these results are equally simple only in abstract formulation. In a more down-to-earth formulation they become rather complicated. For example, a central result of the finite-dimensional linear-quadratic theory is that under suitable assumptions the optimal feedback regulator is linear with respect to the state, and constant with respect to time. This result holds also for time delay systems. It must be remembered, however, that in this case the state involves a function of time and the constant linear regulator is no longer of proportional action; it involves integration.

We shall consider a control system described by the equation

$$\sum_{i=0}^{N} [A_i(t)\dot{x}(t-h_i) + B_i(t)x(t-h_i) + C_i(t)u(t-h_i)] +$$

$$+ \int_{-h}^{0} B(t, \tau)x(t+\tau)\,d\tau = f(t) \qquad t \geqslant 0 \qquad (10.1.1)$$

with the initial condition

$$x(t) = \varphi(t) \quad t \in [-h, 0]$$
$$\dot{x}(t) = \mu(t) \quad t \in (-h, 0) \qquad (10.1.2)$$
$$u(t) = v(t) \quad t \in [-h, 0).$$

The assumptions on coefficient matrices $A_i(t)$, $B_i(t)$, $C_i(t)$, $B(t, \tau)$, function f, control u, and initial functions are the same as in Section 3.2.1. We have discussed this equation in Section 3.2; the appropriate state space theory is given in Section 3.3.

In the construction of quadratic performance indexes for time-delay systems there is one specific feature that may make a difference with the finite-dimensional case. Let us consider the following example. A typical quadratic performance index has the form

$$J(u) = \int_0^T [y(t)^2 + u(t)^2] dt \qquad (10.1.3)$$

where u is a scalar control, and y is the output of a linear system. Assume that the state equation is delay-free,

$$\dot{x}(t) = Ax(t) + Bu(t)$$
$$x(t) \in R^n \quad u(t) \in R^1 \quad x(0) = x^0 \qquad (10.1.4)$$

but there are delays in the output equation,

$$y(t) = \sum_{i=0}^{N} [D_i x(t-h_i) + F_i u(t-h_i)] \quad y(t) \in R^1 \qquad (10.1.5)$$
$$0 = h_0 < h_1 < \ldots < h_N.$$

Obviously enough, the delays in the output are very likely to occur in real systems. They may be caused by the time necessary to take the measurements, or to process the measurement data, or even by the fact that the information signals have finite velocity. If we substitute (10.1.5) into (10.1.3), we obtain

$$J(u) = \sum_{i,j=0}^{N} \int_0^T [x(t-h_i)^T D_i^T D_j x(t-h_j) + 2u(t-h_i) F_i D_j x(t-h_j) +$$
$$+ u(t-h_i) F_i F_j u(t-h_j)] dt + \int_0^T u(t)^2 dt.$$

We can see that if we want to account for the delays in the output, we have to introduce non-trivial argument shifts under the integral sign.

Because of the differences in methods of analysis, problem formulation and the form of results, we strongly distinguish the linear-quadratic problem

0.1 Problem Formulation

with finite control time from that with infinite control time. Let us first consider the linear-quadratic problem of optimal control with finite time. The system equations are (10.1.1) and (10.1.2). The quadratic performance index has the form

$$J(u) = \sum_{i,j=0}^{N} \left[J_{ij}^1(u) + \int_0^T J_{ij}^2(t,u)\,dt \right] + 2 \sum_{i=0}^{N} \left[J_i^3(u) + \int_0^T J_i^4(t,u)\,dt \right] \quad (10.1.6)$$

where

$$J_{ij}^1(u) = x(T-h_i)^T Q_{ij} x(T-h_j)$$

$$J_{ij}^2(t,u) = x(t-h_i)^T W_{ij}(t) x(t-h_j) + u(t-h_i)^T Z_{ij}(t) x(t-h_j) + \\ + u(t-h_i)^T R_{ij}(t) u(t-h_j)$$

$$J_i^3(u) = q_i^T x(T-h_i)$$

$$J_i^4(t,u) = w_i(t)^T x(t-h_i) + r_i(t)^T u(t-h_i).$$

T is the (finite) control time; x is the solution of (10.1.1), (10.1.2) corresponding to control u. We impose assumptions on the coefficients in (10.1.6) such that J is well-defined on $L^2(0, T; R^m)$, strictly convex, and there are reals $a > 0$ and $b \geqslant 0$ such that

$$J(u) \geqslant a\|u\|^2 - b \quad \forall u. \quad (10.1.7)$$

The optimal control problem consists in the determination of a control $u_0 \in L^2(0, T; R^m)$ such that

$$J(u_0) \leqslant J(u) \quad \forall u \in L^2(0, T; R^m). \quad (10.1.8)$$

The control u_0 is called optimal. There are two variants for the formulation of this problem. Either the optimal control u_0 is searched for as an explicit function of time, for fixed initial conditions, or the optimal regulator is synthesized. The first case is called the open-loop optimal control problem, and the second the closed-loop optimal control problem. In this context, the regulator is a mapping $(S_0(t), t) \mapsto u_0(t)$ where $S_0(t)$ is the state of the optimally controlled system at moment t.

In applications we frequently encounter the situation when the termination moment of the control process is so far away that it does not affect the current control actions. The infinite-time optimal control problem is then posed. In the infinite-time problem, synthesis of the optimal regulator is of special importance. We shall study this problem in a simplified formulation, assuming that all the coefficients in the system equations and in the functional are constant, removing linear terms and terms outside the integral in the functional, and assuming $f = 0$ in the system equation. Thus, the control system is described by

$$\sum_{i=0}^{N} [A_i \dot{x}(t-h_i) + B_i x(t-h_i) + C_i u(t-h_i)] +$$

$$+ \int_{-h}^{0} B(\tau) x(t+\tau) d\tau = 0 \quad t \geq 0 \quad (10.1.9)$$

with the initial conditions (10.1.2). The performance functional to be minimized subject to the equations of the system has the form

$$J(u) = \sum_{i,j=0}^{N} \int_{0}^{\infty} [x(t-h_i)^T W_{ij} x(t-h_j) +$$

$$+ u(t-h_i)^T Z_{ij} x(t-h_j) + u(t-h_i)^T R_{ij} u(t-h_j)] dt. \quad (10.1.10)$$

The coefficient matrices are such that J is strictly convex and satisfies (10.1.7) for some a, b, and for every u from its domain in $L^2_{loc}(0, \infty; R^m)$.

In the linear-quadratic problem with infinite time it is important to determine the set in the control space $L^2_{loc}(0, \infty; R^m)$ on which functional (10.1.10) attains finite values. It is obvious that this set is a subset of $L^2(0, \infty; R^m)$. A simple necessary and sufficient condition that the domain of J in (10.1.10) be equal to $L^2(0, \infty; R^m)$ is the exponential stability of system (10.1.9) which in the case where there are no delays in the derivatives ($A_i = 0$, $i > 0$) is equivalent to L^2-stability or asymptotic stability (see Chapter 6). A sufficient condition that the domain be non-empty for every initial condition (10.1.2) is the stabilizability of system (10.1.1). Of course the domain depends on the initial condition. We call system (10.1.1) stabilizable if there is a nonnegative integer M, matrices G_i, $i = 0, \ldots, M$, reals $0 = \tau_0 < \tau_1 < \ldots < \tau_M$ and a matrix-valued, bounded and measurable function $G: [-h, 0] \to L(R^n, R^m)$ such that if we substitute into (10.1.1)

$$u(t) = \sum_{i=0}^{M} G_i x(t-\tau_i) + \int_{-h}^{0} G(\tau) x(t+\tau) d\tau \quad (10.1.11)$$

we obtain an exponentially stable closed-loop system.

10.2 Equivalent Linear-quadratic Problems

By a simple and easily invertible transformation of variables, many linear-quadratic problems may be reduced to a much simpler, canonical form. Therefore it is sufficient to carry out a detailed analysis only for the canonical form since the results may be readily transferred to a whole class of problems. At the same time, such an analysis is simpler than in the case of the explicit most general formulation, and yields the results in a shorter and clearer form.

10.2 Equivalent Linear-quadratic Problems

Let us consider a class of finite-time linear-quadratic problems where the controlled system is described by equation (10.1.1) with $B = 0$, and the performance functional has the form

$$J(u) = x(T)^T[Qx(T)+2q]+$$
$$+ \int_0^T \left[\begin{array}{c}x(t)\\u(t)\end{array}\right]^T \left\{\left[\begin{array}{cc}W(t) & Z(t)\\Z(t)^T & R(t)\end{array}\right]\left[\begin{array}{c}x(t)\\u(t)\end{array}\right]+2\left[\begin{array}{c}w(t)\\r(t)\end{array}\right]\right\} dt. \quad (10.2.1)$$

We assume $Q = Q^T \geq 0$, and for every $t \in [0, T]$: $W(t) = W(t)^T$, $R(t) = R(t)^T > 0$, $R(t)$ has a uniformly bounded inverse, and

$$\left[\begin{array}{cc}W(t) & Z(t)\\Z(t)^T & R(t)\end{array}\right] \geq 0.$$

All functions W, Z, R, w and r are integrable and bounded. Due to these assumptions the functional J is well-defined on the whole space $L^2(0, T; R^m)$, is strictly convex, and there exist reals a, b, $a > 0$ such that $J(u) \geq a\|u\|^2 - b$, $u \in L^2(0, T; R^m)$.

We shall examine to what extent this formulation may be simplified by the following transformation of variables:

$$\begin{aligned}x(t) &= \bar{x}(t)+\beta_1(t)\\u(t) &= \alpha_1(t)\bar{x}(t)+\alpha_2(t)\bar{u}(t)+\beta_2(t).\end{aligned} \quad (10.2.2)$$

We omit the calculations which are lengthy but straightforward. It will be helpful for the reader to verify that maximal simplification is obtained if the matrices $\alpha_1(t)$, $\alpha_2(t)$ and vectors $\beta_1(t)$, $\beta_2(t)$ are chosen as below. Let

$$\begin{aligned}\alpha_1(t) &= -R(t)^{-1}Z(t)^T & t \in [0, T]\\ \alpha_1(t) &= 0 & t \in [-h, 0).\end{aligned} \quad (10.2.3)$$

Due to this assumption all 'mixed' terms of the form $Z_{ij}(t)x_i(t)u_j(t)$ are removed from the expression under the integral sign. Further, let α_2 be a locally integrable matrix function such that

$$\alpha_2(t)^T R(t)\alpha_2(t) = I \quad t \in [0, T] \quad (10.2.4)$$
$$\alpha_2(t) = I \quad t \in [-h, 0). \quad (10.2.5)$$

For example, we may require that $\alpha_2(t)$ be the unique positively definite symmetric matrix satisfying (10.2.4). Such a matrix is denoted by $R(t)^{-1/2}$. Then

$$\alpha_2(t) = R(t)^{-1/2} \quad t \in [0, T]. \quad (10.2.6)$$

By virtue of (10.2.4) the matrix of the quadratic form with respect to control is transformed into a unit matrix. To remove the term that is linear in control we set

$$\beta_2(t) = -R(t)^{-1}[Z(t)^T\beta_1(t)+r(t)] \quad t \in [0, T]$$
$$\beta_2(t) = 0 \quad t \in [-h, 0). \quad (10.2.7)$$

The function β_1 is chosen so that the nonhomogeneity f may be removed from the system equation. We require that β_1 satisfy the equation

$$\sum_{i=0}^{N}[A_i(t)\dot\beta_1(t-h_i)+B_i(t)\beta_1(t-h_i)-C_i(t)\beta_2(t-h_i)] = f(t) \quad t \in [0, T]$$

$$\beta_1(t) = 0 \quad t \in [-h, 0] \quad \dot\beta_1(t) = 0 \quad t \in (-h, 0). \quad (10.2.8)$$

Let us introduce the following notation:

$$\bar B_i(t) = B_i(t)-C_i(t)R(t-h_i)^{-1}Z(t-h_i)^T \quad t-h_i \geq 0$$
$$\bar B_i(t) = B_i(t) \quad t-h_i < 0$$
$$\bar C_i(t) = C_i(t)R(t-h_i)^{-1/2} \quad t-h_i \geq 0$$
$$\bar C_i(t) = C_i(t) \quad t-h_i < 0$$
$$\bar q = Q\beta_1(T)+q$$
$$\bar W(t) = W(t)-Z(t)R(t)^{-1}Z(t)^T$$
$$\bar w(t) = w(t)+\bar W(t)\beta_1(t)-Z(t)R(t)^{-1}r(t)$$
$$\bar J(\bar u) = J(u)-\beta_1(T)^T[Q\beta_1(T)+2q]-$$
$$-\int_0^T \left[\begin{matrix}\beta_1(t)\\ \beta_2(t)\end{matrix}\right]^T\left\{\left[\begin{matrix}W(t) & Z(t)\\ Z(t)^T & R(t)\end{matrix}\right]\left[\begin{matrix}\beta_1(t)\\ \beta_2(t)\end{matrix}\right]+2\left[\begin{matrix}w(t)\\ r(t)\end{matrix}\right]\right\}dt. \quad (10.2.9)$$

The linear-quadratic problem defined by equation (10.1.1) with $B = 0$ and performance functional (10.2.1) then takes the form

$$\sum_{i=0}^{N}[A_i(t)\dot{\bar x}(t-h_i)+\bar B_i(t)\bar x(t-h_i)+\bar C_i(t)\bar u(t-h_i)] = 0 \quad t \in [0, T] \quad (10.2.10)$$

$$\bar x(t) = \varphi(t) \quad t \in [-h, 0]$$
$$\dot{\bar x}(t) = \mu(t) \quad t \in (-h, 0)$$
$$\bar u(t) = v(t) \quad t \in [-h, 0)$$

$$\bar J(\bar u) = \bar x(T)^T[Q\bar x(T)+2\bar q]+\int_0^T [\bar x(t)^T\bar W(t)\bar x(t)+2\bar w(t)^T\bar x(t)+\bar u(t)^T\bar u(t)]dt.$$

$$(10.2.11)$$

10.3 Step Method

Evidently, $\overline{W}$ and $\overline{w}$ are bounded and integrable,
$$\overline{W}(t) = \overline{W}(t)^{\mathrm{T}} \geqslant 0 \quad t \in [0, T].$$
Finally, the transformation (10.2.2) for $t \in [0, T]$ is
$$x(t) = \bar{x}(t) + \beta_1(t)$$
$$u(t) = -R(t)^{-1} Z(t)^{\mathrm{T}} \bar{x}(t) + R(t)^{-1/2} \bar{u}(t) - R(t)^{-1} [Z(t)^{\mathrm{T}} \beta_1(t) + r(t)] \quad (10.2.12)$$
with β_1 determined by (10.2.8). For $t < 0$ we have $\bar{x}(t) = x(t)$, $\dot{\bar{x}}(t) = \dot{x}(t)$, and $\bar{u}(t) = u(t)$. Transformation (10.2.12) is easily invertible:
$$\bar{x}(t) = x(t) - \beta_1(t)$$
$$\bar{u}(t) = R(t)^{-1/2} Z(t)^{\mathrm{T}} x(t) + R(t)^{1/2} u(t) + R(t)^{-1/2} r(t) \quad (10.2.13)$$
for $t \in [0, T]$. Thus, in conclusion we may consider (10.2.10), (10.2.11) as a canonical form for a wide class of linear-quadratic problems of the form (10.1.1), (10.2.1).

In the performance functional (10.1.6), terms with deviated argument appear under the integral. In some cases these deviations can be easily removed. For example let the functional be
$$J(u) = \int_0^2 [x(t-1)^2 + u(t)^2] \mathrm{d}t. \quad (10.2.14)$$
Then of course
$$J(u) = \int_{-1}^0 \varphi(t)^2 \mathrm{d}t + \int_0^2 [W(t) x(t)^2 + u(t)^2] \mathrm{d}t$$
where $\varphi(t)$, $t \in [-1, 0]$ is a given initial function, and
$$W(t) = \begin{cases} 1 & t \in [0, 1] \\ 0 & t \in (1, 2]. \end{cases}$$
Thus, if there are deviations of argument under the integral as in (10.2.14), the problem still may be equivalent to (10.1.1), (10.2.1).

10.3 Step Method

10.3.1 Problem formulation

There are several approaches to the linear-quadratic problem with delays, all of which differ in method of solution, range, generality, form of results and applicability to numerical computations. Below we present an elementary method, (Korytowski, 1973, 1976), called the step method by analogy with the well-known method of solution of differential equations with delays (see

Section 3.2.2, the second variant). The step method approach to more general problems of optimal control with commensurable delays was suggested by Olbrot (1973a). For the finite-time linear-quadratic problem, the step method leads to formulas that permit the calculation of the optimal solution by means of a finite, non-iterative computational method. On the basis of the finite-time results we shall next formulate an iterative algorithm for the determination of the optimal regulator in the case of infinite control time.

Let us consider the system

$$\sum_{i=0}^{N} [A_i(t)\dot{x}(t-i\varrho) + B_i(t)x(t-i\varrho) + C_i(t)u(t-i\varrho)] = f(t) \quad t \in [0, T] \quad (10.3.1)$$

with initial conditions (10.1.2). The assumptions are as for (10.1.1), ϱ is a positive real. On the trajectories of (10.3.1) the performance functional (10.1.6) is minimized, with $h_i = i\varrho$, $i = 0, \ldots, N$.

The chief point to notice in the step method is that problem (10.3.1), (10.1.6) is replaced by an equivalent linear-quadratic problem with increased dimension and split boundary conditions, but without deviations of argument, considered in a time interval of length ϱ.

We define an integer σ such that $T + \varrho > \sigma\varrho \geq T$, and a real $\theta = \sigma\varrho - T$. Note that if $\sigma = 1$, our problem is reduced to a trivial problem without delays. Therefore we shall always assume $\sigma > 1$, pointing out the changes that must be made in the case $\sigma = 1$. Introducing notation (3.2.36)–(3.2.42) we proceed according to the second variant of the step method of Section 3.2.2, and obtain the system equations in the form (3.2.39). Using notation (3.2.42) and

$$I^\sigma_{\sigma-1}(\theta) = [0, \theta) \quad I^\sigma_\sigma(\theta) = [\theta, \varrho] \quad (10.3.2)$$

we may write the equations of the control system in the following way:

$$\tilde{x}^i(t) = \hat{B}_i(t)\tilde{x}^i(t) + \gamma_i(t) \quad t \in I^\sigma_i(\theta) \quad (i = \sigma-1, \sigma) \quad (10.3.3)$$

$$\tilde{x}^{\sigma-1}(\theta-) = e_\sigma \tilde{x}^\sigma(\theta+)$$

$$\tilde{x}^{\sigma-1}(0+) = d_\sigma \tilde{x}^\sigma(\varrho-) \quad (10.3.4)$$

$$x^\sigma(\theta) = x(0).$$

Now we shall rewrite the performance functional (10.1.6) in step notation. By straightforward, but rather lengthy and tedious calculations it can be shown that

$$J(u) = \tilde{J}(\tilde{u}) + c$$

where c is a constant independent of control u, $\tilde{u}$ denotes the pair $(\tilde{u}^\sigma, \tilde{u}^{\sigma-1})$ if $\sigma > 1$, and $\tilde{u}^\sigma$ if $\sigma = 1$; $\tilde{u}^{\sigma-1}$, $\tilde{u}^\sigma$ are connected with u by (3.2.36). Furthermore,

10.3 Step Method

$$\tilde{J}(\tilde{u}) = \tilde{x}^\sigma(\varrho)^T[\hat{Q}_\sigma \tilde{x}^\sigma(\varrho) + 2\hat{q}] +$$

$$+ \sum_{i=\sigma-1}^{\sigma} \int_{I_i^\sigma(\theta)} \left[\begin{pmatrix} \tilde{x}^i(t) \\ \tilde{u}^i(t) \end{pmatrix}^T \begin{bmatrix} \begin{pmatrix} \hat{W}_i(t) & \hat{Z}_i(t) \\ \hat{Z}_i(t)^T & \hat{R}_i(t) \end{pmatrix} \begin{pmatrix} \tilde{x}^i(t) \\ \tilde{u}^i(t) \end{pmatrix} + 2 \begin{pmatrix} \hat{w}_i(t) \\ \hat{r}_i(t) \end{pmatrix} \end{bmatrix} \right] dt. \quad (10.3.5)$$

Here $\tilde{x}^{\sigma-1}$, $\tilde{x}^\sigma$ are the solutions of (10.3.3), (10.3.4) corresponding to the control $\tilde{u}$. We shall not here give explicit formulas for the coefficient matrices and vectors in (10.3.5), that is, we shall not express them by the coefficients in (10.1.6) in the most general case. Such expressions will be given later for some special cases.

At the moment it is important that the coefficients exist and the performance functional can be written in the form (10.3.5). To simplify further analysis let us impose the usual assumptions

$$\hat{Q}_\sigma = \hat{Q}_\sigma^T \geqslant 0,$$

for $i = \sigma-1, \sigma$ and every t from the respective interval

$$\hat{R}_i(t) = \hat{R}_i(t)^T > 0$$

and $\hat{R}_i(t)$ has a uniformly bounded inverse,

$$\hat{W}_i(t) = \hat{W}_i(t)^T,$$

and

$$\begin{pmatrix} \hat{W}_i(t) & \hat{Z}_i(t) \\ \hat{Z}_i(t)^T & \hat{R}_i(t) \end{pmatrix} \geqslant 0.$$

Moreover, we assume that all the coefficients are integrable and bounded functions of time. Of course, with these assumptions the functional satisfies the general requirements of Section 10.1.1.

By virtue of Section 10.1.2 it is evident that without loss of generality we may assume that the performance functional (10.3.5) has the form

$$\tilde{J}(\tilde{u}) = \tilde{x}^\sigma(\varrho)^T[\hat{Q}_\sigma \tilde{x}^\sigma(\varrho) + 2\hat{q}] +$$

$$+ \sum_{i=\sigma-1}^{\sigma} \int_{I_i^\sigma(\theta)} [\tilde{x}^i(t)^T \hat{W}_i(t) \tilde{x}^i(t) + 2\hat{w}_i(t)^T \tilde{x}^i(t) + \tilde{u}^i(t)^T \tilde{u}^i(t)] dt. \quad (10.3.6)$$

Similarly, we may assume that the function f in the system equation (10.3.1) is identically equal to zero. Consider, namely, the transformation of variables

$$\tilde{x}^i(t) = \bar{x}^i(t) + \tilde{\beta}_1^i(t)$$
$$\tilde{u}^i(t) = \alpha_1^i(t)\bar{x}^i(t) + \alpha_2^i(t)\bar{u}^i(t) + \tilde{\beta}_2^i(t) \qquad i = \sigma-1, \sigma \quad t \in I_i^\sigma(\theta) \quad (10.3.7)$$

with
$$\alpha_1^i(t) = -\hat{R}_i(t)^{-1}\hat{Z}_i(t)^T$$
$$\alpha_2^i(t) = \hat{R}_i(t)^{-1/2} \qquad (10.3.8)$$
$$\tilde{\beta}_2^i(t) = -\hat{R}_i(t)^{-1}[\hat{Z}_i(t)^T\tilde{\beta}_1^i(t)+\hat{r}_i(t)]$$

and $\tilde{\beta}_1^i$ determined by

$$\dot{\tilde{\beta}}_1^i(t) = \hat{B}_i(t)\tilde{\beta}_1^i(t)+\hat{f}^i(t) \quad t \in I_i^\sigma(\theta) \quad (i = \sigma-1, \sigma)$$
$$\tilde{\beta}_1^{\sigma-1}(\theta-) = e_\sigma\tilde{\beta}_1^\sigma(\theta+)$$
$$\tilde{\beta}_1^{\sigma-1}(0+) = d_\sigma\tilde{\beta}_1^\sigma(\varrho-) \qquad (10.3.9)$$
$$\beta_1^\sigma(\theta) = 0.$$

A substitution of (10.3.7) into (10.3.5) after the following change of notation

$$\bar{B}_i(t) = \hat{B}_i(t)-\hat{C}_i(t)\hat{R}_i(t)^{-1}\hat{Z}_i(t)^T \quad t \in I_i^\sigma(\theta) \quad (i = \sigma-1, \sigma)$$
$$\bar{C}_i(t) = \hat{C}_i(t)\hat{R}_i(t)^{-1/2}$$
$$\bar{q}_\sigma = \hat{Q}_\sigma\tilde{\beta}_1^\sigma(\varrho)+\hat{q}_\sigma$$
$$\bar{W}_i(t) = \hat{W}_i(t)-\hat{Z}_i(t)\hat{R}_i(t)^{-1}\hat{Z}_i(t)^T$$
$$\bar{w}_i(t) = \hat{w}_i(t)+\hat{W}_i(t)\tilde{\beta}_1^i(t)-\hat{Z}_i(t)\hat{R}_i(t)^{-1}\hat{r}_i(t)$$
$$\bar{J}(\bar{u}) = \hat{J}(\tilde{u})-\tilde{\beta}_1^\sigma(\varrho)^T[\hat{Q}_\sigma\tilde{\beta}_1^\sigma(\varrho)+2\hat{q}_\sigma]-$$
$$-\sum_{i=\sigma-1}^{\sigma}\int_{I_i^\sigma(\theta)}\left[\begin{bmatrix}\tilde{\beta}_1^i(t)\\ \tilde{\beta}_2^i(t)\end{bmatrix}^T\begin{bmatrix}\hat{W}_i(t) & \hat{Z}_i(t)\\ \hat{Z}_i(t)^T & \hat{R}_i(t)\end{bmatrix}\begin{bmatrix}\tilde{\beta}_1^i(t)\\ \tilde{\beta}_2^i(t)\end{bmatrix}+2\begin{bmatrix}\hat{w}_i(t)\\ \hat{r}_i(t)\end{bmatrix}\right]dt \quad (10.3.10)$$

yields a linear-quadratic problem with a homogeneous system equation ($f = 0$) and with a performance functional in which there are no mixed terms or terms linear in control under the integral, and quadratic forms with respect to control occur with unit matrices.

In this way we have succeeded without a loss of generality in transforming the original linear quadratic problem (10.3.1), (10.3.6) into the following one. On the trajectories of the system

$$\dot{\tilde{x}}^{\sigma-1}(t) = \hat{B}_{\sigma-1}(t)\tilde{x}^{\sigma-1}(t)+\hat{C}_{\sigma-1}(t)\tilde{u}^{\sigma-1}(t)+\delta_{\sigma-1}(t)\eta(t+\varrho) \quad t \in [0, \theta]$$
$$\dot{\tilde{x}}^\sigma(t) = \hat{B}_\sigma(t)\tilde{x}^\sigma(t)+\hat{C}_\sigma(t)\tilde{u}^\sigma(t)+\delta_\sigma(t)\eta(t) \quad t \in [\theta, \varrho]$$
$$\tilde{x}^{\sigma-1}(\theta-) = e_\sigma\tilde{x}^\sigma(\theta+) \qquad (10.3.11)$$
$$\tilde{x}^{\sigma-1}(0+) = d_\sigma\tilde{x}^\sigma(\varrho-)$$
$$x^\sigma(\theta) = x(0)$$

10.3 Step Method

minimize the performance functional

$$\tilde{J}(\tilde{u}) = \tilde{x}^\sigma(\varrho)^T[\hat{Q}_\sigma \tilde{x}^\sigma(\varrho) + 2\hat{q}_\sigma] + \sum_{i=\sigma-1}^{\sigma} \int_{I_i^\sigma(\theta)} [\tilde{x}^i(t)^T \hat{W}_i(t) \tilde{x}^i(t) +$$
$$+ 2\hat{w}_i(t)^T \tilde{x}^i(t) + \tilde{u}^i(t)^T \tilde{u}^i(t)] dt. \qquad (10.3.12)$$

Due to the previous assumptions function η is square integrable, and functions $\hat{B}_i$, $\hat{C}_i$, δ_i, $\hat{W}_i$, $\hat{w}_i$ are bounded and integrable. Moreover, $\hat{Q}_\sigma = \hat{Q}_\sigma^T \geqslant 0$ and $\hat{W}_i(t) = \hat{W}_i(t)^T \geqslant 0$, $t \in I_i^\sigma(\theta)$, $i = \sigma-1, \sigma$. If $\sigma = 1$, the problem takes a much simpler form. Only the second equation and the condition $x^\sigma(\theta) = x(0)$ in (10.3.11) remain valid, and in (10.3.12) the first integral term should be dropped.

10.3.2 Variation of the performance functional and the adjoint equation

The classic variational approach will be applied to the problem of optimal control (10.3.11), (10.3.12) and this way the adjoint equation will be derived. Let x be a solution of equation (10.3.1) (with $f = 0$) corresponding to an admissible control u. We consider control variations δu, $\delta u(t) = 0$, $t < s$ $\in [0, T]$. Define a positive integer l such that $T-s+\varrho > l\varrho \geqslant T-s$ and a real ϑ, $\vartheta = l\varrho - T + s$. Let also

$$I_{l-1}^l(\vartheta) = [0, \vartheta] \qquad I_l^l(\vartheta) = [\vartheta, \varrho].$$

To the functions x, u, δu and the corresponding variation of the trajectory δx we assign by means of (3.2.36) functions $\tilde{x}^i$, $\tilde{u}^i$, $i = \sigma-1, \sigma$, and $\delta \tilde{u}^i$, $\delta \tilde{x}^i$, $i = l-1, l$. We shall carry on the derivations for $l > 1$, explaining where necessary the differences in the case $l = 1$.

It is easy to show that the variation of trajectory satisfies the set of equations

$$\delta \dot{\tilde{x}}^i(t) = \hat{B}_i(t) \delta \tilde{x}^i(t) + \hat{C}_i(t) \delta \tilde{u}^i(t) \qquad t \in I_i^l(\vartheta) \quad (i = l-1, l) \qquad (10.3.13)$$

with boundary conditions

$$\delta x^l(\vartheta) = 0$$
$$\delta x^{i+1}(\varrho) = \delta x^{i+1}(\varrho-) = \delta x^i(0+) = \delta x^i(0) \qquad (i = 1, \ldots, l-1) \qquad (10.3.14)$$
$$\delta x^i(\vartheta-) = \delta x^i(\vartheta+) = \delta x^i(\vartheta).$$

To this boundary problem we apply the variation-of-constants formulas of Section 3.2.2 and obtain

$$\delta \tilde{x}^{l-1}(t) = \Pi_{l-1}(t, \vartheta) \xi + \int_\vartheta^t \Pi_{l-1}(t, \tau) \gamma_{l-1}(\tau) d\tau \qquad t \in I_{l-1}^l(\vartheta)$$
$$\delta \tilde{x}^l(t) = \Pi_l(t, \vartheta) e_l^T \xi + \int_\vartheta^t \Pi_l(t, \tau) \gamma_l(\tau) d\tau \qquad t \in I_l^l(\vartheta) \qquad (10.3.15)$$

as the solution, where

$$\gamma_l(t) = \hat{C}_l(t)\,\delta\tilde{u}^l(t)$$
$$\xi = [\Pi_{l-1}(0,\vartheta) - d_l\Pi_l(\varrho,\vartheta)e_l^T]^{-1} \times$$
$$\times \left[d_l\int_\vartheta^\varrho \Pi_l(\varrho,\tau)\gamma_l(\tau)\,\mathrm{d}\tau + \int_0^\vartheta \Pi_{l-1}(0,\tau)\gamma_{l-1}(\tau)\,\mathrm{d}\tau \right].$$

If $l = 1$, we only retain the second equation of system (10.3.13) with the initial condition $\delta x^1(\vartheta) = 0$, and with obvious consequences in (10.3.15).

Now we shall calculate the variation $\delta\tilde{J}(\tilde{u},\delta\tilde{u})$ of the performance functional (10.3.12)

$$\tfrac{1}{2}\delta\tilde{J}(\tilde{u},\delta\tilde{u}) = \delta\tilde{x}^\sigma(\varrho)^T[\hat{Q}_\sigma\tilde{x}^\sigma(\varrho) + \hat{q}_\sigma] +$$
$$+ \sum_{i=\sigma-1}^{\sigma} \int_{I_i^\sigma(\vartheta)} [\delta\tilde{x}^i(t)^T\hat{W}_i(t)\tilde{x}^i(t) + \hat{w}_i(t)^T\delta\tilde{x}^i(t) + \tilde{u}^i(t)^T\delta\tilde{u}^i(t)]\,\mathrm{d}t.$$
(10.3.16)

Let us denote by $\hat{Q}_{k\sigma}$ the $kn \times \sigma n$ submatrix of $\hat{Q}_\sigma$ consisting of its kn first rows. Similarly, let $\hat{W}_{ki}(t)$ be the $(kn \times in)$ submatrix of $\hat{W}_i(t)$ consisting of its kn first rows. Since so far we have not even implicitly defined $\hat{W}_\sigma(t)$ and $\hat{w}_\sigma(t)$, $t \in [0,\theta)$, we may assume that

$$\hat{W}_\sigma(t) = \begin{bmatrix} \hat{W}_{\sigma-1}(t) & 0 \\ 0 & 0 \end{bmatrix} \quad \hat{w}_\sigma(t) = \begin{bmatrix} \hat{w}_{\sigma-1}(t) \\ 0 \end{bmatrix} \quad t \in [0,\theta). \quad (10.3.17)$$

Let, further, $\hat{q}_k$ be a subvector of $\hat{q}_\sigma$ consisting of its kn first components; $\hat{w}_k(t)$ similarly consists of the first kn components of $\hat{w}_\sigma(t)$. Moreover, denote $M_i = \hat{Q}_{i\sigma}\tilde{x}^\sigma(\varrho) + \hat{q}_i$, $V_i(t) = \hat{W}_{i\sigma}(t)\tilde{x}^\sigma(t) + \hat{w}_i(t)$. Then

$$\tfrac{1}{2}\delta\tilde{J}(\tilde{u},\delta\tilde{u}) = M_l^T\,\delta\tilde{x}^l(\varrho) + \sum_{i=l-1}^{l} \int_{I_i^l(\vartheta)} [V_i(t)^T\delta\tilde{x}^i(t) + \tilde{u}^i(t)^T\delta\tilde{u}^i(t)]\,\mathrm{d}t. \quad (10.3.18)$$

The trajectory variation $\delta\tilde{x}^{l-1}$, $\delta\tilde{x}^l$ is a linear continuous function of the control variation $\delta\tilde{u}^{l-1}$, $\delta\tilde{u}^l$, determined by (10.3.15). In order to obtain a simple condition on optimal control, we shall explicitly express the right-hand side of (10.3.18) by $\delta\tilde{u}^{l-1}$, $\delta\tilde{u}^l$, using (10.3.15). To this end we introduce the adjoint functions $\tilde{p}^i\colon I_i^l(\vartheta) \to R^{in}$, $i = l-1, l$. These are absolutely continuous functions of time, satisfying almost everywhere the following system of differential equations, adjoint with respect to (10.3.13):

$$\dot{\tilde{p}}^i(t) = -\hat{B}_i(t)^T\tilde{p}^i(t) - \hat{W}_{i\sigma}(t)\tilde{x}^\sigma(t) - \hat{w}_i(t) \quad t \in I_i^l(\vartheta) \quad (i = l-1, l) \quad (10.3.19)$$

10.3 Step Method

with boundary conditions
$$p^1(\varrho) = p^1(\varrho-) = M_1$$
$$d_l \tilde{p}^l(\varrho-) = \tilde{p}^{l-1}(0+) + d_l M_l \tag{10.3.20}$$
$$p^i(\vartheta-) = p^i(\vartheta+) = p^i(\vartheta), \vartheta > 0 \quad (i = 1, \ldots, l-1).$$

The functions p^i are connected with $\tilde{p}^i$ by (3.2.36).

Let us define the absolutely continuous matrix-valued functions $\Psi_i: I_i^l(\vartheta) \times I_i^l(\vartheta) \to L(R^{in})$, $i = l-1, l$,
$$\frac{\partial}{\partial t}\Psi_i(t, \tau) = -\hat{B}_i(t)^T \Psi_i(t, \tau), \tag{10.3.21}$$
$$\Psi_i(\tau, \tau) = I.$$

We may then write the solution of the adjoint system (10.3.19), (10.3.20) in the form
$$\tilde{p}^{l-1}(t) = \Psi_{l-1}(t, 0)\eta - \int_0^t \Psi_{l-1}(t, \tau)V_{l-1}(\tau)d\tau \quad t \in I_{l-1}^l(\vartheta)$$
$$\tag{10.3.22}$$
$$\tilde{p}^l(t) = \Psi_l(t, \varrho)d_l^T \eta + \Psi_l(t, \varrho)M_l - \int_0^t \Psi_l(t, \tau)V_l(\tau)d\tau \quad t \in I_l^l(\vartheta)$$

where
$$\eta = [\Psi_{l-1}(\vartheta, 0) - e_l \Psi_l(\vartheta, \varrho)d_l^T]^{-1} \times$$
$$\times \left[e_l \Psi_l(\vartheta, \varrho)M_l + e_l \int_\vartheta^\varrho \Psi_l(\vartheta, \tau)V_l(\tau)d\tau + \int_0^\vartheta \Psi_{l-1}(\vartheta, \tau)V_{l-1}(\tau)d\tau \right].$$

To prove that formulas (10.3.22) are valid, it is enough to apply the variation-of-constants formulas to system (10.3.19), taking the condition for $p^1(\varrho)$ and the equality $\eta = \tilde{p}^{l-1}(0)$ into account. Hence
$$\tilde{p}^{l-1}(\vartheta) = \Psi_{l-1}(\vartheta, 0)\tilde{p}^{l-1}(0) - \int_0^\vartheta \Psi_{l-1}(\vartheta, \tau)V_{l-1}(\tau)d\tau$$

$$\tilde{p}^l(\vartheta) = \Psi_l(\vartheta, \varrho)d_l^T \tilde{p}^{l-1}(0) + \Psi_l(\vartheta, \varrho)M_l - \int_\varrho^\vartheta \Psi_l(\vartheta, \tau)V_l(\tau)d\tau.$$

The multiplication of the second equality by e_l and subtraction from the first one yields η, and later (10.3.22).

In the case $l = 1$, only the equation with $i = 1$, with boundary condition $p^1(\varrho) = \hat{Q}_{1\sigma}\tilde{x}^\sigma(\varrho) + \hat{q}_1$ is retained in the system of adjoint equations (10.3.19). The respective changes that must be made in formulas (10.3.22) are obvious.

Below we shall need the relationship between Π_i and Ψ_i, the fundamental solutions of the system equations and the adjoint equations. Obviously, $\Pi_i(\tau, \tau) = \Psi_i(\tau, \tau)^T = I$, and

$$\Psi_i(t, \tau)\Psi_i(\tau, t) = I.$$

Hence, differentiating the last identity with respect to τ, we get

$$\left[\frac{\partial}{\partial \tau}\Psi_i(t, \tau)\right]\Psi_i(\tau, t) - \Psi_i(t, \tau)\hat{B}_i(\tau)^T\Psi_i(\tau, t) = 0$$

and

$$\frac{\partial}{\partial \tau}\Psi_i(t, \tau)^T = \hat{B}_i(\tau)\Psi_i(t, \tau)^T.$$

We see that $\Psi_i(t, \tau)^T$ and $\Pi_i(\tau, t)$ satisfy the same differential equation with identical boundary condition. Therefore

$$\Pi_i(\tau, t) = \Psi_i(t, \tau)^T \quad t, \tau \in I_i^l(\vartheta) \quad (i = l-1, l). \tag{10.3.23}$$

Now we are in a position to express the variation of the performance functional explicitly by $\delta\tilde{u}$, using the adjoint solution $\tilde{p}^{l-1}, \tilde{p}^l$. That is to say, for any admissible $\delta\tilde{u}^{l-1}, \delta\tilde{u}^l$ we have

$$\delta\tilde{J}(\tilde{u}, \delta\tilde{u}) = 2\sum_{i=l-1}^{l} \int_{J_i^l(\vartheta)} [u^i(t) + \hat{C}_i(t)^T\tilde{p}^i(t)]^T \delta u^i(t)\,dt \tag{10.3.24}$$

where $\tilde{p}^i$, $i = l-1, l$ is the solution of (10.3.19), (10.3.20).

To prove (10.3.24), denote

$$\delta J_1 = M_l^T \delta\tilde{x}^l(\varrho) + \sum_{i=l-1}^{l} \int_{I_i^l(\vartheta)} V_i(t)^T \delta\tilde{x}^i(t)\,dt.$$

We have to show that

$$\delta J_1 = \sum_{i=l-1}^{l} \int_{I_i^l(\vartheta)} \tilde{p}^i(t)^T \gamma_i(t)\,dt.$$

The following identity will prove useful below. For every function F

$$\int_a^b\int_s^b F(s, \sigma)\,d\sigma\,ds = \int_a^b\int_a^s F(\sigma, s)\,d\sigma\,ds.$$

By virtue of (10.3.15) and (10.3.22) we may write

$$R = \sum_{i=l-1}^{l} \left[\int_{I_i^l(\vartheta)} V_i(t)^T \delta\tilde{x}^i(t)\,dt - \int_{I_i^l(\vartheta)} \tilde{p}^i(t)^T \gamma_i(t)\,dt\right] + M_l^T \delta\tilde{x}^l(\varrho).$$

10.3 Step Method

$$= \int_0^\vartheta V_{l-1}(t)^T \Pi_{l-1}(t,\vartheta)\delta\tilde{x}^{l-1}(\vartheta)\,dt -$$

$$- \int_0^\vartheta \gamma_{l-1}(t)^T \Psi_{l-1}(t,0)\tilde{p}^{l-1}(0)\,dt +$$

$$+ \int_\vartheta^\varrho V_l(t)^T \Pi_l(t,\vartheta)e_l^T \delta\tilde{x}^{l-1}(\vartheta)\,dt - \int_\vartheta^\varrho \gamma_l(t)^T \Psi_l(t,\varrho)d_l^T \tilde{p}^{l-1}(0)\,dt +$$

$$+ M_l^T \left[\delta\tilde{x}^l(\varrho) - \int_\vartheta^\varrho \Psi_l(t,\varrho)^T \gamma_l(t)\,dt\right]$$

$$= \left[\int_0^\vartheta V_{l-1}(t)^T \Pi_{l-1}(t,\vartheta)\,dt + \int_\vartheta^\varrho V_l(t)^T \Pi_l(t,\vartheta)e_l^T \,dt + \right.$$

$$\left. + M_l^T \Pi_l(\varrho,\vartheta)e_l^T\right]\delta\tilde{x}^{l-1}(\vartheta) -$$

$$- \left[\int_0^\vartheta \gamma_{l-1}(t)^T \Psi_{l-1}(t,0)\,dt + \int_\vartheta^\varrho \gamma_l(t)^T \Psi_l(t,\varrho)d_l^T \,dt\right]\tilde{p}^{l-1}(0).$$

Further, we substitute ξ from (10.3.15) for $\delta\tilde{x}^{l-1}(\vartheta)$, and η from (10.3.22) or $\tilde{p}^{l-1}(0)$, and use (10.3.23). Then we get $R = 0$ which ends the proof.

In the case $l = 1$, we have to drop the term with $i = l-1$ in (10.3.24).

10.3.3 Optimal control and canonical system of equations

In this section we present the fundamental results regarding optimal control and its representation by means of the adjoint variable.

Theorem 10.3.1 *There is exactly one control $u_0 \in L^2(0, T; R^m)$ that minimizes functional (10.1.6) with $h_i = i\varrho$, $i = 0, 1, \ldots, N$, $\varrho > 0$ subject to equation (10.3.1).*

We shall not give the proof which is standard but requires some subtle notions like weak semicontinuity. We shall only go so far as to mention that the uniqueness of optimal control follows from the strict convexity of the functional (which can be shown by elementary methods), and the existence—from (10.1.7) and weak lower semicontinuity of J.

In a similar way one can show the following result.

Theorem 10.3.2 *There is exactly one square integrable control $\tilde{u}_0^{\sigma-1}$, $\tilde{u}_0^\sigma$ that minimizes functional (10.3.12) subject to equations (10.3.11).*

It is evident that if problem (10.3.11), (10.3.12) is equivalent to problem (10.3.1), (10.1.6) that is, if we may obtain one of them from the other by the change of variables (3.2.36) and possibly by the change of notation of Section 10.3.1, then the same relationship holds between the respective optimal controls. Thus, it is enough to solve the linear-quadratic problem (10.3.11), (10.3.12) in order to obtain the optimal solution of the original problem (10.3.1), (10.1.6).

Theorem 10.3.3 *The unique optimal control of problem* (10.3.11), (10.3.12) *is determined by the equalities*

$$\tilde{u}_0^i(t) = -\hat{C}_i(t)^T \tilde{p}_0^i(t) \quad t \in I_i^l(\vartheta) \quad i = l-1, l \quad \text{for } l > 1$$
$$\quad i = l \quad \text{for } l = 1 \quad (10.3.25)$$

where $l = 1, \ldots, \sigma$ *and* $\vartheta \in [0, \varrho)$ *for* $l < \sigma$, $\vartheta \in [\theta, \varrho)$ *for* $l = \sigma$.

The adjoint functions $\tilde{p}_0^i$ *satisfy the canonical system of equations*

$$\dot{\tilde{x}}_0^i(t) = \hat{B}_i(t)\tilde{x}_0^i(t) - \hat{C}_i(t)\hat{C}_i(t)^T \tilde{p}_0^i(t) + \delta_i(t)\eta_i(t),$$
$$t \in I_i^l(\vartheta) \quad i = l-1, l \quad \text{for } l > 1, \quad i = l \quad \text{for } l = 1 \quad (10.3.26)$$

$$\eta_i = \text{col}(\dot{x}_0^{i+1}, \ldots, \dot{x}_0^{i+N}, x_0^{i+1}, \ldots, x_0^{i+N}, u_0^{i+1}, \ldots, u_0^{i+N})$$

$$\tilde{x}_0^{l-1}(\vartheta -) = e_l \tilde{x}_0^l(\vartheta +) \quad \tilde{x}_0^{l-1}(0+) = d_l \tilde{x}_0^l(\varrho -) \quad (10.3.27)$$

$$x_0^l(\vartheta) = x_0(s)$$

$$\dot{\tilde{p}}_0^i(t) = -\hat{B}_i(t)^T \tilde{p}_0^i(t) - \hat{W}_{l\sigma}(t) x_0^\sigma(t) - \hat{w}_i(t) \quad t \in I_i^l(\vartheta)$$
$$i = l-1, l \quad \text{for } l > 1, \quad i = l \quad \text{for } l = 1 \quad (10.3.28)$$

$$p_0^1(\varrho) = \hat{Q}_{1\sigma}\tilde{x}_0^\sigma(\varrho) + \hat{q}_1$$

$$d_l \tilde{p}_0^l(\varrho -) = \tilde{p}_0^{l-1}(0+) + d_l[\hat{Q}_{l\sigma}\tilde{x}_0^\sigma(\varrho) + \hat{q}_l] \quad (10.3.29)$$

$$\tilde{p}_0^{l-1}(\vartheta -) = e_l \tilde{p}_0^l(\vartheta +).$$

If $l = 1$, *the boundary conditions* (10.3.27) *and* (10.3.29) *are dropped.*

The canonical set of equations has a unique solution $\tilde{x}_0^i, \tilde{p}_0^i, i = l-1, l$ *for* $l > 1$ *and* $i = l$ *for* $l = 1$.

Proof The optimal control satisfies (10.3.25) by virtue of (10.3.24). The canonical equations are easily recognized to be the system equations with the substitution of (10.3.25) and the adjoint equations. The fact that the canonical system has a unique solution follows immediately from the existence and uniqueness of optimal control. □

Note that it is possible to fully characterize the optimal control more simply than in Theorem 10.3.3. Indeed, we can do it by taking fixed values for l and ϑ in (10.3.25)–(10.3.29), $l = \sigma$ and $\vartheta = \theta$. However, the approach presented above is more convenient in a discussion of the optimal regulator

10.3.4 Basic algebraic equation and solution of optimal control problem

We shall apply the well-known variation-of-constants formulas to the canonical system. Let us decompose the matrix $\hat{W}_{i\sigma}(t)$ in the following way:

$$\hat{W}_{i\sigma}(t)\tilde{x}_0^\sigma(t) = v_{i1}(t)\tilde{x}_0^l(t) + v_{i2}(t)\mathrm{col}(x_0^{l+1}(t), \ldots, x_0^\sigma(t))$$

and define

$$z_i(t) = \mathrm{col}(\tilde{x}_0^i(t), \tilde{p}_0^i(t))$$

$$\alpha_i(t) = \begin{bmatrix} \hat{B}_i(t) & -\hat{C}_i(t)\hat{C}_i(t)^T \\ -v_{i1}(t) & -\hat{B}_i(t)^T \end{bmatrix}$$

$$\beta_i(t) = \mathrm{col}(0, -\hat{w}_i(t)).$$

Let also $p = \max(\sigma, l+N)$. We define y_i by

$$y_i(t) = \mathrm{col}(\delta_i(t)\eta_i(t), -v_{i2}(t)\mathrm{col}(x_0^{l+1}(t), \ldots, x_0^\sigma(t))).$$

It is noteworthy that $y_i(t)$ depends linearly on $\mathrm{col}(x_0^{l+1}(t), \ldots, x_0^p(t))$. Let also $\mu_i = \mathrm{col}(0, \hat{Q}_{i\sigma}\tilde{x}_0^\sigma(\varrho) + \hat{q}_i)$.

We rewrite the canonical equations (10.3.26)–(10.3.29) in the new notation

$$\dot{z}_i(t) = \alpha_i(t)z_i(t) + y_i(t) + \beta_i(t) \quad t \in I_i^l(\vartheta)$$

$$i = l-1, l \quad \text{for} \quad l > 1, \quad i = l \quad \text{for} \quad l = 1$$

$$z_{l-1}(\vartheta-) = \mathrm{diag}(e_l, e_l)z_l(\vartheta+) \quad \text{for} \quad l > 1$$

$$x_0^l(\vartheta) = x_0(s)$$

$$z_{l-1}(0+) = \mathrm{diag}(d_l, d_l)[z_l(\varrho-) - \mu_l] \quad \text{for} \quad l > 1$$

$$p_0^l(\varrho) = \hat{Q}_{1\sigma}\tilde{x}_0^\sigma(\varrho) + \hat{q}_1. \tag{10.3.30}$$

The fundamental solutions are defined as the absolutely continuous solutions of the equations

$$\frac{\partial}{\partial t}\Phi_i(t, \tau) = \alpha_i(t)\Phi_i(t, \tau)$$

$$\Phi_i(\tau, \tau) = I. \tag{10.3.31}$$

$t, \tau \in I_i^l(\vartheta) \quad i = l-1, l \quad \text{for} \quad l > 1, \quad i = l \quad \text{for} \quad l = 1$

The variation-of-constants formulas yield

$$z_{l-1}(0) = \Phi_{l-1}(0, \vartheta)z_{l-1}(\vartheta) + \int_\vartheta^0 \Phi_{l-1}(0, \tau)[y_{l-1}(\tau) + \beta_{l-1}(\tau)]d\tau$$

$$z_l(\varrho) = \Phi_l(\varrho, \vartheta)z_l(\vartheta) + \int_\vartheta^\varrho \Phi_l(\varrho, \tau)[y_l(\tau) + \beta_l(\tau)]d\tau. \tag{10.3.32}$$

The first of these equations is valid only in the case $l > 1$.

Equations (10.3.32) together with the boundary conditions of (10.3.30) constitute a linear algebraic set of equations. From the solution of this set we can easily obtain the solution of the optimal control problem. The algebraic set of equations can be slightly simplified without matrix inversion. That is, from the second equation (10.3.32) and the terminal condition on $\tilde{p}_0^l$ we get

$$\begin{bmatrix} \tilde{x}_0^l(\varrho) \\ \begin{bmatrix} 0 \\ \tilde{p}_0^{l-1}(0+) \end{bmatrix} + \hat{Q}_{l\sigma}\hat{x}_0^\sigma(\varrho) + \hat{q}_l \end{bmatrix} = \Phi' \begin{bmatrix} \tilde{x}_0^{l-1}(\vartheta) \\ \tilde{p}_0^l(\vartheta) \end{bmatrix} + \Phi'' x_0^l(\vartheta) +$$

$$+ \int_\vartheta^\varrho \Phi_l(\varrho, \tau)[y_l(\tau) + \beta_l(\tau)] d\tau \qquad (10.3.33)$$

with appropriately defined Φ', Φ''.

Using the first ln scalar equations we eliminate $\tilde{x}_0^l(\varrho)$ from the remaining ln equations in (10.3.33). Further, using the boundary conditions we eliminate $\tilde{x}_0^{l-1}(0)$ from the first of equations (10.3.32), and substitute the expression for $\tilde{p}_0^{l-1}(0+)$, given by the first equation of (10.3.32), into the last $(l-1)n$ scalar equations of (10.3.33). Finally, we obtain the so-called basic algebraic equation

$$\Delta_l(\vartheta) \begin{bmatrix} \tilde{x}_0^{l-1}(\vartheta) \\ \tilde{p}_0^l(\vartheta) \end{bmatrix} = P_l(\vartheta) \qquad (10.3.34)$$

where P_l is an affine function of $\{x_0^l(\vartheta), x_0^{l+1}(\varrho-), ..., x_0^\sigma(\varrho-)\}$, and of $\{x_0^l(t), ..., x_0^{p-1}(t), t \in [0, \vartheta)\}$, $\{x_0^{l+1}(t), ..., x_0^p(t), t \in (\vartheta, \varrho)\}$. The dependence on the two last sets of arguments is by an integral operation.

In the case $l = 1$ an analogous derivation gives the basic algebraic equation in the form

$$\Delta_1(\vartheta)\tilde{p}_0^1(\vartheta) = P_1(\vartheta). \qquad (10.3.35)$$

The matrices $\Delta_l(\vartheta)$ are nonsingular for every $\vartheta \in [0, \varrho)$ and every l, $1 \leq l \leq \sigma$. To prove this, suppose for a moment that $\Delta_l(\vartheta)$ is singular. Then the solution of equation (10.3.34) (resp. (10.3.35)) either does not exist or is not unique. Hence the optimal control either does not exist or is not unique, contrary to Theorem 10.3.2.

In the step method the basic equation is a fundamental result. It is a starting point for the construction of solutions, both analytic and numerical, to different problems of optimal control. From equation (10.3.34) or (10.3.35), from the boundary conditions in (10.3.30), and from (10.3.25) we obtain the optimal solution, that is, the optimal control and optimal trajectory, as an affine continuous function of the past trajectory up to the moment s. This affine function is the optimal regulator and a solution of the optimal closed-loop control problem.

10.3.5 Step method solution in case of a delay-free performance functional

We shall apply the step method to a linear-quadratic optimal control problem with the system equation

$$\sum_{i=0}^{N} [A_i(t)\dot{x}(t-i\varrho) + B_i(t)x(t-i\varrho) + C_i(t)u(t-i\varrho)] = 0 \quad t \in [0, T] \tag{10.3.36}$$

and the performance functional

$$J(u) = x(T)^{\mathrm{T}}[Qx(T) + 2q] + \int_0^T [x(t)^{\mathrm{T}} W(t)x(t) + u(t)^{\mathrm{T}} u(t) + 2w(t)^{\mathrm{T}} x(t)] \, dt. \tag{10.3.37}$$

Again the assumptions regarding the system equation are the same as for (10.1.1). The initial conditions are given by (10.1.2). The performance functional is a special case of (10.1.6); by the arguments of Section 10.2 our results may be generalized to cover functionals of the form (10.2.1). It is assumed that

$$Q = Q^{\mathrm{T}} \geqslant 0$$
$$W(t) = W(t)^{\mathrm{T}} \geqslant 0 \quad \forall t \tag{10.3.38}$$

W and w are piecewise continuous functions of time. If we define

$$\tilde{W}_i = \mathrm{diag}(W^1, \ldots, W^i)$$

we obtain the following step-method formulation of the linear-quadratic problem (10.3.36), (10.3.37). Minimize

$$\tilde{J}(\tilde{u}) = x^1(\varrho)^{\mathrm{T}}[Qx^1(\varrho) + 2q] +$$

$$+ \sum_{i=\sigma-1}^{\sigma} \int_{I_i^\sigma(\theta)} [\tilde{x}^i(t)^{\mathrm{T}} \tilde{W}_i(t)\tilde{x}^i(t) + \tilde{u}^i(t)^{\mathrm{T}} \tilde{u}^i(t) + 2\tilde{w}^i(t)^{\mathrm{T}} \tilde{x}^i(t)] \, dt \tag{10.3.39}$$

subject to

$$\dot{\tilde{x}}^i(t) = \hat{B}_i(t)\tilde{x}^i(t) + \hat{C}_i(t)\tilde{u}^i(t) + \delta_i(t)\eta_i(t) \quad t \in I_i^i(\theta)$$
$$i = \sigma-1, \sigma \quad \text{for} \quad \sigma > 1, \quad i = \sigma \quad \text{for} \quad \sigma = 1$$
$$\tilde{x}^{\sigma-1}(\theta-) = e_\sigma \tilde{x}^\sigma(\theta+) \quad \tilde{x}^{\sigma-1}(0+) = d_\sigma \tilde{x}^\sigma(\varrho-) \tag{10.3.40}$$
$$x^\sigma(\theta) = x(0).$$

If $\sigma = 1$, we drop the term with $i = \sigma - 1$ in (10.3.39).

The unique optimal control satisfies the equalities (10.3.25) with the adjoint functions $\tilde{p}_0^i$ determined by the canonical set of equations (10.3.30). However for our special case this set may be written in a simpler and clearer form. Let us denote

$$\alpha_i = \begin{bmatrix} \hat{B}_i & -\hat{C}_i\hat{C}_i^T \\ -\tilde{W}_i & -\hat{B}_i^T \end{bmatrix}$$

$$\beta_i = \mathrm{col}(0, -\tilde{w}^i) \qquad \delta_i' = \mathrm{col}(\delta_i, 0).$$

Then the canonical set of equations takes the form

$$\dot{z}_i(t) = \alpha_i(t)z_i(t) + \delta_i'(t)\eta_i(t) + \beta_i(t) \qquad t \in I_i^l(\vartheta)$$
$$i = l-1, l \quad \text{for} \quad l > 1, \quad i = l \quad \text{for} \quad l = 1$$
$$z_{l-1}(\vartheta-) = \mathrm{diag}(e_l, e_l)z_l(\vartheta+) \qquad (10.3.41)$$
$$z_{l-1}(0+) = \mathrm{diag}(d_l, d_l)z_l(\varrho-)$$
$$x_0^l(\vartheta) = x_0(s) \qquad p_0^1(\varrho) = Qx_0^1(\varrho) + q.$$

In order to construct the basic algebraic equation we partition the fundamental matrices Φ_i into submatrices

$$\Phi_i = \begin{bmatrix} \Phi_i^1 & \Phi_i^2 & \Phi_i^3 \\ \Phi_i^4 & \Phi_i^5 & \Phi_i^6 \\ \Phi_i^7 & \Phi_i^8 & \Phi_i^9 \\ \Phi_i^{10} & \Phi_i^{11} & \Phi_i^{12} \end{bmatrix} \begin{matrix} n \\ (i-1)n \\ n \\ (i-1)n \end{matrix} = [\Phi_i^A \ \Phi_i^A] \atop {in \quad in}$$

$$\qquad \qquad (i-1)n \quad n \quad in \qquad \qquad \qquad (10.3.42)$$

$$\Phi_i^A = \begin{bmatrix} \Phi_i^{A1} \\ \Phi_i^{A2} \\ \Phi_i^{A3} \\ \Phi_i^{A4} \end{bmatrix} \begin{matrix} n \\ (i-1)n \\ n \\ (i-1)n \end{matrix}.$$

In the case $i = 1$

$$\Phi_1 = \begin{bmatrix} \Phi_1^2 & \Phi_1^3 \\ \Phi_1^8 & \Phi_1^9 \end{bmatrix} \begin{matrix} n \\ n \end{matrix} \qquad \Phi_1^A = \begin{bmatrix} \Phi_1^{A1} \\ \Phi_1^{A3} \end{bmatrix}. \qquad (10.3.43)$$
$$\quad n \quad n$$

The basic equations have the form (10.3.34) and (10.3.35). For $l > 1$

$$\Delta_l(\vartheta) = \begin{bmatrix} \Phi_l^7(\varrho,\vartheta) - Q\Phi_l^1(\varrho,\vartheta) & \Phi_l^9(\varrho,\vartheta) - Q\Phi_l^3(\varrho,\vartheta) \\ \begin{bmatrix} \Phi_l^4(\varrho,\vartheta) & \Phi_l^6(\varrho,\vartheta) \\ \Phi_l^{10}(\varrho,\vartheta) & \Phi_l^{12}(\varrho,\vartheta) \end{bmatrix} - \Phi_{l-1}(0,\vartheta)e_{2l-1} \end{bmatrix} \qquad (10.3.44)$$

$$P_l(\vartheta) = \begin{bmatrix} q \\ 0 \end{bmatrix} - \begin{bmatrix} \Phi_l^8(\varrho,\vartheta) - Q\Phi_l^2(\varrho,\vartheta) \\ \Phi_l^5(\varrho,\vartheta) \\ \Phi_l^{11}(\varrho,\vartheta) \end{bmatrix} x_0(s) -$$

10.3 Step Method

$$-\int_0^\theta \left\{ \begin{bmatrix} 0 \\ \Phi_{l-1}^A(0,t) \end{bmatrix} \delta_{l-1}(t)\eta_{l-1}(t) - \begin{bmatrix} 0 \\ \Phi_{l-1}^B(0,t) \end{bmatrix} \tilde{w}^{l-1}(t) \right\} dt -$$

$$-\int_\vartheta^\varrho \left\{ \begin{bmatrix} \Phi_l^{A3}(\varrho,t) - Q\Phi_l^{A1}(\varrho,t) \\ \Phi_l^{A2}(\varrho,t) \\ \Phi_l^{A4}(\varrho,t) \end{bmatrix} \delta_l(t)\eta_l(t) - \begin{bmatrix} \Phi_l^9(\varrho,t) - Q\Phi_l^3(\varrho,t) \\ \Phi_l^6(\varrho,t) \\ \Phi_l^{12}(\varrho,t) \end{bmatrix} \tilde{w}^l(t) \right\} dt.$$

(10.3.45)

for $l = 1$

$$\Delta_1(\vartheta) = \Phi_1^9(\varrho,\vartheta) - Q\Phi_1^3(\varrho,\vartheta) \qquad (10.3.46)$$

$$P_1(\vartheta) = q - [\Phi_1^8(\varrho,\vartheta) - Q\Phi_1^2(\varrho,\vartheta)]x_0(s) -$$

$$-\int_\vartheta^\varrho \{[\Phi_1^8(\varrho,t) - Q\Phi_1^2(\varrho,t)]\delta_1(t)\eta_1(t) -$$

$$-[\Phi_1^9(\varrho,t) - Q\Phi_1^3(\varrho,t)]\tilde{w}^1(t)\}dt. \qquad (10.3.47)$$

10.3.6 Open-loop and closed-loop optimal control

The formulas given above allow us to determine optimal control both as an explicit function of time, that is, in open-loop form, and as a function of the current state, that is in closed-loop or regulator form. As an explicit function of time, optimal control is determined according to the following algorithm. We assume for a moment $\sigma > 1$.

1. Determine an integer σ, $(\sigma-1)\varrho < T \leq \sigma\varrho$, and a real $\theta = \sigma\varrho - T$.
2. Compute $\Phi_\sigma(\varrho, \theta)$ and $\Phi_{\sigma-1}(0, \theta)$ using equations (10.3.31).
3. Compute the matrix $\Delta_\sigma(\theta)$ (10.3.44).
4. Compute

$$\gamma_{\sigma-1}(0) = \int_0^\theta \Phi_{\sigma-1}(0,t)[\delta'_{\sigma-1}(t)\eta_{\sigma-1}(t) + \beta_{\sigma-1}(t)]dt,$$

$$\gamma_\sigma(\varrho) = \int_\theta^\varrho \Phi_\sigma(\varrho,t)[\delta'_\sigma(t)\eta_\sigma(t) + \beta_\sigma(t)]dt,$$

by solving the equations

$$\dot{\gamma}_{\sigma-1}(t) = \alpha_{\sigma-1}(t)\gamma_{\sigma-1}(t) - \delta'_{\sigma-1}(t)\eta_{\sigma-1}(t) - \beta_{\sigma-1}(t), \qquad \gamma_{\sigma-1}(\theta) = 0$$

$$\dot{\gamma}_\sigma(t) = \alpha_\sigma(t)\gamma_\sigma(t) + \delta'_\sigma(t)\eta_\sigma(t) + \beta_\sigma(t), \qquad \gamma_\sigma(\theta) = 0.$$

5. Determine $P_\sigma(\theta)$ (10.3.45).
6. Compute $\tilde{x}_0^{\sigma-1}(\theta)$ and $\tilde{p}_0^\sigma(\theta+)$ using the basic equation (10.3.34).

7. Compute $z_\sigma(t)$, $t \in [\theta, \varrho]$ from (10.3.41) with $i = l$, setting $l = \sigma$. The initial condition is
$$z_\sigma(\theta+) = \mathrm{col}(\tilde{x}_0^{\sigma-1}(\theta), x(0), \tilde{p}_0^\sigma(\theta+)).$$

8. Compute $z_{\sigma-1}(t)$, $t \in [0, \theta]$ from (10.3.41) with $i = l-1$, setting $l = \sigma$. The terminal condition is
$$z_{\sigma-1}(\theta-) = \mathrm{col}(\tilde{x}_0^{\sigma-1}(\theta), \tilde{p}_0^{\sigma-1}(\theta-)).$$

9. Determine $\tilde{u}_0^{\sigma-1}(t)$, $t \in [0, \theta)$, and $\tilde{u}_0^\sigma(t)$, $t \in [\theta, \varrho]$ using (10.3.25).

Finally, the solution of the original problem (10.3.36), (10.3.37) is obtained by means of formulas (3.2.36).

If $\sigma = 1$, the same algorithm is used with the following small changes. In step 2 only $\Phi_1(\varrho, \theta)$ is computed; in step 4 only $\gamma_1(\varrho)$; in step 6 only $\tilde{p}_0^1(\theta+)$; in step 7 $z_1(\theta+) = \mathrm{col}(x(0), \tilde{p}_0^1(\theta+))$; step 8 is cancelled; and in step 9 only $u_0^\sigma(t)$ is determined. Formulas (10.3.44) and (10.3.45) are replaced by (10.3.46) and (10.3.47).

Let us pass on to the synthesis of optimal regulator. Denote

$$\Delta_l(\vartheta)^{-1} = \begin{bmatrix} \Gamma_l^1(\vartheta) & \Gamma_l^2(\vartheta) \\ \Gamma_l^3(\vartheta) & \Gamma_l^4(\vartheta) \end{bmatrix} \begin{matrix} (l-1)n \\ ln \end{matrix} \quad \text{for } l > 1 \tag{10.3.48}$$
$$\phantom{\Delta_l(\vartheta)^{-1} = }\begin{matrix} n & 2(l-1)n \end{matrix}$$

$$\Delta_1(\vartheta)^{-1} = \Gamma_1^3(\vartheta).$$

Our purpose here is to present the functional dependence of the optimal control in problem (10.3.36), (10.3.37) on the optimal state trajectory. We shall also give the conditions under which it is possible to synthesize the optimal control on the state without the derivative component $\dot{\hat{x}}(t)$. We denote by $U(s)$ the $ln \times m$ submatrix of $\hat{C}_l(\vartheta)$ consisting of its m last columns.

Theorem 10.3.4 *For every $s \in [0, T)$ the value of optimal control in problem (10.3.36), (10.3.37) at time s, $u_0(s)$ is expressed by the optimal state in the following way*:

$$u_0(s) = -U(s)^\mathrm{T}\Big\{G(s)x_0(s) + \sum_{i=0}^{N-1} \int_{-\varrho}^{0} [G_i^a(s,\tau)\dot{x}_0(s-i\varrho+\tau) +$$
$$+ G_i^b(s,\tau)x_0(s-i\varrho+\tau) + G_i^c(s,\tau)u_0(s-i\varrho+\tau)]\,\mathrm{d}\tau +$$
$$+ \int_s^T G^d(s,\tau)w(\tau)\,\mathrm{d}\tau + G^e(s)q\Big\} = -U(s)^\mathrm{T}\Big\{G(s)x_0(s) +$$
$$+ \int_{s-N\varrho}^{s} [G^a(s,\tau)\dot{x}_0(\tau) + G^b(s,\tau)x_0(\tau) + G^c(s,\tau)u_0(\tau)]\,\mathrm{d}\tau +$$
$$+ \int_s^T G^d(s,\tau)w(\tau)\,\mathrm{d}\tau + G^e(s)q\Big\}. \tag{10.3.49}$$

0.3 Step Method

If $s < T - \varrho$, we set

$$G(s) = -\Gamma_l^4(\vartheta)\begin{bmatrix}\Phi_l^5(\varrho, \vartheta)\\ \Phi_l^{11}(\varrho, \vartheta)\end{bmatrix} - \Gamma_l^3(\vartheta)[\Phi_l^8(\varrho, \vartheta) - Q\Phi_l^2(\varrho, \vartheta)] \quad (10.3.50)$$

$$G_i^a(s, \tau) = -\Gamma_l^4(\vartheta)\Phi_{l-1}^A(0, \tau+\vartheta)a_{l-1}^i(\tau+\vartheta) \quad \tau \in (-\vartheta, 0] \quad (10.3.51)$$

$$G_i^a(s, \tau) = -\left\{\Gamma_l^4(\vartheta)\begin{bmatrix}\Phi_l^{A2}(\varrho, \tau+\vartheta+\varrho)\\ \Phi_l^{A4}(\varrho, \tau+\vartheta+\varrho)\end{bmatrix}+\right.$$

$$\left. +\Gamma_l^3(\vartheta)[\Phi_l^{A3}(\varrho, \tau+\vartheta+\varrho) - Q\Phi_l^{A1}(\varrho, \tau+\vartheta+\varrho)]\right\} a_l^i(\tau+\vartheta+\varrho)$$

$$\tau \in [-\varrho, -\vartheta]. \quad (10.3.52)$$

Replacing letter a by b and c we obtain G_i^b and G_i^c, respectively.

$$G^d(s, \tau) = \Gamma_l^4(\vartheta)[\Phi_{i-1}^B(0, \tau-T+i\varrho)]_i \quad \tau \in [T-i\varrho, T-i\varrho+\vartheta)$$
$$(i = 1, \ldots, l-1) \quad (10.3.53)$$

$$G^d(s, \tau) = \Gamma_l^4(\vartheta)\left[\begin{pmatrix}\Phi_i^6(\varrho, \tau-T+i\varrho)\\ \Phi_i^{12}(\varrho, \tau-T+i\varrho)\end{pmatrix}\right]_i +$$

$$+ \Gamma_l^3(\vartheta)[\Phi_i^9(\varrho, \tau-T+i\varrho) - Q\Phi_i^3(\varrho, \tau-T+i\varrho)]_i$$

$$\tau \in [T-i\varrho+\vartheta, T-(i-1)\varrho] \quad (i = 1, \ldots, l). \quad (10.3.54)$$

$[\cdot]_i$ denotes here the submatrix consisting of columns with numbers from $(i-1)n+1$ through in,

$$G^e(s) = \Gamma_l^3(\vartheta) \quad (10.3.55)$$

$$G^a(s, \tau) = G_i^a(s, \tau-s+i\varrho) \quad \tau \in [s-(i+1)\varrho, s-i\varrho]$$
$$(i = 0, \ldots, N-1). \quad (10.3.56)$$

G^b and G^c are obtained by replacing a by b and c in this formula.
For $s \geq T-\varrho$ formulas (10.3.55), (10.3.56) are valid, too, and

$$U(s) = -A_0(s)^{-1}C_0(s)$$
$$G(s) = -\Gamma_1^3(\vartheta)[\Phi_1^8(\varrho, \vartheta) - Q\Phi_1^2(\varrho, \vartheta)]$$
$$G_i^a(s, \tau) = -\Gamma_1^3(\vartheta)[\Phi_1^8(\varrho, \tau+\varrho+\vartheta) - Q\Phi_1^2(\varrho, \tau+\varrho+\vartheta)]a_1^i(\tau+\varrho+\vartheta)$$
$$\tau \in [-\varrho, -\vartheta) \quad (10.3.57)$$
$$G_i^a(s, \tau) = 0 \quad \tau \in [-\vartheta, 0]$$
$$G^d(s, \tau) = \Gamma_1^3(\vartheta)[\Phi_1^9(\varrho, \tau+\varrho-T) - Q\Phi_1^3(\varrho, \tau+\varrho-T)].$$

Proof Let $l > 1$. From Theorem 10.3.3 and the basic equation (10.3.34) it follows that

$$u_0(s) = -U(s)^T\left(\Gamma_1^3(\vartheta) \quad \Gamma_l^4(\vartheta)\right)P_l(\vartheta).$$

Using (10.3.45) and (3.2.36) we obtain (10.3.49)–(10.3.56). The proof for $l = 1$ is similar. □

A simplification of the optimal regulator synthesis is possible if there are no delays in the derivative of the momentary state in the system equation (10.3.36). Of course $A_i = 0$, $i > 0$ and without loss of generality we may assume $A_0(t) \equiv I$. Then $\tilde{A}_i = I$, $\hat{C}_i = -\tilde{C}_i$ and by virtue of (10.3.25)

$$\tilde{u}_0^l(t) = \tilde{C}_i(t)^T \tilde{p}_0^l(t). \tag{10.3.58}$$

Let $r_1 = \max(i: C_i \neq 0)$ and $\varkappa = \min(r_1, l-1)$. Hence

$$u_0(s) = u_0^l(\vartheta) = \bar{U}(s)^T \bar{p}_0^l(\vartheta), \tag{10.3.59}$$

where $\bar{U}(s)$ is a $((\varkappa+1)n \times m)$-matrix

$$\bar{U}(s) = \mathrm{col}\big(C_\varkappa(s+\varkappa\varrho), \ldots, C_1(s+\varrho), C_0(s)\big) \tag{10.3.60}$$

and $\bar{p}_0^l(\vartheta)$ is a vector of $(\varkappa+1)n$ last components of $\tilde{p}_0^l(\vartheta)$. Let $\bar{\Gamma}_l^3(\vartheta)$ and $\bar{\Gamma}_l^4(\vartheta)$ be matrices constructed of the last $(\varkappa+1)n$ last rows of $\Gamma_l^3(\vartheta)$ and $\Gamma_l^4(\vartheta)$. The optimal control is determined by formula (10.3.49) with U, Γ_l^3 and Γ_l^4 replaced by $\bar{U}$, $\bar{\Gamma}_l^3$ and $\bar{\Gamma}_l^4$, respectively.

An interesting question is under what conditions it is possible to remove the derivatives of the instantaneous state from the right-hand side of formula (10.3.49) or, in other words, when the optimal regulator can be synthesized on the state that does not include derivatives. A trivial condition is $A_i = 0$, $i > 0$. To give a non-trivial answer let us formulate the following.

Assumption 10.3.1 (i) *In the initial conditions* (10.1.2) *the function φ is absolutely continuous and $\mu = d\varphi/dt$.*

(ii) *The matrix-valued functions A_i, $i = 0, \ldots, N$, are absolutely continuous.*

It is easy to verify that if Assumption 10.3.1 (ii) is satisfied, then $G^a(s, \cdot)$ is piecewise absolutely continuous for every s. Discontinuities may occur only at points $\tau = s-i\varrho$, $\tau = s-\vartheta-i\varrho$, $i = 1, \ldots, N-1$, and at $\tau = s-\vartheta$ if $\vartheta > 0$. Moreover,

$$G^a(s, s-i\varrho-) - G^a(s, s-i\varrho+) = G_i^a(s, 0-) - G_{i-1}^a(s, -\varrho+)$$
$$(i = 1, \ldots, N-1)$$

$$G^a(s, s-\vartheta-i\varrho-) - G^a(s, s-\vartheta-i\varrho+) = \Gamma_i^3(\vartheta) Q \check{a}_i^i \quad \text{for} \quad \vartheta > 0$$
$$(i = 0, \ldots, N-1) \tag{10.3.61}$$

$\check{a}_i^i$ denotes the matrix consisting of the first n rows of $a_i^i(\varrho)$.

Now, if we notice that under Assumption 10.3.1 the term

$$\int_{s-N\varrho}^{s} G^a(s, \tau) \dot{x}_0(\tau) \, d\tau$$

in (10.3.49) can be integrated by parts, we easily obtain the following theorem.

Theorem 10.3.5 *Let Assumption 10.3.1 hold. Then for every $s \in [0, T)$ the value of optimal control in problem* (10.3.36), (10.3.37) *at time s, $u_0(s)$ is expressed by the optimal state in the following way:*

$$u_0(s) = -U(s)^T \Big\{ \sum_{i=0}^{N} g_i^1(s) x_0(s - i\varrho) + \sum_{i=0}^{N-1} g_i^2(s) x_0(s - i\varrho - \vartheta) + $$
$$+ \sum_{i=0}^{N-1} \int_{-\varrho}^{0} [g_i^b(s, \tau) x_0(s - i\varrho + \tau) + G_i^c(s, \tau) u_0(s - i\varrho + \tau)] d\tau + $$
$$+ \int_{s}^{T} G^d(s, \tau) w(\tau) d\tau + G^e(s) q \Big\} \qquad (10.3.62)$$

where

$$g_0^1(s) = G(s) + G_0^a(s, 0-)$$
$$g_i^1(s) = G_i^a(s, 0-) - G_{i-1}^a(s, -\varrho+) \quad (i = 1, \ldots, N-1)$$
$$g_N^1(s) = -G_{N-1}^a(s, -\varrho+)$$
$$g_i^2(s) = \Gamma_i^3(\vartheta) Q \check{a}_i^i \quad \vartheta \neq 0 \quad (i = 0, \ldots, N-1)$$
$$g_i^2(s) = 0 \quad \vartheta = 0$$
$$g_i^b(s, \tau) = G_i^b(s, \tau) - \frac{\partial}{\partial \tau} G_i^a(s, \tau) \quad \tau \neq -\vartheta \quad (i = 0, \ldots, N-1). \quad (10.3.63)$$

10.3.7 Other variants of the step method

In some cases results can be obtained in a simpler form using a variant of the step method, different from the one described above. The differences between the variants lie in the construction of the vectors z_i—more precisely, in the order of succession of the vectors x_0^j and p_0^j in the vector z_i. This order, of course, has consequences for the construction of matrices α_i, β_i and δ_i'.

The details will be given for the linear-quadratic problem (10.3.36), (10.3.37) with $Q = 0$, $q = 0$, and $w = 0$. For any matrix S with appropriate dimensions we denote by $[S]_j^i$ its submatrix consisting of the elements situated at the intersections of rows $(i-1)n+1$ through in and columns $(j-1)n$ through jn, and by $[S]^i$ its submatrix consisting of the rows $(i-1)n+1$ through in.

The definitions of α_i, δ_i' of Section 10.3.5, and z_i of Section 10.3.4 are replaced by

$$[\alpha_i]_k^j = \begin{bmatrix} [\hat{B}_i]_k^j & -[\hat{C}_i \hat{C}_i^T]_k^j \\ -[\tilde{W}_i]_k^j & -[\hat{B}_i^T]_k^j \end{bmatrix}$$
$$\delta_i' = \text{col}([\delta_i]^1, 0, \ldots, [\delta_i]^i, 0)$$
$$z_i = \text{col}(x_0^1, p_0^1, \ldots, x_0^i, p_0^i).$$

As before, we denote the fundamental solutions of the canonical equations by Φ_i. This time, however, we partition them differently,

$$\Phi_i = \left[\begin{matrix} \Phi_i^1 \\ \Phi_i^2 & \Phi_i^3 & \Phi_i^4 \end{matrix} \right]_{\substack{2(i-1)n \ n \ n}}^{n} \quad (i > 1).$$

The basic algebraic equation for $l > 1$ has the form

$$\Delta_l(\vartheta) \begin{bmatrix} z_{l-1}(\vartheta) \\ p_0^l(\vartheta) \end{bmatrix} = P_l(\vartheta)$$

where

$$\Delta_l(\vartheta) = [\Phi_l^2(\varrho, \vartheta) - \pi_l(\vartheta) \quad \Phi_l^4(\varrho, \vartheta)] \qquad (10.3.64)$$

$$\pi_l(t) = \operatorname{col}(0, \Phi_{l-1}(0, t))$$

$$P_l(\vartheta) = -\Phi_l^3(\varrho, \vartheta) x_0(s) - \int_0^\vartheta \pi_l(t) \delta_{l-1}'(t) \eta_{l-1}(t) \mathrm{d}t -$$

$$- \int_\vartheta^\varrho [\Phi_l^2(\varrho, t) \quad \Phi_l^3(\varrho, t) \quad \Phi_l^4(\varrho, t)] \delta_l'(t) \eta_l(t) \mathrm{d}t.$$

As in the previous variant, the basic equation is a basis for the construction of the optimal solution in the form of an explicit function of time or an optimal regulator. Equation (10.3.64) is a particularly convenient starting point for optimal regulator synthesis if there are no delays in the derivative of instantaneous state or in control in the system equation.

10.3.8 Linear-quadratic problem with infinite control time

The step method does not give an elegant solution of the linear-quadratic problem with infinite control time. Nevertheless, it permits the construction of a suboptimal regulator and an iterative algorithm for searching for the optimal solution. We now consider a stationary problem, with constant coefficients in the system equation and performance functional. The system is described by

$$\sum_{i=0}^{N} [A_i \dot{x}(t - i\varrho) + B_i x(t - i\varrho) + C_i u(t - i\varrho)] = 0 \quad t \in [0, \infty) \quad (10.3.65)$$

with initial conditions (10.1.2) and assumptions as for (10.1.1). Of course $\det A_0 \neq 0$. The space of admissible controls is $L^2_{\mathrm{loc}}(0, \infty; R^m)$.

We begin with a suboptimal conception. Let us suppose that the control time is unbounded, but the planning horizon is finite. This horizon is denoted

10.3 Step Method

by T, $T > \varrho$. At every moment s the control $u(s)$ is chosen so that the system behaviour and control cost is optimized in the interval $[s, s+T]$. In such a situation it is advisable to use a suboptimal control u^*, whose value at every moment of time s is equal to the initial value of the control $u_T:[s, s+T] \to R^m$, minimizing the functional

$$J_T(u) = \int_s^{s+T} [x(t)^T W x(t) + u(t)^T u(t) + 2w^T x(t)] dt +$$
$$+ x(s+T)^T [Q x(s+T) + 2q] \qquad (10.3.66)$$

where W and Q are $(n \times n)$-matrices, $W = W^T \geqslant 0$, $Q = Q^T \geqslant 0$.

Let us denote by $S(s)$ the state of the system at time s,

$$S(s) = (\tilde{x}(s), \dot{\tilde{x}}(s), \tilde{u}(s)) \in R^n \times L^2(-\varrho N, 0; R^n) \times L^2(-\varrho N, 0; R^n) \times$$
$$\times L^2(-\varrho N, 0; R^m). \qquad (10.3.67)$$

According to Theorem 10.3.4, control u^* may be represented in the form

$$u^*(s) = K_T S(s) + k_T \qquad (10.3.68)$$

where K_T and k_T are determined by (10.3.49)–(10.3.57). Of course, as the problem is stationary the regulator (10.3.68) is constant.

Now we shall consider the optimal control in the case of infinite control time. We are dealing with system (10.3.65); the performance functional to be minimized is

$$J(u) = \lim_{T \to \infty} J_T(u) \qquad (10.3.69)$$

$$J_T(u) = \int_0^T [x(t)^T W x(t) + u(t)^T u(t)] dt \qquad (10.3.70)$$

$$W = W^T \geqslant 0.$$

Theorem 10.3.6 *Assume that there is an optimal control u_0 in problem* (10.3.65), (10.3.69). *Of course this optimal control is unique. Denote by S_0 the corresponding state trajectory. Then*:

(i) *There exists a constant in time linear continuous operator K, such that*

$$u_0(s) = K S_0(s). \qquad (10.3.71)$$

Moreover,

$$K = \lim_{T \to \infty} K_T. \qquad (10.3.72)$$

(ii) *There is a constant $(m \times m)$-matrix G and matrix-valued functions G^a, G^b, G^c such that*

$$KS_0(s) = Gx_0(s) + \int_{-\varrho N}^{0} [G^a(t)\dot{x}_0(s+t) + G^b(t)x_0(s+t) +$$
$$+ G^c(t)u_0(s+t)]dt. \tag{10.3.73}$$

(iii) *If Assumption 10.3.1 holds, then there are matrices* G_i, $i = 0, \ldots, N$, *and matrix-valued functions* G^b, G^c *such that*

$$KS_0(s) = \sum_{i=0}^{N} G_i x_0(s - i\varrho) + \int_{-\varrho N}^{0} [G^b(t)x_0(s+t) + G^c(t)u_0(s+t)]dt. \tag{10.3.74}$$

This theorem can be proved by methods presented in the work by Delfour et al. (1975), but we shall omit the details.

Theorem 10.3.7 *Stabilizability of system* (10.3.65) *is a sufficient condition for the existence of optimal control in problem* (10.3.65), (10.3.69).

Proof Denote by u_T the restriction of a stabilizing control (see Section 10.1) to the interval $[0, T]$. Let u_{0T} denote the optimal control in problem (10.3.65), (10.3.70). It is evident that $J_T(u_{0T})$ and $J_T(u_T)$ increase when $T \to \infty$ and are bounded because $J_T(u_{0T}) \leq J_T(u_T)$. Then of course zero extensions of u_{0T} to the whole real half-line are convergent to the optimal control. □

Following Pandolfi (1976) and Olbrot (1977) we shall give conditions of stabilizability for system (10.3.65). Denote

$$\Delta_0(\lambda) = \sum_{i=0}^{N} (\lambda A_i + B_i) e^{-i\varrho\lambda}$$

$$C(\lambda) = \sum_{i=0}^{N} e^{-i\varrho\lambda} C_i. \tag{10.3.75}$$

Theorem 10.3.8 *Assume that there is a real* $a < 0$ *such that the equation*

$$\det \Delta_0(\lambda) = 0 \tag{10.3.76}$$

has at most a finite number of roots with real parts greater than a. *Then system* (10.3.65) *is stabilizable if and only if*

$$\operatorname{rank}[\Delta_0(\lambda) \quad C(\lambda)] = n \quad \forall \lambda: \operatorname{Re}\lambda \geq 0. \tag{10.3.77}$$

The results presented above make it possible to construct a numerical algorithm for the determination of the optimal regulator in an infinite-time problem, of course provided the optimal solution exists. For a sufficiently large T in (10.3.70), the respective regulator K_T is arbitrarily close to K.

10.3 Step Method

In subsequent steps of the algorithm the values of K_T are calculated for increasing integer values of σ with $T = \sigma\varrho$. The algorithm stops when K_T may be considered constant with respect to T within a given accuracy. In every step the formulas of Sections 10.3.6 or 10.3.7 are used. For a fixed $\sigma > 1$ the algorithm is as follows:

1. Determine $\Phi_\sigma(\varrho, t)$ $t \in [0, \varrho]$.
2. Determine $\Delta_\sigma(0)$.
3. Determine $\bar{p}_0^\sigma(0)$ (10.3.59).
4. Determine the regulator.

10.3.9 Examples

Example 10.3.1 We shall determine the optimal solution in the problem

$$\dot{x}(t) - x(t-1) - u(t) = 0 \quad t \in [0, 1.5]$$
$$x(t) = 1 \quad t \in [-1, 0]$$
$$J(u) = \int_0^{1.5} [x(t)^2 + u(t)^2]dt.$$

Of course $\varrho = 1$, $\sigma = 2$, and $\theta = 0.5$. We apply the version of the step method described in Section 10.3.7. According to the formulas given there

$$\alpha_1 = \begin{bmatrix} 0 & -1 \\ -1 & 0 \end{bmatrix} \quad \delta_1' = \begin{bmatrix} 1 \\ 0 \end{bmatrix}$$

$$\alpha_2 = \begin{bmatrix} 0 & -1 & 1 & 0 \\ -1 & 0 & 0 & 0 \\ 0 & 0 & 0 & -1 \\ 0 & -1 & -1 & 0 \end{bmatrix} \quad \delta_2' = \begin{bmatrix} 0 \\ 0 \\ 1 \\ 0 \end{bmatrix}$$

$$\Phi_1(0, t) = \begin{bmatrix} \cosh t & \sinh t \\ \sinh t & \cosh t \end{bmatrix}.$$

Let

$$a = \sqrt{\frac{\sqrt{2}+1}{2}} \quad b = \sqrt{\frac{\sqrt{2}-1}{2}}$$

$$\text{cc}(t) = \cosh at \cos bt \quad \text{cs}(t) = \cosh at \sin bt$$
$$\text{sc}(t) = \sinh at \cos bt \quad \text{ss}(t) = \sinh at \sin bt.$$

Then

$$\Phi_2(t, 0) = \begin{bmatrix} \text{cc}(t) & r_3(t) & r_4(t) & -\text{ss}(t) \\ r_1(t) & \text{cc}(t) & -\text{ss}(t) & -r_2(t) \\ r_2(t) & \text{ss}(t) & \text{cc}(t) & r_1(t) \\ \text{ss}(t) & -r_4(t) & r_3(t) & \text{cc}(t) \end{bmatrix}$$

where

$$r_1(t) = -\frac{a}{\sqrt{2}}\operatorname{sc}(t) - \frac{b}{\sqrt{2}}\operatorname{cs}(t)$$

$$r_2(t) = \frac{b}{\sqrt{2}}\operatorname{sc}(t) - \frac{a}{\sqrt{2}}\operatorname{cs}(t)$$

$$r_3(t) = b\operatorname{cs}(t) - a\operatorname{sc}(t)$$

$$r_4(t) = b\operatorname{sc}(t) + a\operatorname{cs}(t).$$

Finally,

$$u_0(t) = -0.6\operatorname{cc}(t) + 1.35\operatorname{sc}(t) + 0.15\operatorname{cs}(t) - 0.97\operatorname{ss}(t) - 1 \quad t \in [0, 0.5]$$
$$u_0(t) = 0.45\cosh(t-1) + 0.97\sinh(t-1) - 1 \quad t \in [0.5, 1]$$
$$u_0(t) = -0.55\operatorname{cc}(t-1) + 0.95\operatorname{sc}(t-1) - 0.15\operatorname{cs}(t-1) + \operatorname{ss}(t-1) \quad t \in [1, 1.5]$$
$$x_0(t) = \operatorname{cc}(t) - 0.15\operatorname{sc}(t) - 0.95\operatorname{cs}(t) + 0.55\operatorname{ss}(t) \quad t \in [0, 0.5]$$
$$x_0(t) = 0.97\cosh(t-1) + 0.45\sinh(t-1) \quad t \in [0.5, 1]$$
$$x_0(t) = 0.97\operatorname{cc}(t-1) - 0.15\operatorname{sc}(t-1) + 1.35\operatorname{cs}(t-1) - 0.6\operatorname{ss}(t-1)$$
$$\quad t \in [1, 1.5].$$

Example 10.3.2 In the problem

$$\dot{x}(t) - \dot{x}(t-2) - u(t) = 0 \quad t \in [0, 3]$$

$$J(u) = x(3)^2 + \int_0^3 u(t)^2 dt,$$

we determine the optimal control as a function of the initial condition. It is required that Assumption 10.3.1 (i) be satisfied. We have $\sigma = 2$, $\theta = 1$, and $\varrho = 2$. The basic version of the method is used (see Section 10.3.5):

$$\alpha_1 = \begin{bmatrix} 0 & -1 \\ 0 & 0 \end{bmatrix} \quad \delta_1' = \begin{bmatrix} 1 \\ 0 \end{bmatrix}$$

$$\alpha_2 = \begin{bmatrix} 0 & 0 & -2 & -1 \\ 0 & 0 & -1 & -1 \\ 0 & 0 & 0 & 0 \\ 0 & 0 & 0 & 0 \end{bmatrix} \quad \delta_2' = \begin{bmatrix} 1 \\ 1 \\ 0 \\ 0 \end{bmatrix}$$

$$\Phi_1(0, t) = \begin{bmatrix} 1 & t \\ 0 & 1 \end{bmatrix}$$

$$\Phi_2(2, t) = \begin{bmatrix} 1 & 0 & 2(t-2) & t-2 \\ 0 & 1 & t-2 & t-2 \\ 0 & 0 & 1 & 0 \\ 0 & 0 & 0 & 1 \end{bmatrix}$$

10.4 Finite-time Problem with Arbitrary Delays

$$\Delta_2(1) = \begin{bmatrix} -1 & 3 & 1 \\ -1 & -2 & -1 \\ 0 & -1 & 1 \end{bmatrix}$$

$$P_2(1) = \begin{bmatrix} \int_1^2 \dot{x}_0^3(t)\,dt \\ -x_0(0) - \int_0^1 \dot{x}_0^3(t+2)\,dt - \int_1^2 \dot{x}_0^3(t)\,dt \\ 0 \end{bmatrix}.$$

Denoting $V = \frac{1}{7}[2x_0(0) + x_0(-1) - 2x_0(-2)]$, we obtain

$$u_0(t) = \begin{cases} -2V & t \in [0, 1) \\ -V & t \in [1, 3]. \end{cases}$$

10.4 Finite-time Problem with Arbitrary Delays

10.4.1 Problem formulation

In Sections 10.4 and 10.5 we present a fairly general theory of the linear-quadratic problem with arbitrary delays. Unless otherwise stated, all theorems and proofs of Section 10.4 are a reformulation, in some cases corrected and generalized, of results due to Delfour and Mitter (1972), and Delfour (1977a). We consider system (10.1.1) with $C_i = 0$, $i > 0$, together with the performance functional (10.2.1). Using the formulas of Section 10.2 we may reduce this kind of problem without a loss of generality to the following. The system is described by

$$\dot{x}(t) + \sum_{i=0}^N B_i(t)x(t-h_i) + \int_{-h}^0 B(t,\tau)x(t+\tau)\,d\tau + C(t)u(t) = 0 \quad t \geq 0 \tag{10.4.1}$$

with the initial condition

$$x(0) = \varphi^0 \quad x(t) = \varphi^1(t) \quad t \in [-h, 0) \tag{10.4.2}$$

$$\varphi = (\varphi^0, \varphi^1) \in H$$

where $H = R^n \times L^2(-h, 0; R^n)$ is the state space discussed in Section 3.3.2. The assumptions on coefficients are the same as for (3.2.5). The performance functional to be minimized on the trajectories of (10.4.1) is $J: L^2(0, T; R^m) \to R$,

$$J(u) = x(T)^T[Qx(T)+2q]+ \int_0^T [x(t)^T W(t)x(t)+2w(t)^T x(t)+u(t)^T u(t)]dt \qquad (10.4.3)$$

$Q = Q^T \geqslant 0 \quad W(t) = W(t)^T \geqslant 0 \quad$ for every $\quad t \in [0, T]$,

W is bounded and absolutely integrable, and w is square integrable.

On the one hand the problem considered here is more general than that of Section 10.3; the delays may be non-commensurate and distributed. On the other hand, there are no delays in control, derivative of the momentary state or in the performance index.

The theory of equation (10.4.1) is presented in Section 3.2.1, and the corresponding state equation is discussed in Section 3.3.2.

10.4.2 Optimal control and canonical set of equations

We shall formulate the main result of the theory, the existence and uniqueness theorem which also gives a full characterization of optimal control by means of canonical equations.

Theorem 10.4.1 *For every initial condition* $\varphi \in H$ *of* (10.4.2) *in the problem* (10.4.1), (10.4.3), *there is a unique optimal control* $u_0 \in L^2(0, T; R^m)$. *The optimal control is fully characterized by the following set of canonical equations satisfied almost everywhere in* $[0, T]$:

$$\dot{x}_0(t) + \sum_{i=0}^N B_i(t)x_0(t-h_i) + \int_{-h}^0 B(t, \tau)x_0(t+\tau)d\tau + C(t)u_0(t) = 0$$

$$x_0(0) = \varphi^0, \quad x_0(t) = \varphi^1(t) \quad t \in [-h, 0) \qquad (10.4.4)$$

$$\dot{p}_0(t) - \sum_{i=0}^N B_i(t+h_i)^T p_0(t+h_i) - \int_{-h}^0 B(t-\tau, \tau)^T p_0(t-\tau)d\tau +$$

$$+ W(t)x_0(t)+w(t) = 0 \qquad (10.4.5)$$

$$p_0(T) = Qx_0(T)+q, \quad p_0(t) = 0 \quad t > T$$

$$u_0(t) = C(t)^T p_0(t). \qquad (10.4.6)$$

We shall only sketch the proof. For details the reader is referred to the literature (Delfour and Mitter, 1972). Existence follows from the fact that the functional J (10.4.3) is weakly lower semicontinuous and coercive, that is there are two reals $a > 0$ and b such that $J(u) \geqslant a||u||^2 - b$ for every $u \in L^2$. The optimal control is unique because of the strict convexity of the functional J. The relation (10.4.6) and canonical equations follow from the first-order

10.4 Finite-time Problem with Arbitrary Delays

necessary condition of optimality. This necessary condition, which is obtained in the standard way, has the form

$$\int_0^T [y(t)^T W(t) x_0(t) + w(t)^T y(t) + v(t)^T u_0(t)] dt + y(T)^T [Q x_0(T) + q] = 0$$

$$\forall v \in L^2(0, T; R^m)$$

where y is the solution of (10.4.1) generated by the control v from a zero initial condition. If we introduce the adjoint function p_0 satisfying the adjoint equation (10.4.5), by elementary transformations the necessary condition takes the form

$$\int_0^T v(t)^T [u_0(t) - C(t)^T p_0(t)] dt = 0 \quad \forall v \in L^2(0, T; R^m).$$

Hence we easily obtain (10.4.6).

It will be helpful, especially for the discussion of the optimal regulator, to present parallel considerations in the state space notation of Section 3.3.2. The canonical system (10.4.4)–(10.4.6) can be rewritten in the form

$$\dot{\tilde{x}}_0(t) + \tilde{B}(t) \tilde{x}_0(t) + \tilde{C}(t) u_0(t) = 0 \qquad (10.4.7)$$
$$\tilde{x}_0(0) = \varphi \in V$$

$$\dot{\tilde{p}}_0(t) - \tilde{B}(t)^* \tilde{p}_0(t) + \Lambda^* [\tilde{W}(t) \Lambda \tilde{x}_0(t) + \tilde{w}(t)] = 0 \qquad (10.4.8)$$
$$\tilde{p}_0(T) = \tilde{Q} \Lambda \tilde{x}_0(T) + \tilde{q}$$

$$u_0(t) = \tilde{C}(t)^* \tilde{p}_0(t) \qquad (10.4.9)$$

where $\tilde{W}(t), \tilde{Q} \in L(H)$, $\tilde{w}(t), \tilde{q} \in H$,

$$\tilde{W}(t) \varphi = (W(t) \varphi^0, 0) \quad \tilde{Q} \varphi = (Q \varphi^0, 0) \quad \forall \varphi \in H$$

$$\tilde{w}(t) = (w(t), 0) \quad \tilde{q} = (q, 0).$$

Let us recall that the canonical system is valid only for initial conditions φ taken from V, a dense subspace of H. The interconnection between equations (10.4.7) and (10.4.4) has been clarified in Section 3.3.2. The adjoint state equation (10.4.8) requires some explanation. Let V' denote the dual of V and let us identify H with its dual H'. Then the adjoint of Λ, $\Lambda^*: H \to V'$ is a dense and continuous injection. $\tilde{B}(t)^*: H \to V'$ is the adjoint of $\tilde{B}(t)$, and $\tilde{C}(t)^*: H \to R^m$ the adjoint of $\tilde{C}(t)$. The adjoint trajectory $\tilde{p}_0$ is treated as an element of

$$W^*(0, T) = \{z \in L^\infty(0, T; H) : \dot{z} \in L^2(0, T; V')\}.$$

It can be shown that (10.4.8) has a unique solution that satisfies for every t

$$\tilde{p}_0(t) = \tilde{\Phi}(T, t)^* \tilde{p}_0(T) + \int_t^T \tilde{\Phi}(\tau, t)^* [\tilde{W}(\tau) \Lambda \tilde{x}_0(\tau) + \tilde{w}(\tau)] d\tau \qquad (10.4.10)$$

$\tilde{\Phi}(t, s)^*$ is the adjoint of $\tilde{\Phi}(t, s)$.

Formula (10.4.10) is easily obtained from the variation-of-constants formula for equation (10.4.5)

$$p_0(t) = \Phi^0(T, t)^T p_0(T) + \int_t^T \Phi^0(\tau, t)^T [W(\tau) x_0(\tau) + w(\tau)] d\tau \qquad (10.4.11)$$

if we take into account the fact that $\tilde{p}_0$ and p_0 are connected by the following relationship:

$$\tilde{p}_0(t) = (\tilde{p}_0(t)^0, \tilde{p}_0(t)^1) \in H$$
$$\tilde{p}_0(t)^0 = p_0(t) \in R^n \qquad (10.4.12)$$
$$\tilde{p}_0(t)^1 = \{p_0(t-\tau),\ \tau \in [-h, 0)\} \in L^2(-h, 0; R^n).$$

10.4.3 Optimal regulator

In order to express the optimal control value $u_0(s)$ by the state $\tilde{x}_0(s)$ for some $s \in [0, T]$, we have to consider the optimization problem in the interval $[s, T]$. On the trajectories of system (10.4.1), considered in $[s, T]$ with the initial condition φ at time s, we minimize the functional

$$J_s(u) = x(T)^T [Q x(T) + 2q] +$$
$$+ \int_s^T [x(t)^T W(t) x(t) + 2 w(t)^T x(t) + u(t)^T u(t)] dt. \qquad (10.4.13)$$

The set of admissible controls is $L^2(s, T; R^m)$.

Because of the relationships (10.4.6), (10.4.9) it is evident that it is enough to find the dependence between $p_0(s)$ and $\tilde{x}_0(s)$.

Theorem 10.4.2 Let x_0, p_0 be the solution of (10.4.4)–(10.4.6) and let $\tilde{x}_0$ be the state trajectory corresponding to the trajectory of instantaneous state x_0. For every $s, t \in [0, T], s \leqslant t$ there is a linear continuous operator $D(t, s) : H \to R^n$ and a vector $d(t, s) \in R^n$ such that

$$p_0(t) = D(t, s) \tilde{x}_0(s) + d(t, s). \qquad (10.4.14)$$

$D(t, s)$ and $d(t, s)$ are determined by the following rules.

10.4 Finite-time Problem with Arbitrary Delays

(i) *Solve the system*

$$\dot{\beta}(t)+ \sum_{i=0}^{N} B_i(t)\beta(t-h_i)+ \int_{-h}^{0} B(t,\tau)\beta(t+\tau)\mathrm{d}\tau +$$
$$+ C(t)C(t)^{\mathrm{T}}\gamma(t) = 0 \quad t \in [s, T]$$
$$\beta(s) = \varphi^0 \quad \beta(s+t) = \varphi^1(t) \quad t \in [-h, 0) \quad (10.4.15)$$

where

$$\varphi = (\varphi^0, \varphi^1) \in H$$

$$\dot{\gamma}(t) - \sum_{i=0}^{N} B_i(t+h_i)^{\mathrm{T}}\gamma(t+h_i) - \int_{-h}^{0} B(t-\tau, \tau)^{\mathrm{T}}\gamma(t-\tau)\mathrm{d}\tau +$$
$$+ W(t)\beta(t) = 0 \quad t \in [s, T]$$
$$\gamma(T) = Q\beta(T) \quad \gamma(t) = 0 \quad t > T. \quad (10.4.16)$$

Hence we obtain

$$D(t, s)\varphi = \gamma(t) \quad t \in [s, T]. \quad (10.4.17)$$

(ii) *Solve the system*

$$\dot{\eta}(t)+ \sum_{i=0}^{N} B_i(t)\eta(t-h_i)+ \int_{-h}^{0} B(t,\tau)\eta(t+\tau)\mathrm{d}\tau +$$
$$+ C(t)C(t)^{\mathrm{T}}\xi(t) = 0 \quad t \in [s, T] \quad (10.4.18)$$
$$\eta(t) = 0 \quad t \leqslant s$$

$$\dot{\xi}(t) - \sum_{i=0}^{N} B_i(t+h_i)^{\mathrm{T}}\xi(t+h_i) - \int_{-h}^{0} B(t-\tau, \tau)^{\mathrm{T}}\xi(t-\tau)\mathrm{d}\tau +$$
$$+ W(t)\eta(t)+w(t) = 0 \quad t \in [s, T]$$
$$\xi(T) = Q\eta(T)+q \quad \xi(t) = 0 \quad t > T. \quad (10.4.19)$$

Hence we obtain

$$d(t, s) = \xi(t) \quad t \in [s, T]. \quad (10.4.20)$$

This theorem is an obvious consequence of Theorem 10.4.1. Let us note that for fixed $s \in [0, T)$ and $\varphi \in H$, the functions $t \mapsto D(t, s)\varphi$ and $t \mapsto d(t, s)$ belong to $W^{1,2}(s, T; R^n)$.

Now we shall reformulate the results using state space notation.

Theorem 10.4.3 *Let $\tilde{x}_0$ and $\tilde{p}_0$ be the solution of system* (10.4.7)–(10.4.9). *For almost every $s \in [0, T)$ there is a unique $K(s) \in L(H)$ and a unique $k(s) \in H$ such that*

$$\tilde{p}_0(s) = K(s)\Lambda\tilde{x}_0(s)+k(s). \quad (10.4.21)$$

$K(s)$ and $k(s)$ are obtained in the following way.
   (i) Solve the system
$$\dot{\beta}(t)+\tilde{B}(t)\beta(t)+\tilde{C}(t)\tilde{C}(t)^*\gamma(t) = 0 \quad t \in [s, T] \quad (10.4.22)$$
$$\beta(s) = \varphi \in V$$
$$\dot{\gamma}(t)-\tilde{B}(t)^*\gamma(t)+\Lambda^*\tilde{W}(t)\Lambda\beta(t) = 0 \quad t \in [s, T]$$
$$\gamma(T) = \tilde{Q}\Lambda\beta(T). \quad (10.4.23)$$

Hence
$$K(s)\Lambda\varphi = \gamma(s). \quad (10.4.24)$$

(ii) Solve the system
$$\dot{\eta}(t)+\tilde{B}(t)\eta(t)+\tilde{C}(t)\tilde{C}(t)^*\xi(t) = 0 \quad t \in [s, T]$$
$$\eta(s) = 0 \in V \quad (10.4.25)$$
$$\dot{\xi}(t)-\tilde{B}(t)^*\xi(t)+\Lambda^*[\tilde{W}(t)\Lambda\eta(t)+\tilde{w}(t)] = 0 \quad t \in [s, T]$$
$$\xi(T) = \tilde{Q}\Lambda\eta(T)+\tilde{q}. \quad (10.4.26)$$

Hence
$$k(s) = \xi(s). \quad (10.4.27)$$

The operator $K(s)$ and relationship (10.4.21) play a central role in optimal regulator theory.

10.4.4 General characterization of operator $K(s)$

We shall study the properties of the linear part of the optimal regulator, $K(s)$. In order to do this we assume $w = 0$ and $q = 0$ in the functional (10.4.3). Then $d = 0$ in (10.4.14) and $k = 0$ in (10.4.21), so the optimal regulator is a linear one.

Theorem 10.4.4
   (i) *Denote by β (resp. $\bar{\beta}$) the solution of the equation*
$$\dot{\beta}(t)+[\tilde{B}(t)+\tilde{C}(t)\tilde{C}(t)^*K(t)\Lambda]\beta(t) = 0 \quad t \in [s, T] \quad (10.4.28)$$
$$\beta(s) = \varphi \ (resp. \ \bar{\varphi}) \ in \ V.$$

Then
$$(K(s)\Lambda\varphi|\Lambda\bar{\varphi}) = (\tilde{Q}\Lambda\beta(T)|\Lambda\bar{\beta}(T))+$$
$$+\int_s^T ([\tilde{W}(t)+K(t)\tilde{C}(t)\tilde{C}(t)^*K(t)]\Lambda\beta(t)|\Lambda\bar{\beta}(t))dt \quad (10.4.29)$$

and
$$J_s(u_0) = (K(s)\Lambda\varphi|\Lambda\varphi). \quad (10.4.30)$$

$J_s(u_0)$ *is the optimal value of the functional with initial condition φ.*

10.4 Finite-time Problem with Arbitrary Delays

(ii) *The operator $K(s)$ is a selfadjoint nonnegative definite element of $L(H)$, and there is a constant $c > 0$ (independent of s and φ) such that*

$$\forall s \in [0, T] \quad \forall \varphi \in H \quad \|K(s)\varphi\| \leq c\|\varphi\|. \tag{10.4.31}$$

The function

$$s \mapsto K(s): [0, T] \to L(H) \tag{10.4.32}$$

is weakly continuous, then strongly measurable and bounded. The operator $K(s)$ can be decomposed in a unique way into four operators

$$K(s)(\varphi^0, \varphi^1) = \big(K_{00}(s)\varphi^0 + K_{01}(s)\varphi^1, K_{10}(s)\varphi^0 + K_{11}(s)\varphi^1\big) \quad \forall (\varphi^0, \varphi^1) \in H$$

$$K_{00}(s) \in L(R^n) \quad K_{01}(s) \in L\big(L^2(-h, 0; R^n), R^n\big)$$
$$K_{10}(s) \in L\big(R^n, L^2(-h, 0; R^n)\big) \tag{10.4.33}$$
$$K_{11}(s) \in L\big(L^2(-h, 0; R^n)\big).$$

Moreover,

$$K_{00}(s) = K_{00}(s)^* \geq 0 \quad K_{01}(s) = K_{10}(s)^*$$
$$K_{11}(s) = K_{11}(s)^* \geq 0. \tag{10.4.34}$$

There is a function $K_{01}(s, \cdot) \in L^2(-h, 0; L(R^n))$ such that

$$K_{01}(s)\varphi^1 = \int_{-h}^{0} K_{01}(s, \alpha)\varphi^1(\alpha) \, d\alpha \tag{10.4.35}$$

and there is a function $K_{11}(s, \cdot, \cdot) \in L^2(-h, 0; -h, 0; L(R^n))$ such that

$$[K_{11}(s)\varphi^1](\tau) = \int_{-h}^{0} K_{11}(s, \tau, \alpha)\varphi^1(\alpha) \, d\alpha. \tag{10.4.36}$$

(iii) *If we denote by $\tilde{\Psi}(t, s) \in L(H)$ the evolution operator (in the sense of (3.3.24), (3.3.25)) generated by $\tilde{B}(t) + \tilde{C}(t)\tilde{C}(t)^*K(t)\Lambda$, then equation (10.4.29) may be rewritten in the form*

$$(K(s)\varphi|\bar{\varphi}) = (\tilde{Q}\tilde{\Psi}(t, s)\varphi|\tilde{\Psi}(t, s)\bar{\varphi}) +$$
$$+ \int_{s}^{T} ([\tilde{W}(t) + K(t)\tilde{C}(t)\tilde{C}(t)^*K(t)]\tilde{\Psi}(t, s)\varphi|\tilde{\Psi}(t, s)\bar{\varphi}) \, dt \quad \forall \varphi, \bar{\varphi} \in H.$$
$$\tag{10.4.37}$$

This equation (or more precisely, set of equations, after adding the equation defining $\tilde{\Psi}$) has a unique solution with respect to K.

(iv) *For all $\varphi, \bar{\varphi} \in H$ the function $s \mapsto (K(s)\varphi|\bar{\varphi}): [0, T] \to R$ is continuous; for all $\varphi, \bar{\varphi} \in V$ the function $s \mapsto (K(s)\Lambda\varphi|\Lambda\bar{\varphi}): [0, T] \to R$ is differentiable; the function $s \mapsto K(s)$ has a distributional derivative $[0, T] \to L(V, V')$ strongly measurable and bounded. Moreover, the following operator Riccati equation is satisfied:*

$$\dot{K}(t) - \tilde{B}(t)^*K(t)\Lambda - \Lambda^*K(t)\tilde{B}(t) + \Lambda^*[\tilde{W}(t) - K(t)\tilde{C}(t)\tilde{C}(t)^*K(t)]\Lambda = 0$$
(10.4.38)

$$K(T) = \tilde{Q}.$$
(10.4.39)

(v) *Equation* (10.4.38) *with final condition* (10.4.39) *has a unique solution in* $W(0, T; H, V, V') = \{P: [0, T] \to L(H): \forall \varphi, \bar{\varphi} \in H$ (*resp.* V) *the function* $t \mapsto (P(t)\varphi|\bar{\varphi})$ *is continuous* (*resp.* $t \mapsto (P(t)\Lambda\varphi|\Lambda\bar{\varphi})$ *has a distributional derivative in* $L^\infty(0, T; R))\}$.

Proof (i) By virtue of Theorem 10.4.3 $K(s)\Lambda\varphi = \gamma(s)$. Then

$$(K(s)\Lambda\varphi|\Lambda\bar{\varphi}) = (\gamma(s)|\Lambda\bar{\varphi})$$

$$= \left(\tilde{\Phi}(T, s)^*\tilde{Q}\Lambda\beta(T) + \int_s^T \tilde{\Phi}(t, s)^*\tilde{W}(t)\Lambda\beta(t)\,dt \Big| \Lambda\bar{\varphi}\right)$$

$$= (\tilde{Q}\Lambda\beta(T)|\tilde{\Phi}(T, s)\Lambda\bar{\varphi}) + \int_s^T (\tilde{W}(t)\Lambda\beta(t)|\tilde{\Phi}(t, s)\Lambda\bar{\varphi})\,dt.$$

Since

$$\tilde{\Phi}(t, s)\Lambda\bar{\varphi} = \Lambda\bar{\beta}(t) + \int_s^t \tilde{\Phi}(t, \tau)\tilde{C}(\tau)\tilde{C}(\tau)^*\bar{\gamma}(\tau)\,d\tau,$$

we have

$$(K(s)\Lambda\varphi|\Lambda\bar{\varphi}) = (\tilde{Q}\Lambda\beta(T)|\Lambda\bar{\beta}(T)) + \int_s^T (\tilde{W}(t)\Lambda\beta(t)|\Lambda\bar{\beta}(t))\,dt +$$

$$+ \int_s^T \Big[(\tilde{Q}\Lambda\beta(T)|\tilde{\Phi}(T, t)\tilde{C}(t)\tilde{C}(t)^*\bar{\gamma}(t)) +$$

$$+ \int_s^t (\tilde{W}(t)\Lambda\beta(t)|\tilde{\Phi}(t, \tau)\tilde{C}(\tau)\tilde{C}(\tau)^*\bar{\gamma}(\tau))\,d\tau\Big]\,dt$$

$$= (\tilde{Q}\Lambda\beta(T)|\Lambda\bar{\beta}(T)) + \int_s^T (\tilde{W}(t)\Lambda\bar{\beta}(t)|\Lambda\tilde{\beta}(t))\,dt + \int_s^T (\tilde{\Phi}(T, t)^*\tilde{Q}\Lambda\beta(T) +$$

$$+ \int_t^T \tilde{\Phi}(\tau, t)^*\tilde{W}(\tau)\Lambda\beta(\tau)\,d\tau | C(t)C(t)^*\bar{\gamma}(t))\,dt$$

$$= (\tilde{Q}\Lambda\beta(T)|\Lambda\bar{\beta}(T)) +$$

$$+ \int_s^T [(\tilde{W}(t)\Lambda\beta(t)|\Lambda\bar{\beta}(t)) + (\gamma(t)|\tilde{C}(t)\tilde{C}(t)^*\bar{\gamma}(t))]\,dt.$$

(10.4.40)

10.4 Finite-time Problem with Arbitrary Delays

Hence we obtain (10.4.29) and (10.4.30).

(ii) First we shall prove (10.4.31). For every $\varphi \in V$

$$(K(s)\Lambda\varphi|\Lambda\varphi) = J_s(u_0, \Lambda\varphi) \leqslant J_s(0, \Lambda\varphi)$$

where $J_s(u, \Lambda\varphi)$ is the value of the functional (10.4.13) with control u and initial condition $\tilde{x}(s) = \Lambda\varphi$. There exists a constant $c > 0$ (independent of s and φ) such that

$$J_s(0, \Lambda\varphi) \leqslant c\|\varphi\|_H^2.$$

As Λ is a continuous and dense injection, we obtain from these two inequalities

$$(K(s)\varphi|\varphi) \leqslant c\|\varphi\|_H^2 \quad \forall \varphi \in H$$

hence (10.4.31).

Since $\tilde{Q}, \tilde{W}(t)$, and $\tilde{C}(t)\tilde{C}(t)^*$ are selfadjoint and nonnegative definite, by virtue of (10.4.40) $K(s)$ is selfadjoint and nonnegative definite. The weak continuity of K is understood as the continuity of $s \mapsto (K(s)\varphi|\bar{\varphi})$, $\forall \varphi, \bar{\varphi} \in H$, $s \in [0, T]$. This follows easily from (10.4.29).

The decomposition of $K(s)$ in (10.4.33) is obvious, the symmetry and nonnegative definiteness follow immediately from $K(s) = K(s)^* \geqslant 0$. The function $K_{01}(s, \cdot)$ exists by virtue of the theorem on the general form of a linear bounded functional in L^2. The existence of the function $K_{11}(s, \cdot, \cdot)$ follows easily from (10.4.37) after substitution $\varphi = (0, \varphi^1)$, $\bar{\varphi} = (0, \bar{\varphi}^1)$.

(iii) Since K is measurable and bounded, the operator $\tilde{B}(t) + \tilde{C}(t)\tilde{C}(t)^*K(t)\Lambda$ is the infinitesimal generator of a strongly continuous evolution operator defined by means of the solution of equation (10.4.28) $\tilde{\Psi}(t, s)\varphi = \beta(t)$. The uniqueness of the solution of (10.4.37) follows from the fact that the optimal control is unique.

(iv) For every $\varphi \in H$ the function $(t, s) \mapsto \tilde{\Psi}(t, s)\varphi$: $\{(t, s): s \in [0, T], s \leqslant t \leqslant T\} \to H$ is continuous, for every $\varphi \in V$

$$\tilde{\Psi}(t, s)\Lambda\varphi = \Lambda\varphi + \int_s^t \tilde{\Psi}(t, \tau)[\tilde{B}(\tau) + \tilde{C}(\tau)\tilde{C}(\tau)^*K(\tau)\Lambda]\varphi\,d\tau$$

and the function $s \mapsto \tilde{\Psi}(t, s)\Lambda\varphi$ has a distributional derivative in $L^\infty(0, t; H)$. From this it follows that for all $\varphi, \bar{\varphi} \in V$ the function $s \mapsto (K(s)\Lambda\varphi|\Lambda\bar{\varphi})$ has a distributional derivative which may be calculated by the differentiation of (10.4.37) with respect to s. Hence we obtain (10.4.38).

(v) Obviously $K \in W(0, T; H, V, V')$. Suppose that (10.4.38) has two solutions K_1 and K_2. Let $D = K_1 - K_2$. Then $D(T) = 0$ and

$$\dot{D}(t) - [\tilde{B}(t) + \tilde{C}(t)\tilde{C}(t)^*K_2(t)\Lambda]^*D(t)\Lambda - \Lambda^*D(t)[\tilde{B}(t) + \\ + \tilde{C}(t)\tilde{C}(t)^*K_2(t)\Lambda] - \Lambda^*D(t)\tilde{C}(t)\tilde{C}(t)^*D(t)\Lambda = 0 \quad (10.4.41)$$

and
$$\dot{D}(t) - [\tilde{B}(t) + \tilde{C}(t)\tilde{C}(t)^*K_1(t)\Lambda]^*D(t)\Lambda - \Lambda^*D(t)[\tilde{B}(t) + \\ + \tilde{C}(t)\tilde{C}(t)^*K_1(t)\Lambda] + \Lambda^*D(t)\tilde{C}(t)\tilde{C}(t)^*D(t)\Lambda = 0. \quad (10.4.42)$$

Hence
$$(\varphi|D(s)\varphi) = -\int_s^T \left(D(t)\tilde{\Psi}_1(t,s)\varphi|\tilde{C}(t)\tilde{C}(t)^*D(t)\tilde{\Psi}_1(t,s)\varphi\right)dt$$
$$= \int_s^T \left(D(t)\tilde{\Psi}_2(t,s)\varphi|\tilde{C}(t)\tilde{C}(t)^*D(t)\tilde{\Psi}_2(t,s)\varphi\right)dt.$$

where $\tilde{\Psi}_i(t,s)$ is the evolution operator of the equation
$$\dot{D}(t) - [\tilde{B}(t) + \tilde{C}(t)\tilde{C}(t)^*K_i(t)\Lambda]^*D(t)\Lambda - \\ - \Lambda^*D(t)[\tilde{B}(t) + \tilde{C}(t)\tilde{C}(t)^*K_i(t)\Lambda] = 0.$$

Therefore for any φ, $(\varphi|D(s)\varphi)$ is at the same time nonnegative and nonpositive, hence $D(s) = 0$ and $K_1 = K_2$. □

10.4.5 Characterization of components of $K(s)$ by means of differential equations

The functions K_{00}, K_{01}, K_{10} and K_{11} can be determined by means of a set of differential equations. In a non-abstract formulation, this set is an analogue of the Riccati equation for the optimal regulator in the finite-dimensional linear-quadratic problem, and therefore it is also sometimes called 'the Riccati set of equations'. As we shall see, the equations for the time-delay case are rather complicated.

Theorem 10.4.5 Let the functions B, B_i, $i = 1, ..., N$, be continuously differentiable. Then:

(i) *The function $t \mapsto K_{00}(t)$ is continuous and for almost all $t \in [0, T]$ satisfies the equation*
$$\dot{K}_{00}(t) - K_{00}(t)C(t)C(t)^T K_{00}(t) - K_{00}(t)B_0(t) - B_0(t)^T K_{00}(t) + \\ + K_{01}(t, 0) + K_{01}(t, 0)^T + W(t) = 0 \quad (10.4.43)$$
$$K(T) = Q. \quad (10.4.44)$$

(ii) *The function $(t, \xi) \mapsto K_{01}(t, \xi): [0, T] \times [-h, 0] \to L(R^n)$ may be discontinuous only at points of the set $\{(t, \xi): \xi = -h_i \text{ or } \xi = T - t - h_j, i = 1, ..., N-1, j = 1, ..., N\}$ and almost everywhere satisfies the equation*
$$\left(\frac{\partial}{\partial t} - \frac{\partial}{\partial \xi}\right)K_{01}(t, \xi) - [B_0(t) + C(t)C(t)^T K_{00}(t)]^T K_{01}(t, \xi) - \\ - K_{00}(t)B(t, \xi) + K_{11}(t, 0, \xi) = 0 \quad (10.4.45)$$

10.4 Finite-time Problem with Arbitrary Delays

with boundary conditions
$$K_{01}(T, \xi) = 0 \quad K_{01}(t, -h) = -K_{00}(t)B_N(t) \quad \xi \in [-h, 0]$$
$$t \in [0, T). \tag{10.4.46}$$

The jumps at $\xi = -h_i$ are determined by
$$K_{01}(t, -h_i+) - K_{01}(t, -h_i-) = -K_{00}(t)B_i(t) \quad t \in [0, T)$$
$$i = 1, \ldots, N-1 \tag{10.4.47}$$

and at $\xi = T-t-h_i$
$$K_{01}(t, T-t-h_i+) - K_{01}(t, T-t-h_i-) = 0 \quad \text{if} \quad QB_i(T) = 0$$
$$t \in [\max(0, T-h_i), T] \quad (i = 1, \ldots, N). \tag{10.4.48}$$

(iii) *The function* $(t, \xi, \zeta) \mapsto K_{11}(t, \xi, \zeta)$: $[0, T] \times [-h, 0] \times [-h, 0]$
$\to L(R^n)$ *may be discontinuous only in the set* $\{(t, \xi, \zeta): \xi = -h_i \text{ or } \zeta = -h_i,$
or $\xi = T-t-h_j$, *or* $\zeta = T-t-h_j$, $i = 1, \ldots, N-1$, $j = 1, \ldots, N\}$ *and almost everywhere satisfies the equation*

$$\left(\frac{\partial}{\partial t} - \frac{\partial}{\partial \xi} - \frac{\partial}{\partial \zeta}\right) K_{11}(t, \xi, \zeta) - K_{01}(t, \xi)^T B(t, \zeta) -$$
$$- B(t, \xi)^T K_{01}(t, \zeta) - K_{01}(t, \xi) C(t) C(t)^T K_{01}(t, \zeta) = 0 \tag{10.4.49}$$

with boundary conditions
$$K_{11}(t, \xi, -h) = -K_{01}(t, \xi)^T B_N(t) \quad t \in [0, T) \quad \xi \in [-h, 0]$$
$$K_{11}(T, \xi, \zeta) = 0 \quad \xi, \zeta \in [-h, 0]. \tag{10.4.50}$$

Moreover
$$K_{11}(t, \xi, \zeta) = K_{11}(t, \zeta, \xi)^T \tag{10.4.51}$$

hence we easily obtain the remaining boundary conditions. The jumps are determined by
$$K_{11}(t, -h_i+, \zeta) - K_{11}(t, -h_i-, \zeta) = -B_i(t)^T K_{01}(t, \zeta) \quad t \in [0, T)$$
$$\zeta \in [-h, 0] \quad (i = 1, \ldots, N-1) \tag{10.4.52}$$
$$K_{11}(t, T-t-h_i+, \zeta) - K_{11}(t, T-t-h_i-, \zeta) = 0 \quad \text{if} \quad QB_i(T) = 0$$
$$t \in [\max(0, T-h_i), T] \quad \zeta \in [-h, 0] \quad (i = 1, \ldots, N).$$

The jumps at $\zeta = -h_i$, $\zeta = T-t-h_i$ are obtained by means of (10.4.51).

(iv) *The system of equations* (10.4.43)–(10.4.52) *with appropriate boundary conditions and jump values has a unique solution, continuous outside the described discontinuity sets.*

Proof First we determine the boundary conditions and the jumps at the discontinuity points. From (10.4.29) it follows easily that the function $t \mapsto K_{00}(t)$ is continuous and (10.4.44) holds. Further, we use the identity (cf. Theorem 3.2.1, formula (3.2.9))

$$\Psi^1(t, s, \tau) = \sum_{i=1}^{N} \begin{Bmatrix} \Psi(t, s+\tau+h_i) B_i(s+\tau+h_i) & s+\tau-t < -h_i \leq \tau \\ 0 & \text{otherwise} \end{Bmatrix} +$$

$$+ \int_{\max(-h, s+\tau-t)}^{\tau} \Psi(t, s+\tau-\xi)[B(s+\tau-\xi, \xi) +$$

$$+ C(s+\tau-\xi)C(s+\tau-\xi)^T K_{01}(s+\tau-\xi, \xi)]d\xi. \qquad (10.4.53)$$

From this and from careful analysis of (10.4.29) it follows that $K_{01}(\cdot, \cdot)$ may have discontinuities only at $\xi = -h_i$, $i = 1, \ldots, N-1$, or $\xi = T-t-h_i$, $i = 1, \ldots, N$. (10.4.46)–(10.4.48) are obtained also from (10.4.29) and from the following equality:

$$\Psi^1(t, s, -h_i+) - \Psi^1(t, s, -h_i-) = \Psi(t, s)B_i(s). \qquad (10.4.54)$$

Similarly, we obtain that the function $(s, \xi, \zeta) \mapsto K_{11}(s, \xi, \zeta)$ may have discontinuities only at $\xi = T-s-h_i$, $\zeta = T-s-h_i$, $\xi = -h_i$, $\zeta = -h_i$ and that conditions (10.4.50) and (10.4.52) hold. Condition (10.4.51) follows immediately from (10.4.34).

To derive the differential equations (10.4.43), (10.4.45) and (10.4.49) we write the abstract Riccati differential equation in the form

$$\frac{d}{dt}(\Lambda\varphi|K(t)\Lambda\bar{\varphi}) - (\tilde{B}(t)\varphi|K(t)\Lambda\bar{\varphi}) - (K(t)\Lambda\varphi|\tilde{B}(t)\bar{\varphi}) +$$

$$+ ([\tilde{W}(t) - K(t)\tilde{C}(t)\tilde{C}(t)^*K(t)]\Lambda\varphi|\Lambda\bar{\varphi}) = 0 \quad \forall \varphi, \bar{\varphi} \in V. \qquad (10.4.55)$$

Let us define

$$\varphi_a(t) = \begin{cases} \varphi^0 \left(1 + a \dfrac{t}{h}\right) & -\dfrac{h}{a} \leq t \leq 0 \\ 0 & \text{otherwise} \end{cases}$$

$$\varphi_b(t) = \begin{cases} \varphi^0 \left(1 + b \dfrac{t}{h}\right) & -\dfrac{h}{b} \leq t \leq 0 \\ 0 & \text{otherwise}. \end{cases}$$

If we substitute $\varphi = \varphi_a$ and $\bar{\varphi} = \varphi_b$ into (10.4.55), by standard calculations involving the limit procedures $a \to \infty$, $b \to \infty$ we obtain (10.4.43). To obtain (10.4.45) we substitute $\varphi = \varphi_a$ and take a $\bar{\varphi} \in V$ such that supp $\bar{\varphi} \subset (-h_i, -h_{i-1})$, for an arbitrary i.

Standard arguments, with $a \to \infty$, yield (10.4.45). The third equation of the system, (10.4.49) is obtained after the substitution of $\varphi, \bar{\varphi} \in V$ such that supp $\varphi \subset (-h_i, -h_{i-1})$, supp $\bar{\varphi} \subset (-h_j, -h_{j-1})$. $\square$

10.4 Finite-time Problem with Arbitrary Delays

10.4.6 Characterization of the free term of the optimal regulator

We shall derive a differential equation for the free term k of the optimal regulator (10.4.21). To this end we consider the linear-quadratic problem 10.4.1), (10.4.13) with a zero initial condition $\varphi = 0$.

Theorem 10.4.6 *The free term* $k: [0, T] \to H$ *in formula* (10.4.21) *is the unique solution in* $W^*(0, T)$ *of the equation*

$$\dot{k}(t) = [\tilde{B}(t)^* + \Lambda^* K(t) \tilde{C}(t) \tilde{C}(t)^*] k(t) - \Lambda^* \tilde{w}(t) \quad (10.4.56)$$

$$k(T) = \tilde{q}. \quad (10.4.57)$$

Proof We differentiate the equality (10.4.21). This can be done as $\tilde{p}_0$ and $K(\cdot)\Lambda\tilde{x}_0(\cdot)$ belong to $W^*(0, T)$. $\dot{K}(t)$ is determined by the operator Riccati equation (10.4.38), $\dot{\tilde{x}}_0(t)$ by the state equation and $\dot{\tilde{p}}_0(t)$ by the adjoint state equation. Assuming $\tilde{x}_0(s) = 0$ in problem (10.4.1), (10.4.13) we obtain

$$\dot{k}(s) = \dot{\tilde{p}}_0(s) - \Lambda^* K(s) \dot{\tilde{x}}_0(s)$$

for any $s \in [0, T)$. Hence (10.4.56) follows easily. The final condition for k results from the final condition for $\tilde{p}_0$. □

10.4.7 Generalizations

First we shall show how the theory presented in the previous sections can be applied to problems with delays in the control in the system equation. Let us consider a linear-quadratic problem in which the system is described by the equation

$$\dot{x}(t) + \sum_{i=0}^{N} [B_i(t) x(t-h_i) + C_i(t) u(t-h_i)] +$$

$$+ \int_{-h}^{0} [B(t, \tau) x(t+\tau) + C(t, \tau) u(t+\tau)] d\tau = 0 \quad t \geq 0 \quad (10.4.58)$$

with the initial condition

$$x(0) = \varphi^0 \quad x(t) = \varphi^1(t) \quad t \in [-h, 0)$$
$$\varphi = (\varphi^0, \varphi^1) \in H \quad (10.4.59)$$
$$u(t) = v(t) \quad t \in [-h, 0) \quad v \in L^2(-h, 0; R^m).$$

The function C is bounded and measurable with respect to both arguments; the remaining assumptions are as for (10.1.1). The performance functional to be minimized is (10.4.3) or (10.4.13), respectively. The necessary and sufficient condition of optimality which in the proof of Theorem 10.4.1 has the form

$$\int_0^T v(t)^T[u_0(t) - C(t)^T p_0(t)]\,dt = 0 \quad \forall v \in L^2(0, T; R^m) \quad (10.4.60)$$

is replaced in our problem by

$$\int_0^T v(t)^T u_0(t)\,dt = \int_0^T p_0(t)^T \Big[\sum_{i=0}^N C_i(t)v(t-h_i) +$$

$$+ \int_{-h}^0 C(t, \tau)v(t+\tau)\,d\tau\Big]dt \quad \forall v \in L^2(-h, T; R^m)$$

(10.4.61)

where $v(t) = 0$, $t < 0$.

A simple derivation leads to the conclusion that Theorem 10.4.1 remains valid if only we replace formula (10.4.6) by

$$u_0(t) = \sum_{i=0}^N C_i(t+h_i)^T p_0(t+h_i) + \int_t^{t+h} C(\tau, t-\tau)^T p_0(\tau)\,d\tau$$

$$t \in [0, T] \quad (10.4.62)$$

with zero extensions of the functions C_i, and C beyond the interval $[0, T]$ and $[0, T] \times [-h, 0]$.

The theory of optimal regulator can be developed as before. However, as the notation becomes complicated we shall not go into that here.

Let us consider in turn a system described by an integro-differential equation

$$\dot{x}(t) + B_0(t)x(t) + \int_0^t B_1(t, \tau)x(\tau)\,d\tau + C(t)u(t) = f(t)$$

$$t \in [0, T] \quad (10.4.63)$$

were B_0 and B_1 are bounded and measurable. The performance functional is given by (10.4.3). As shown by Delfour (1977a), this problem is covered by the present theory. To demonstrate this, let us define

$$B(t, \tau) = \begin{cases} B_1(t, t+\tau) & -t \leqslant \tau \leqslant 0 \\ 0 & -T \leqslant \tau \leqslant -t \end{cases} \quad (10.4.64)$$

and rewrite (10.4.63) in the form

$$\dot{x}(t) + B_0(t) + \int_{-T}^0 B(t, \tau)x(t+\tau)\,d\tau + C(t)u(t) = f(t)$$

$$t \in [0, T] \quad (10.4.65)$$

$$x(0) = \varphi^0, \quad x(t) = \varphi^1(t) \quad t \in [-T, 0), \quad \varphi^1 \in L^2.$$

As could have been expected the system is independent of φ^1, but the theory can be applied immediately.

10.5 Infinite-time Problem with Arbitrary Delays

10.5.1 Problem formulation

Following Delfour, McCalla and Mitter (1975) we shall present below the general theory of linear-quadratic problem with infinite control time. We consider a stationary system described by

$$\dot{x}(t) + \sum_{i=0}^{N} B_i x(t-h_i) + \int_{-h}^{0} B(\tau) x(t+\tau) d\tau + Cu(t) = 0 \quad t \geq 0 \quad (10.5.1)$$

with initial condition

$$x(0) = \varphi^0 \quad x(t) = \varphi^1(t) \quad t \in [-h, 0) \quad (10.5.2)$$

$$\varphi = (\varphi^0, \varphi^1) \in H.$$

As before, $x(t)$ is an n-vector, $u(t)$ is an m-vector, B_i and C are matrices of compatible dimensions and B is a bounded measurable function with matrix values. For the space of admissible controls we take $L^2_{loc}(0, \infty; R^m)$.

The performance functional to be minimized on the trajectories of (10.5.1) is

$$J(u) = \int_0^{\infty} [x(t)^T W x(t) + u(t)^T u(t)] dt \quad (10.5.3)$$

where $W = W^T \geq 0$. It is evident that the functional J takes finite values only on some subset of $L^2_{loc}(0, \infty; R^m)$. If system (10.5.1) is asymptotically stable, then the domain of J is the whole of $L^2(0, \infty; R^m)$, independently of the initial condition φ. A sufficient condition that the domain of J be nonempty for any initial condition is the stabilizability of system (10.5.1). Stability was discussed at length in Chapter 6; a sufficient and necessary condition of stabilizability is given by the following theorem.

Theorem 10.5.1 *System* (10.5.1) *is stabilizable if and only if*

$$\text{rank}[\Delta_0(\lambda) \quad C] = n \quad \forall \lambda \quad \text{Re } \lambda \geq 0 \quad (10.5.4)$$

where

$$\Delta_0(\lambda) = \lambda I + \sum_{i=0}^{N} e^{-\lambda h_i} B_i + \int_{-h}^{0} e^{\lambda \tau} B(\tau) d\tau. \quad (10.5.5)$$

The proof can be found in Pandolfi (1975).

10.5.2 Optimal regulator

An examination of the asymptotic behaviour (as $T \to \infty$) of the optimal regulator in the problem with finite control time yields important results pertaining to the infinite-time linear-quadratic problem. Let us consider an auxiliary linear-quadratic problem with system equation (10.5.1) and performance functional

$$J_T(u) = \int_0^T [x(t)^T W x(t) + u(t)^T u(t)] dt. \qquad (10.5.6)$$

It follows from the previous section that the optimal control in problem (10.5.1), (10.5.6) has the form

$$u_0(t) = \tilde{C}^* K_T(t) \Lambda \tilde{x}_0(t) \qquad t \in [0, T] \qquad (10.5.7)$$

where K_T is characterized in Theorem 10.4.3. The optimal value of the performance functional is

$$J_T(u_0) = (\Lambda\varphi | K_T(0) \Lambda\varphi)_H. \qquad (10.5.8)$$

Theorem 10.5.2 *Assume that system (10.5.1) is stabilizable. Then*:
(i) *There is an operator* $K \in L(H)$, $K = K^* \geq 0$ *such that*

$$\lim_{t < T \to \infty} K_T(t)\varphi = K\varphi \qquad \forall \varphi \in H \quad \forall t \geq 0. \qquad (10.5.9)$$

(ii) *For every* $\varphi \in V$,

$$(K\Lambda\varphi | \Lambda\varphi) = \int_0^\infty ((\tilde{W} + K\tilde{C}\tilde{C}^* K)\Lambda\tilde{x}(t) | \Lambda\tilde{x}(t)) dt \qquad (10.5.10)$$

where $\tilde{x}$ is the solution of the equation

$$\dot{z}(t) + (\tilde{B} + \tilde{C}\tilde{C}^* K\Lambda) z(t) = 0 \qquad t \in [0, \infty) \qquad (10.5.11)$$
$$z(0) = \varphi \in V.$$

(iii) *For every* $\varphi \in V$

$$(K\Lambda\varphi | \Lambda\varphi) = J(\tilde{C}^* K\Lambda\tilde{x}(\cdot)). \qquad (10.5.12)$$

Proof Consider the problem of optimal control in the interval $[s, T]$. Let J_T^s denote a functional created from (10.5.3) by replacing the upper and lower integration limits by T and s, respectively. By virtue of the stabilizability assumption there is a feedback operator $\tilde{G}$ of the form (10.1.11) such that the operator $\tilde{B} + \tilde{C}\tilde{G}$ is L^2-stable. It is evident that for $T > s \geq 0$

$$(K_T(s)\Lambda\varphi | \Lambda\varphi)_H = \inf\{J_T^s(u): u \in L^2(s, T; R^m)\}$$
$$\leq \|W\| \int_0^T \|z(t)\|^2 dt + \int_0^T \|\tilde{G}\tilde{z}(t)\|^2 dt \qquad (10.5.13)$$

10.5 Infinite-time Problem with Arbitrary Delays

where $\tilde{z}$ is the solution of
$$\dot{\tilde{z}}(t)+(\tilde{B}+\tilde{C}\tilde{G})\tilde{z}(t) = 0 \quad t \geqslant 0 \tag{10.5.14}$$
$$\tilde{z}(0) = \varphi.$$

The state trajectory $\tilde{z}$ is connected with z by (3.3.16). Further,

$$\left[\int_0^T ||\tilde{G}\tilde{z}(t)||^2 dt\right]^{1/2} \leqslant \sum_{i=0}^M ||G_i|| \left[\int_{-\tau_i}^0 ||\varphi(t)||^2 dt + \int_0^T ||z(t)||^2 dt\right]^{1/2} + $$
$$+ ||G|| \sqrt{h} \left[\int_{-h}^0 ||\varphi(t)||^2 dt + \int_0^T ||z(t)||^2 dt\right]^{1/2}.$$
(10.5.15)

Since the closed-loop system is L^2-stable, there is a real $c > 0$ (independent of φ, T, and s) such that

$$(K_T(s)\Lambda\varphi|\Lambda\varphi)_H \leqslant c||\Lambda\varphi||_H^2 \quad \forall \varphi, T, s, \quad T \geqslant s \geqslant 0. \tag{10.5.16}$$

It is easy to show that if $T_2 \geqslant T_1 \geqslant s$, then $K_{T_2}(s) \geqslant K_{T_1}(s)$ where $\geqslant$ denotes the natural partial ordering of nonnegative definite operators. Moreover, there is a real $c > 0$ such that

$$||K_T(s)||_{L(H)} \leqslant c \quad \forall T \geqslant s.$$

This follows immediately from (10.5.16). On the basis of these two facts and the well-known theorem on nonnegative definite operators (Kantorovich and Akilov, 1964) we conclude that there is a selfadjoint nonnegative definite operator $K(s) \in L(H)$ such that

$$\forall \varphi \in H, \quad K_T(s)\varphi \to K(s)\varphi \quad \text{for} \quad T \to \infty.$$

Since the problem is stationary, the limit operator $K(s)$ does not depend on time s, and can be denoted simply by K.

(ii) Consider the equations

$$\dot{\tilde{x}}_T(t)+[\tilde{B}+\tilde{C}\tilde{C}^*K_T(t)\Lambda]\tilde{x}_T(t) = 0 \quad t \in [0, T]$$
$$\tilde{x}_T(0) = \varphi$$
$$\dot{\tilde{x}}(t)+[\tilde{B}+\tilde{C}\tilde{C}^*K\Lambda]\tilde{x}(t) = 0 \quad t \in [0, T]$$
$$\tilde{x}(0) = \varphi.$$

Denoting $\tilde{y}_T = \tilde{x}_T - \tilde{x}$ we have

$$\dot{\tilde{y}}_T(t)+[\tilde{B}+\tilde{C}\tilde{C}^*K_T(t)\Lambda]\tilde{y}_T(t)+\tilde{C}\tilde{C}^*[K_T(t)-K]\Lambda\tilde{x}(t) = 0 \quad t \in [0, T]$$
$$\tilde{y}_T(0) = 0.$$

From this and (10.5.9) we obtain for $T \to \infty$

$$\tilde{x}_T(t) \to \tilde{x}(t) \quad \forall t$$
$$K_T(t)\Lambda\tilde{x}_T(t) \to K\Lambda\tilde{x}(t) \quad \forall t.$$

By virtue of Theorem 10.4.4 (i)

$$(K_T(0)\Lambda\varphi|\Lambda\varphi) = \int_0^T ([\tilde{W}+K_T(t)\tilde{C}\tilde{C}^*K_T(t)]\Lambda\tilde{x}_T(t)|\Lambda\tilde{x}_T(t))dt,$$

hence (10.5.10).
(iii) is an obvious consequence of (ii). □

Theorem 10.5.3 *Assume that system* (10.5.1) *is stabilizable. Then for every* $\varphi \in H$ *there exists an optimal control* $u_0 \in L^2(0, \infty; R^m)$, *and*

$$J(u_0) = \inf\{J(u): u \in L^2(0, \infty; R^m)\} = (\varphi|K\varphi). \tag{10.5.17}$$

The optimal feedback regulator is determined by

$$\begin{aligned} u_0(t) &= \tilde{C}^*K\tilde{x}(t) \\ \tilde{x}(t) &= \tilde{\Psi}(t)\varphi \end{aligned} \tag{10.5.18}$$

where $\tilde{\Psi}$ *is the strongly continuous semigroup of evolution operators generated by the operator* $\tilde{B}+\tilde{C}\tilde{C}^*K\Lambda$ *which means that* $\forall \psi \in V$, $\forall t \geqslant 0$ *the solution of*

$$\dot{\tilde{y}}(t)+(\tilde{B}+\tilde{C}\tilde{C}^*K\Lambda)\tilde{y}(t) = 0 \quad \tilde{y}(0) = \psi \tag{10.5.19}$$

satisfies

$$\Lambda\tilde{y}(t) = \tilde{\Psi}(t)\Lambda\psi. \tag{10.5.20}$$

Proof For an arbitrary $u \in L^2(0, \infty; R^m)$ denote by x the solution of (10.5.1) generated by the control u from the initial condition φ. Since

$$(\varphi|K_T(0)\varphi) \leqslant \int_0^T [x(t)^T W x(t)+u(t)^T u(t)]dt$$

we have

$$(\varphi|K\varphi) \leqslant \int_0^\infty [x(t)^T W x(t)+u(t)^T u(t)]dt.$$

Finally, we obtain the result using Theorem 10.5.2 (iii). □

10.5.3 Operator Riccati equation and stability of the closed-loop system

In this section we shall study the interconnection between stabilizability and the operator Riccati equation.

Theorem 10.5.4 *Assume* $W > 0$. *Then*:
(i) *System* (10.5.1) *is stabilizable if and only if there is a selfadjoint operator* $K \geqslant 0$ *in* $L(H)$ *satisfying the operator Riccati equation*

10.5 Infinite-time Problem with Arbitrary Delays

$$(\tilde{B}\varphi|K\Lambda\bar{\varphi}) + (\Lambda\varphi|K\tilde{B}\bar{\varphi}) + (\Lambda\varphi|K\tilde{C}\tilde{C}^*K\Lambda\bar{\varphi}) - (\Lambda\varphi|\tilde{W}\Lambda\bar{\varphi}) = 0$$
$$\forall \varphi, \bar{\varphi} \in V. \qquad (10.5.21)$$

(ii) *If there is a selfadjoint nonnegative definite solution of* (10.5.21), *then it is unique and identical with the operator K of Theorem 10.5.3. The operator $\tilde{B} + \tilde{C}\tilde{C}^*K\Lambda$ is L^2-stable, the operator $\tilde{G} = \tilde{C}^*K\Lambda$ defines an L^2-stable regulator and K is positive definite on R^n, that is,*

$$y^{\mathrm{T}} K_{00} y > 0 \quad \forall y \neq 0 \quad y \in R^n$$

where K_{00} is defined by

$$K_{00} \varphi^0 = (K\varphi)^0 \quad \varphi = (\varphi^0, 0) \quad \forall \varphi^0 \in R^n.$$

Proof By virtue of (10.5.10)

$$(K\Lambda\varphi|\Lambda\bar{\varphi}) = \int_0^\infty ([\tilde{W} + K\tilde{C}\tilde{C}^*K]\Lambda\tilde{x}(t,\varphi)|\Lambda\tilde{x}(t,\bar{\varphi}))\mathrm{d}t$$

where $\tilde{x}(\cdot, \varphi)$ (resp. $\tilde{x}(\cdot, \bar{\varphi})$) is the solution of equation (10.5.11) with initial condition φ (resp. $\bar{\varphi}$). Let $\tilde{\Psi}$ be defined as in Theorem 10.5.3. For any $\varphi, \bar{\varphi} \in V$

$$(K(\tilde{B} + \tilde{C}\tilde{C}^*K\Lambda)\varphi|\Lambda\bar{\varphi}) = \int_0^\infty ((\tilde{W} + K\tilde{C}\tilde{C}^*K)\tilde{\Psi}(t)(\tilde{B} + \tilde{C}\tilde{C}^*K\Lambda)\varphi|\tilde{\Psi}(t)\Lambda\bar{\varphi})\mathrm{d}t$$

$$(K\Lambda\varphi|(\tilde{B} + \tilde{C}\tilde{C}^*K\Lambda)\bar{\varphi}) = \int_0^\infty ((\tilde{W} + K\tilde{C}\tilde{C}^*K)\tilde{\Psi}(t)\Lambda\varphi|\tilde{\Psi}(t)(\tilde{B} + \tilde{C}\tilde{C}^*K\Lambda)\bar{\varphi})\mathrm{d}t.$$

If system (10.5.11) is L^2-stable, then by addition of these two equalities we obtain

$$(K(\tilde{B} + \tilde{C}\tilde{C}^*K\Lambda)\varphi|\Lambda\bar{\varphi}) + (K\Lambda\varphi|(\tilde{B} + \tilde{C}\tilde{C}^*K\Lambda)\bar{\varphi})$$
$$= -\int_0^\infty \frac{\mathrm{d}}{\mathrm{d}t}((\tilde{W} + K\tilde{C}\tilde{C}^*K)\Lambda\tilde{x}(t,\varphi)|\Lambda\tilde{x}(t,\bar{\varphi}))\mathrm{d}t = ((\tilde{W} + K\tilde{C}\tilde{C}^*K)\Lambda\varphi|\Lambda\bar{\varphi}),$$

hence (10.5.21).

Assume now that there is a nonnegative definite selfadjoint solution K of equation (10.5.21). This equation may be written in the form

$$(K(\tilde{B} + \tilde{C}\tilde{C}^*K\Lambda)\varphi|\Lambda\bar{\varphi}) + (K\Lambda\varphi|(\tilde{B} + \tilde{C}\tilde{C}^*K\Lambda)\bar{\varphi}) -$$
$$- (K\tilde{C}\tilde{C}^*K\Lambda\varphi|\Lambda\bar{\varphi}) - (\tilde{W}\Lambda\varphi|\Lambda\bar{\varphi}) = 0 \quad \forall \varphi, \bar{\varphi} \in V.$$

By virtue of Lemma 6.2.8 this means that system (10.5.11) is L^2-stable. It is easy to verify that the stabilizing regulator is $\tilde{G} = \tilde{C}^*K\Lambda$. From the results of Chapter 6 it also follows that K is positive definite on R^n.

To prove uniqueness let us suppose there are two selfadjoint solutions $K_1 \geq 0$ and $K_2 \geq 0$. Let $D = K_1 - K_2$. From (10.5.21)
$$((\tilde{B}+\tilde{C}\tilde{C}^*K_2\Lambda)\varphi|D\Lambda\bar{\varphi}) + (\Lambda\varphi|D(\tilde{B}+\tilde{C}\tilde{C}^*K_1\Lambda)\bar{\varphi}) = 0$$
or
$$\frac{d}{dt}(\tilde{\Psi}_2(t)\Lambda\varphi|D\tilde{\Psi}_1(t)\Lambda\bar{\varphi}) = 0$$

where $\tilde{\Psi}_i$ is the semigroup of evolution operators generated by $\tilde{B}+\tilde{C}\tilde{C}^*K_i\Lambda$. As the closed-loop system is L^2-stable, we have
$$(\Lambda\varphi|D\Lambda\bar{\varphi}) = (\tilde{\Psi}_2(t)\Lambda\varphi|D\tilde{\Psi}_1(t)\Lambda\bar{\varphi}) \to 0 \quad (t \to \infty).$$
Then $D = 0$. □

10.5.4 Characterization of operator K

The following theorem gives a full characterization of the operator K which is the solution of the optimal regulator problem.

Theorem 10.5.5 Let a selfadjoint operator $K \geq 0$, $K \in L(H)$ be the solution of (10.5.21). Then:

(i) $\quad (\varphi|K\bar{\varphi}) = \int_0^\infty (\tilde{\Psi}(t)\varphi|(\tilde{W}+K\tilde{C}\tilde{C}^*K)\tilde{\Psi}(t)\bar{\varphi})dt \quad \forall \varphi, \bar{\varphi} \in H \quad (10.5.22)$

where $\tilde{\Psi}$ is defined in Theorem 10.5.3.

(ii) *The operator K can be decomposed into four operators*
$$K(\varphi^0, \varphi^1) = (K_{00}\varphi^0 + K_{01}\varphi^1, K_{10}\varphi^0 + K_{11}\varphi^1)$$
$$\forall (\varphi^0, \varphi^1) \in H \quad (10.5.23)$$
$$K_{00} \in L(R^n) \quad K_{01} \in L(L^2(-h,0;R^n), R^n)$$
$$K_{10} \in L(R^n, L^2(-h,0;R^n))$$
$$K_{11} \in L(L^2(-h,0;R^n)).$$

Moreover, there is a function $K_{10}(\cdot) \in L^2(-h,0;L(R^n))$ such that
$$(K_{10}\varphi^0)(\xi) = K_{10}(\xi)\varphi^0 \quad \varphi^0 \in R^n \quad \xi \in [-h,0] \quad (10.5.24)$$

and there is a function $K_{11}(\cdot,\cdot) \in L^2(-h,0;-h,0;L(R^n))$ such that
$$(K_{11}\varphi^1)(\alpha) = \int_{-h}^0 K_{11}(\alpha,\beta)\varphi^1(\beta)d\beta \quad \varphi^1 \in L^2(-h,0;R^n)$$
$$\alpha \in [-h,0]. \quad (10.5.25)$$

K_{01} *is obtained from the equality*
$$K_{01}\varphi^1 = \int_{-h}^0 K_{10}(\alpha)^T \varphi^1(\alpha)d\alpha. \quad (10.5.26)$$

10.5 Infinite-time Problem with Arbitrary Delays

(iii) *The operator K_{00} satisfies the equation*
$$K_{00}B_0 + B_0^T K_{00} - K_{10}(0) - K_{10}(0)^T - W + K_{00}CC^T K_{00} = 0, \quad (10.5.27)$$
and
$$K_{00} = K_{00}^T \geq 0.$$

(iv) *The function $K_{10}(\cdot)$ may only have discontinuities of the form*
$$K_{10}(-h_i+) - K_{10}(-h_i-) = -B_i^T K_{00} \quad (i = 1, \ldots, N-1). \quad (10.5.28)$$
Elsewhere it is absolutely continuous and almost everywhere satisfies the equation
$$\dot{K}_{10}(\alpha) + K_{10}(\alpha)(B_0 + CC^T K_{00}) + B_{01}(\alpha)^T K_{00} - K_{11}(\alpha, 0) = 0$$
$$\alpha \in [-h, 0] \quad (10.5.29)$$
with the boundary condition
$$K_{10}(-h) = -B_N^T K_{00}. \quad (10.5.30)$$

(v) *The function $K_{11}(\cdot, \cdot)$ may only have discontinuities of the form*
$$K_{11}(-h_i+, \beta) - K_{11}(-h_i-, \beta) = -B_i^T K_{10}(\beta)^T \quad (i = 1, \ldots, N-1)$$
$$K_{11}(\alpha, -h_i+) - K_{11}(\alpha, -h_i-) = -K_{10}(\alpha)B_i. \quad (10.5.31)$$
Elsewhere it is absolutely continuous and almost everywhere satisfies the equation
$$\left(\frac{\partial}{\partial \alpha} + \frac{\partial}{\partial \beta}\right) K_{11}(\alpha, \beta) + K_{10}(\alpha)B(\beta) + B(\alpha)^T K_{10}(\beta)^T +$$
$$+ K_{10}(\alpha)CC^T K_{10}(\beta)^T = 0. \quad (10.5.32)$$
The boundary conditions are
$$K_{11}(-h, \beta) = -B_N^T K_{10}(\beta)^T$$
$$K_{11}(\alpha, -h) = -K_{10}(\alpha)B_N. \quad (10.5.33)$$
Furthermore, K_{11} satisfies the relationship
$$K_{11}(\alpha, \beta) = K_{11}(\beta, \alpha)^T. \quad (10.5.34)$$

(vi) *The unique solution of equation (10.5.32), satisfying (10.5.31), (10.5.33), and (10.5.34) is*
$$K_{11}(\alpha, \beta) = -\begin{cases} K_{10}(\alpha-\beta-h)B_N & \alpha \geq \beta \\ B_N^T K_{10}(\beta-\alpha-h)^T & \alpha < \beta \end{cases} -$$
$$- \sum_{i=1}^{N-1} \begin{cases} B_i^T K_{10}(\beta-\alpha-h_i)^T & -h \leq \beta-\alpha-h_i \quad -h_i < \alpha \\ 0 & \text{otherwise} \end{cases} -$$
$$- \sum_{j=1}^{N-1} \begin{cases} K_{10}(\alpha-\beta-h_j)B_j & -h \leq \alpha-\beta-h_j \quad -h_j < \beta \\ 0 & \text{otherwise} \end{cases} -$$

$$-\int_{-h}^{\alpha} \begin{cases} B(\xi)^T K_{10}(\xi-\alpha+\beta)^T & \xi \geqslant \alpha-\beta-h \\ 0 & \text{otherwise} \end{cases} d\xi -$$

$$-\int_{-h}^{\beta} \begin{cases} K_{10}(\zeta-\beta+\alpha)B(\zeta) & \zeta \geqslant \beta-\alpha-h \\ 0 & \text{otherwise} \end{cases} d\zeta -$$

$$-\begin{cases} \int_{-h}^{\beta} K_{10}(\alpha-\beta+\zeta)CC^T K_{10}(\zeta)^T d\zeta & \alpha \geqslant \beta \\ \int_{-h}^{\alpha} K_{10}(\xi)CC^T K_{10}(\beta-\alpha+\xi)^T d\xi & \alpha < \beta \end{cases}. \qquad (10.5.35)$$

The proof is similar to that in the case of the finite-time problem and will be omitted.

10.6 Discrete Approximation of the Linear-quadratic Problem with Finite Control Time

Numerical solution of the linear-quadratic problem of optimal control was partly discussed in Section 10.3. The present section is devoted to the two most frequently used approaches to numerical calculation of optimal control: discretization of the original problem (i.e. replacing the original problem by a discrete-time one), and discretization of the Riccati set of equations. Our considerations are based on the results of Delfour (1977a). Some simple cases of discretization of the Riccati equations may be found in the works of Aggarwal (1970), and Eller, Aggarwal and Banks (1969). Below we shall use the notation introduced in Section 4.4.1.

10.6.1 Discrete approximation of the optimal control problem

We begin with the discrete approximation of the optimal control problem (10.4.1), (10.4.3). The system equation will be approximated as described in Section 4.4.1. To approximate the performance index let us define for $i = 0, \ldots, M-1$

$$\overline{W}(i) = \frac{1}{\delta} \int_{i\delta}^{(i+1)\delta} W(t) dt$$

$$\overline{w}(i) = \frac{1}{\delta} \int_{i\delta}^{(i+1)\delta} w(t) dt. \qquad (10.6.1)$$

10.6 Discrete Approximation of the Linear-quadratic Problem

For a given control sequence $\bar{u} \in \{0, M-1; m\}$ and the corresponding solution z of the equation we determine the approximate performance functional by the formula

$$J_\delta(\bar{u}) = (Qz(M)+2q|z(M)) + \delta \sum_{i=0}^{M-1} [(\overline{W}(i)z(i)+2\bar{w}(i)|z(i)) + (\bar{u}(i)|\bar{u}(i))]. \tag{10.6.2}$$

The approximate optimal control problem consists in the minimization of J_δ on the space $\{0, M-1; m\}$ subject to the equation (4.4.15):

$$z(i+1)-z(i) + \delta \left[\sum_{j=0}^{N} \bar{B}_j(i) z(i-L_j) + \delta \sum_{j=-L}^{0} \bar{B}(i,j) z(i+j) + \bar{C}(i)\bar{u}(i) \right] = 0$$

$$(i = 1, \ldots, M-1) \tag{10.6.3}$$

$$z(i) = \bar{\varphi}(i) \quad (i = -L, \ldots, 0) \tag{10.6.4}$$

where $\bar{\varphi}$ is a given element of H^δ.

In order to write the functional J_δ in the discrete state space notation we define $\tilde{Q} \in L(H^\delta)$ and $\tilde{W}(i) \in L(H^\delta)$, $i = 0, \ldots, M-1$

$$(\tilde{Q}y|v)_{H^\delta} = (Qy(0)|v(0))_{R^n} \quad \forall y, v \in H^\delta$$
$$(\tilde{W}(i)y|v)_{H^\delta} = (\overline{W}(i)y(0)|v(0))_{R^n}. \tag{10.6.5}$$

Let further $\tilde{q}, \tilde{w}(i) \in H^\delta$, $i = 0, \ldots, M-1$ be defined by

$$\tilde{q}(0) = q \quad \tilde{q}(i) = 0 \quad i < 0$$
$$[\tilde{w}(i)](0) = \bar{w}(i) \quad [\tilde{w}(i)](j) = 0 \quad j < 0. \tag{10.6.6}$$

Using the state space notation we can now formulate the approximate optimal control problem (10.6.3), (10.6.2) in the following form. On the trajectories of the system

$$\tilde{z}(i+1) - \tilde{z}(i) + \delta[\tilde{B}(i)\tilde{z}(i) + \tilde{C}(i)\bar{u}(i)] = 0 \quad (i = 0, \ldots, M-1)$$
$$\tilde{z}(0) = \bar{\varphi} \tag{10.6.7}$$

minimize the performance functional

$$J_\delta(\bar{u}) = (\tilde{Q}\tilde{z}(M) + 2\tilde{q}|\tilde{z}(M)) + \delta \sum_{i=0}^{M-1} [(\tilde{W}(i)\tilde{z}(i) + 2\tilde{w}(i)|\tilde{z}(i)) + (\bar{u}(i)|\bar{u}(i))]. \tag{10.6.8}$$

This problem has a unique solution characterized by the following theorem.

Theorem 10.6.1 *The optimal control problem* (10.6.7), (10.6.8) *has a unique solution* $\bar{u}$. *It is completely characterized by the following canonical set of equations:*

$$\tilde{z}(i+1) - \tilde{z}(i) + \delta[\tilde{B}(i)\tilde{z}(i) + \tilde{C}(i)\bar{u}(i)] = 0 \quad (i = 0, \ldots, M-1) \quad (10.6.9)$$

$$\tilde{z}(0) = \bar{\varphi}$$

$$p(i+1) - p(i) - \delta[I_\delta^{-1}\tilde{B}(i)^* I_\delta p(i+1) - \tilde{W}(i)\tilde{z}(i) - \tilde{w}(i)] = 0$$
$$(i = 0, \ldots, M-1) \quad (10.6.10)$$

$$p(M) = \tilde{Q}\tilde{z}(M) + \tilde{q}$$

$$\bar{u}(i) = \tilde{C}(i)^* p(i+1) \quad (i = 0, \ldots, M-1) \quad (10.6.11)$$

where

$$I_\delta = q_H^* q_H \colon H^\delta \to H^\delta. \quad (10.6.12)$$

For the proofs of this and the following theorems we refer the reader to Delfour (1977a).

Theorem 10.6.2 *Assume that some* $\varphi \in H$ *is given. The approximate optimal control problem* (10.6.7), (10.6.8) *with initial condition* $\bar{\varphi} = r_H \varphi$ *has a unique solution* $\bar{u} \in \{0, M-1; m\}$. *If* $\delta \to 0$, *then* $q_u \bar{u}$ *tends in* $L^2(0, T; R^m)$ *to the control* u_0, *that is, the optimal solution of problem* (10.4.1), (10.4.2), (10.4.3).

10.6.2 Discrete approximation of the optimal regulator

As in Section 10.4.3 for problem (10.4.1), (10.4.3), we can separate the set of equations (10.6.9), (10.6.10), (10.6.11). To this end we consider the approximate optimal control problem (10.6.7), (10.6.8) on a discrete set of time moments $\sigma, \sigma+1, \ldots, M$ and obtain as a result a new set of equations characterizing the optimal solution.

Theorem 10.6.3 *Let* $\tilde{z}$ *and* p *be the solution of system* (10.6.9), (10.6.10) (10.6.11). *There is a matrix function* $\bar{K} \colon \{0, \ldots, M\} \to L(H^\delta)$ *and a vector function* $\bar{k} \colon \{0, \ldots, M\} \to H^\delta$ *such that*

$$I_\delta p(i) = \bar{K}(i)\tilde{z}(i) + I_\delta \bar{k}(i) \quad (i = 0, \ldots, M). \quad (10.6.13)$$

$\bar{K}$ *and* $\bar{k}$ *can be determined in the following way.*

(i) *Solve the system of equations*

$$\beta(i+1) - \beta(i) + \delta[\tilde{B}(i)\beta(i) + \tilde{C}(i)\tilde{C}(i)^*\gamma(i+1)] = 0 \quad (i = \sigma, \ldots, M-1)$$
$$(10.6.14)$$

$$\beta(\sigma) = \bar{\varphi}$$

$$\gamma(i+1) - \gamma(i) - \delta[I_\delta^{-1}\tilde{B}(i)^* I_\delta \gamma(i+1) - \tilde{W}(i)\beta(i)] = 0 \quad (i = \sigma, \ldots, M-1)$$

$$\gamma(M) = \tilde{Q}\beta(M). \quad (10.6.15)$$

10.6 Discrete Approximation of the Linear-quadratic Problem

Hence
$$\bar{K}(\sigma)\bar{\varphi} = I_\delta \gamma(\sigma). \tag{10.6.16}$$

(ii) *Solve the system of equations*
$$\eta(i+1) - \eta(i) + \delta[\tilde{B}(i)\eta(i) + \tilde{C}(i)\tilde{C}(i)^*\xi(i+1)] = 0$$
$$(i = \sigma, \ldots, M-1) \tag{10.6.17}$$
$$\eta(\sigma) = 0$$
$$\xi(i+1) - \xi(i) - \delta[I_\delta^{-1}\tilde{B}(i)^*I_\delta\xi(i+1) - \tilde{W}(i)\eta(i) - \tilde{w}(i)] = 0$$
$$(i = \sigma, \ldots, M-1) \tag{10.6.18}$$
$$\xi(M) = \tilde{Q}\eta(M) + \tilde{q}.$$

Hence
$$\bar{k}(\sigma) = \xi(\sigma). \tag{10.6.19}$$

The approximation of the differential equation for the operator K of (10.4.38) is our next objective. As before, for the study of $\bar{K}$ we assume $w = 0$, $q = 0$ in the problem with continuous time which results in $\tilde{w} = 0$, $\tilde{q} = 0$ in the discrete-time problem. Thus, we have to consider the following problem with a fixed initial moment σ, $0 \leq \sigma < M$. On the trajectories of

$$z(i+1) - z(i) + \delta[\tilde{B}(i)z(i) + \tilde{C}(i)\bar{u}(i)] = 0 \quad (i = \sigma, \ldots, M-1) \tag{10.6.20}$$
$$\tilde{z}(\sigma) = \bar{\varphi}$$

minimize the functional
$$J_\delta^\sigma(\bar{u}) = (\tilde{Q}\tilde{z}(M)|z(M)) + \delta \sum_{i=\sigma}^{M-1} [(\tilde{W}(i)\tilde{z}(i)|z(i)) + (\bar{u}(i)|\bar{u}(i))]. \tag{10.6.21}$$

This problem has a unique solution fully characterized by the system
$$\tilde{z}(i+1) - \tilde{z}(i) + \delta[\tilde{B}(i)\tilde{z}(i) + \tilde{C}(i)\tilde{C}(i)^*p(i+1)] = 0$$
$$(i = \sigma, \ldots, M-1) \tag{10.6.22}$$
$$\tilde{z}(\sigma) = \bar{\varphi}$$
$$p(i+1) - p(i) - \delta[I_\delta^{-1}\tilde{B}(i)^*I_\delta p(i+1) - \tilde{W}(i)\tilde{z}(i)] = 0$$
$$(i = \sigma, \ldots, M-1) \tag{10.6.23}$$
$$p(M) = \tilde{Q}\tilde{z}(M)$$
$$\bar{u}(i) = \tilde{C}(i)^*p(i+1) \quad (i = \sigma, \ldots, M-1). \tag{10.6.24}$$

The above system may be separated by means of operators $\bar{K}(i)$ determined in Theorem 10.6.3. Namely, we have
$$I_\delta p(\sigma) = \bar{K}(\sigma)\bar{\varphi}. \tag{10.6.25}$$

Lemma 10.6.1 *Let us fix* $s \in [0, T]$ *and let* σ *be an integer such that* $s = \sigma\delta$. *Moreover, let* $\overline{K}(M) = \tilde{Q}$.

(i) *If* $\overline{u}$ *is the optimal control in problem* (10.6.20), (10.6.21) *then*

$$J_\delta^\sigma(\overline{u}) = (\overline{K}(\sigma)\overline{\varphi}|\overline{\varphi})_{H^\delta} \qquad (10.6.26)$$

and

$$\overline{K}(\sigma) = \overline{K}(\sigma)^* \geq 0. \qquad (10.6.27)$$

(ii) *There is a real* c *(independent of* δ *and* φ*) such that*

$$(\overline{K}(\sigma)r_H\varphi|r_H\varphi) \leq c\|\varphi\|^2 \quad \forall \varphi \in H. \qquad (10.6.28)$$

(iii) *If* $\tilde{z}$ *and* p *are the solution of* (10.6.22), (10.6.23), *then*

$$I_\delta p(i) = \overline{K}(i)\tilde{z}(i) \quad (i = \sigma, \ldots, M). \qquad (10.6.29)$$

Theorem 10.6.4 *Define*

$$P(i) = r_H^* \overline{K}(i) r_H \in L(H) \quad (i = 0, \ldots, M). \qquad (10.6.30)$$

(i) *(Stability) There is a real* c *(independent of* δ*) such that*

$$\|P(i)\|_{L(H)} \leq c \quad (i = 0, \ldots, M). \qquad (10.6.31)$$

(ii) *(Convergence) Fix an* $s \in [0, T]$. *Then*

$$(P(i)\varphi|\eta) \to (K(s)\varphi|\eta) \quad \forall \varphi, \eta \in H \qquad (10.6.32)$$

if $\delta \to 0$ *with* $i\delta = s$.

We introduce a decomposition of the operators $P(i)$,

$$P(i)\varphi = (P_{00}(i)\varphi^0 + P_{01}(i)\varphi^1, \ P_{10}(i)\varphi^0 + P_{11}(i)\varphi^1) \quad \forall \varphi \in H. \qquad (10.6.33)$$

It is easy to verify that

$$P_{00}(i) = \overline{K}_{00}(i)$$

$$P_{01}(i)\varphi^1 = \int_{-h}^{0} \sum_{j=-L}^{-1} \delta^{-1} \overline{K}_{0j}(i) \chi_j(t) \varphi^1(t) \, dt$$

$$[P_{11}(i)\varphi^1](\alpha) = \int_{-h}^{0} \sum_{j=-L}^{-1} \sum_{k=-L}^{-1} \delta^{-2} \overline{K}_{jk}(i) \chi_j(\alpha) \chi_k(t) \varphi^1(t) \, dt. \qquad (10.6.34)$$

$\overline{K}_{jk}(i)$ is defined by

$$y = \overline{K}(i)z \Rightarrow y(j) = \sum_{k=-L}^{0} \overline{K}_{jk}(i) z(k) \quad \forall z \in H^\delta. \qquad (10.6.35)$$

10.6 Discrete Approximation of the Linear-quadratic Problem

If we fix an $s \in [0, T]$ and let δ tend to zero so that $i\delta = s$, then

$$P_{00}(i) \to K_{00}(s) \quad \text{in} \quad L(H)$$
$$P_{01}(i)y \to K_{01}(s)y \quad \text{in} \quad R^n \quad \forall y \in L^2(-h, 0; R^n) \quad (10.6.36)$$
$$P_{11}(i)y \to K_{11}(s)y \quad \text{weakly in} \quad L^2(-h, 0; R^n) \quad \forall y \in L^2(-h, 0; R^n)$$

and the norms of $P_{00}(i)$, $P_{01}(i)$ and $P_{11}(i)$ are bounded from above by some constant c.

Theorem 10.6.5 *The function $\overline{K}$ defined by equations (10.6.22)–(10.6.25) is the solution of the equation*

$$\overline{K}(i) = [I - \delta \tilde{B}(i)]^* \overline{K}(i+1)[I + \delta \tilde{C}(i) \tilde{C}(i)^* \overline{K}(i+1)]^{-1}[I - \delta \tilde{B}(i)] +$$
$$+ \delta \tilde{W}(i) \quad (i = 0, \ldots, M-1) \quad (10.6.37)$$
$$\overline{K}(M) = \tilde{Q} \quad (10.6.38)$$

where I stands for the identity operation in $L(H^\delta)$.

Let us note that the inverse $[I + \delta \tilde{C}(i) \tilde{C}(i)^* \overline{K}(i+1)]^{-1}$ exists, and the operation $\overline{K}(i+1)[I + \delta \tilde{C}(i) \tilde{C}(i)^* \overline{K}(i+1)]^{-1}$ is symmetric and nonnegative definite.

In some cases the order of convergence can be estimated.

Theorem 10.6.6 *Assume that $B_1, \ldots, B_N$ and C are constant matrices, and $B = 0$. Then there exists a constant c (independent of δ) such that*

$$\|K(i\delta) - P(i)\| \leq c\delta \quad (i = 0, \ldots, M). \quad (10.6.39)$$

Now the approximation of the differential equation for the free term of the optimal regulator (10.4.56) will be given.

Theorem 10.6.7 *Let $\overline{k}$ be defined as in Theorem 10.6.3. Then:*

(i) *$\overline{k}$ is the solution of*

$$\overline{k}(i) = I_\delta^{-1}[I - \delta \tilde{B}(i)]^* [(I + \delta \tilde{C}(i) \tilde{C}(i)^* \overline{K}(i+1))^{-1}]^* I_\delta \overline{k}(i+1) + \delta \tilde{w}(i)$$
$$(i = 0, \ldots, M-1) \quad (10.6.40)$$
$$\overline{k}(M) = \tilde{q}. \quad (10.6.41)$$

(ii) *There is a constant c, independent of δ, such that*

$$\|q_H \overline{k}(i)\| \leq c \quad (i = 0, \ldots, M). \quad (10.6.42)$$

(iii) *For every $\varphi \in H$ and every $s \in [0, T]$*

$$(q_H \overline{k}(i) | \varphi) \to (k(s) | \varphi) \quad (10.6.43)$$

when $\delta \to 0$ with $i\delta = s$.

References

Aggarwal, J. K. (1970). Computation of optimal control for time-delay systems, *IEEE Trans. Automatic Control* AC-15, 683–685.

Anderson, B. D. O. and Vongpanitlerd, S. (1973). *Network Analysis and Synthesis. A Modern Systems Theory Approach*, Prentice-Hall, Englewood Cliffs, New Jersey.

Balakrishnan, A. V. (1976). *Applied Functional Analysis*, Springer-Verlag, New York.

Banks, H. T. (1969). Representations for solutions of linear functional differential equations, *J. Differential Equations*, **5**, 399–409.

Banks, H. T. and Jacobs, M. (1973). An attainable sets approach to optimal control of functional differential equations with function space side conditions, *J. Differential Equations*, **13**, 127–149.

Banks, H. T. and Kent, G. (1972). Control of functional differential equations to target sets in function space, *SIAM J. Control*, **10**, 567–593.

Banks, H. T. and Manitius, A. (1975). Projection series for retarded functional differential equations with applications to optimal control problems, *J. Differential Equations* **18**, 296–332.

Bellman, R. and Cooke, K. L. (1963). *Differential-Difference Equations*, Academic Press, New York.

Bhatia, N. P. and Szegö, G. P. (1970). *Stability Theory of Dynamical Systems*, Springer-Verlag, Berlin, Heidelberg, New York.

Brierley, S. D., Chiasson, J. N., Lee, E. B. and Żak, S. H. (1982). On stability independent of delay for linear systems, *IEEE Trans. Automatic Control*, AC-27, 252–254.

Calvarho, L. A. V., Walker, J. A. and Infante, E. F. (1980). On the existence of simple Liapunov functions for linear retarded difference-differential equations, *Tôhoku Mathematical Journal*, **32**, 283–297.

Castelan, W. B. (1980). A Liapunov Functional for a Matrix Retarded Difference-Differential Equation with Several Delays, Lect. Notes in Math., **799**, 88–118, Springer-Verlag, Berlin, Heidelberg, New York.

Castelan, W. B. and Infante, E. F. (1977). On a functional equation arising in the stability theory of difference-differential equations, *Quarterly of Appl. Math.*, **35**, 311–319.

Castelan, W. B. and Infante, E. F. (1979). A Liapunov functional for a matrix neutral difference-differential equation with one delay, *J. Math. Anal. Appl.*, **71**, 105–130.

Chow, S. N. (1974). Existence of periodic solutions of autonomous functional differential equations, *J. Differential Equations*, **15**, 350–378.

References

Chukwu, E. N. (1981). Global asymptotic behaviour of functional differential equations of the neutral type, *Nonlinear Analysis Theory Methods and Applications*, **5**, 853–872.

Chyung, D. H. and Lee, E. B. (1966). Linear optimal systems with time delays, *SIAM J. Control*, **4**, 568–575.

Chyung, D. H. and Lee, E. B. (1970). Delayed action control problems, *Automatica*, **6**, 395–400.

Crandall, M. G. (1971). Semigroups of nonlinear transformations, *Proc Symposium Conducted by Math. Res. Center*, April 12–14, 1971, Univ. Wisconsin, Madison. In: Zarantonello, E. (ed.), *Contributions to Nonlinear Functional Analysis*, Academic Press, New York.

Crandall, M. G. and Liggett, T. M. (1971). Generation of semigroups of nonlinear transformations on general Banach spaces, *American J. Math.*, **93**, 265–298.

Datko, R. (1970). Extending a theorem of Liapunov to Hilbert space, *J. Math. Anal. Appl.*, **32**, 610–616.

Datko, R. (1980). Lyapunov functionals for a certain linear delay differential equation in a Hilbert space, *J. Math. Anal. Appl.*, **76**, 37–57.

Delfour, M. C. (1972). Theory of differential delay systems in the space M^2; stability and Lyapunov equation, *Proc. Symposium on Differential Delay and Functional Equations*, June 26–July 7, 1972, Control Theory Centre, University of Warwick, England.

Delfour, M. C. (1974). *Generalisation de resultats de R. Datko sur les fonctions de Lyapunov quadratiques définées sur un espace de Hilbert*, Centre de Recherches Mathématiques, Université de Montreal, CRM-457.

Delfour, M. C. (1977a). The linear quadratic optimal control problem for hereditary differential systems: theory and numerical solution, *Appl. Math. and Optimization*, **3**, 101–162.

Delfour, M. C. (1977b). State theory of linear hereditary differential systems, *J. Math. Anal. Appl.*, **60**, 8–35.

Delfour, M. C. and Manitius, A. (1980a). The structural operator F and its role in the theory of retarded systems, part I, *J. Math. Anal. Appl.*, **73**, 466–490.

Delfour, M. C. and Manitius, A. (1980b). The structural operator F and its role in the theory of retarded systems, part II. *J. Math. Anal. Appl.*, **74**, 359–381.

Delfour, M. C., McCalla, C. and Mitter, S. K. (1975). Stability and the infinite-time quadratic cost problem for linear hereditary differential systems, *SIAM J. Control*, **13**, 48–88.

Delfour, M. C. and Mitter, S. K. (1972). Controllability, observability and optimal feedback control of affine hereditary differential systems, *SIAM J. Control*, **10**, 298–328.

Devyatov, B. N. (1950). On the problem of approximate representation of transfer functions of heat exchangers. In: *Automatic Control of Continuous Processes*, Siberian Division of Acad. Sci. USSR, 29–41 (in Russian).

Dickerson, J. and Gibson, J. (1976). Stability of linear functional differential equations in Banach spaces, *J. Math. Anal. Appl.*, **55**, 150–155.

Doetsch, G. (1956). *Handbuch der Laplace Transformation*, vol. III, Birkhäuser Verlag, Basel, Switzerland.

Dunford, N. and Schwartz, J. T. (1967). *Linear Operators I*, Interscience, New York.

Dyson, J. and Villella-Bresan, R. (1976). Functional differential equations and nonlinear evolution operators, *Proc. Roy. Soc. Edinburgh*, **A75**, 223–234.

Dyson, J. and Villella-Bresan, R. (1979). Semigroups of translations associated with functional differential equations, *Proc. Roy. Soc. Edinburgh*, **A82**, 121–128.

El'sgol'c, L. E. and Norkin S. B. (1971). *Introduction to the Theory of Differential Equations with Deviating Argument*, 2nd ed. Nauka, Moscow (in Russian).

Eller, D. H., Aggarwal, J. K. and Banks, H. T. (1969). Optimal control of linear time-delay systems, *IEEE Trans. Automatic Control*, AC-14, 678–687.

Feldstein, A. and Goodman, R. (1973). Numerical solution of ordinary and retarded differential equations with discontinuous derivatives, *Numer. Math.*, **21**, 1–13.

Flaschka, H. and Leitmann, M. J. (1975). On semigroup of nonlinear operators and the solution of the FDE $\dot{x}(t) = F(\dot{x}_t)$, *J. Math. Anal. Appl.*, **49**, 649–658.

Foiaş, C. (1973). Sur une question de M. Reghis, *An. Univ. Timisoara, ser. sti. mat.*, **11**, 111–114.

Gabasov, R. and Kirillova, F. M. (1971). *Qualitative Theory of Optimal Processes*, Nauka, Moscow (in Russian).

Gayshun, I. V. (1972). Asymptotic stability of systems with retardation, *Differencial'nye Uravneniya*, **8**, 906–908 (in Russian).

Gokhman, E. Kh. (1958). *Stieltjes Integral and Its Applications*, GIFML, Moscow (in Russian).

Gorbunov, V. P. and Shikhov, S. B. (1975). *Nonlinear Dynamics of Nuclear Reactors*, Atomizdat, Moscow (in Russian).

Górecki, H. (1971). *Analasis and Synthesis of Control Systems with Delay*, WNT, Warsaw (in Polish).

Górecki, H. and Popek, L. (1982). Control of the systems with time delay, Prep., III *IFAC Symposium on Control of Distributed Parameter Systems*, June 29–July 2, 1982, session XXII.9, pp. XXII.9–XXII.12, Toulouse, France.

Goryachenko, V. D. (1971). *Methods in the Stability Theory in Dynamics of Nuclear Reactors*, Atomizdat, Moscow (in Russian).

Grabbe, E. M., Ramo, S. and Woolridge, D. E. (1961). *Handbook of Automation, Computation and Control*, Vol. 3, J. Wiley, New York.

Grabowski, P. (1979). *Accretive Operators in the Theory of Semidynamical Systems*, Ph. D. dissertation, AGH, Kraków (in Polish).

Grabowski, P. (1980). Determination of a subset of the domain of attraction of the null-solution of the Lur'e control system, *Proc. 8-th National Conf. on Automatic Control*, Szczecin, September 16–17, 1980, **1**, 50–55 (in Polish).

Grabowski, P. (1983). A Lyapunov functional approach to a parametric optimization of infinite-dimensional control systems, *Elektrotechnika*, **2**, 207–232.

Gromova, P. S. and Pelevina, A. F. (1977). Absolute stability of automatic control systems with delay, *Differencialnye Uravneniya*, **13**, 1375–1383 (in Russian).

Gruber, M. (1969). Path integrals and Lyapunov functionals, *IEEE Trans. Automatic Control*, AC-14, 465–475.

Halanay, A. (1966). *Differential Equations: Stability, Oscillations, Time Lags*, Academic Press, New York.

Hale, J. K. (1977). *Theory of Functional Differential Equations*, Springer-Verlag, New York.

Hale, J. K. and Cruz, M. (1969). Asymptotic behaviour of NFDE's, *Archive for Rational Mechanics and Analysis*, **34**, 331–353.

Hale, J. K. and Cruz, M. (1970). Existence, uniqueness and continuous dependence for hereditary systems, *Ann. Math. Pure Appl.*, **4** (85), 63–82.

Hale, J. K. and Ize, A. F. (1971). On the uniform asymptotic stability of FDE's of neutral type, *Proc. AMS.*, **28**, 100–106.

References

Hale, J. K. and Martinez-Amores, P. (1977). Stability in neutral equations, *Nonlinear Analysis Theory, Methods and Applications*, **1** 161–173.

Hartman, Ph. (1964). *Ordinary Differential Equations*, J. Wiley and Sons, New York.

Hautus, M. L. J. (1969). Controllability and observability conditions of linear systems, *Indag. Math.*, **31**, 443–448.

Heiden, U. (1979). Periodic solutions of a nonlinear second order differential equation with delay, *J. Math. Anal. Appl.*, **70**, 599–609.

Henry, D. (1974). Linear autonomous NFDE's, *J. Diff. Eqs.*, **15**, 106–128.

Hille, E. and Phillips, R. S. (1968). *Functional Analysis and Semi-Groups*, AMS Colloq. Publ., **31**, Providence.

Infante, E. F. and Castelan, W. B. (1978). A Liapunov functional for a matrix difference-differential equation, *J. Diff. Eqs.*, **29**, 439–451.

Infante, E. F. and Slemrod, M. (1972). Asymptotic stability criteria for linear systems of difference-differential equations of neutral type and their discrete analogues, *J. Math. Anal. Appl.*, **38**, 399–415.

Israelson, D. and Johnson, A. (1968). Applications of a theory for circumnutations to geotropic movement, *Physiologia Plantorium*, **21**, 282–291.

Israelson, D. and Johnson, A. (1969). Phase shift in geotropical oscillations—a theoretical and experimental study, *Physiologia Plantorium*, **22**, 1226–1237.

Jacobs, M. Q. (1968). Attainable sets in systems with unbounded controls, *J. Diff. Eqs.*, **4**, 408–423.

Jury, E. I. and Mansour, D. (1982). Stability conditions for class of DDE systems, *Int. J. Control*, **35**, 689–699.

Kamen, E. (1980). On the relationship between zero criteria for two-variable polynomials and asymptotic stability of DDE's, *IEEE Trans. Automatic Control*, AC-25, 983–984.

Kantorovich, L. V. and Akilov G. P. (1964). *Functional Analysis in Normed Spaces*, Pergamon Press, London.

Kaplan, J. and Yorke, J. (1977). On the nonlinear differential-difference equation: $\dot{x}(t) = -f(x(t), x(t-1))$, *J. Diff. Eqs.*, **23**, 293–314.

Kharatishvili, G. L. (1966). *Optimal Processes with Time-Lag*, Izdat. Mecniereba, Tbilisi (in Russian).

Korn, G. A. and Korn, T. M. (1961). *Mathematical Handbook for Scientists and Engineers*, McGraw-Hill Book Company, Inc, New York, Toronto, London.

Korytowski, A. (1973). *Selected Problems of Optimal Control in Time Delay Systems*, Ph.D. dissertation, AGH, Kraków (in Polish).

Korytowski, A. (1976). Optimal control for quadratic problem with neutral system equation, *Control and Cybernetics*, **5**, 53–66.

Krall, A. (1964). Stability criteria for feedback systems with time lag, *SIAM J. Appl. Math.*, A2, 160–170.

Krasovskii, N. N. (1962). On the analytic construction of an optimal control in a system with time lags, *Prikl. Mat. Mekh.*, **26**, 39–51, or *J. Appl. Math. Mech.*, **26**, 50–67.

Kwakernaak, H. and Sivan, R. (1972). *Linear Optimal Control Systems*, Wiley-Interscience, New York, London, Sydney, Toronto.

Lancaster, P. (1969). *Theory of Matrices*, Academic Press, New York.

Lerner, A. Ya. (1958). *Introduction to the Theory of Automatic Control*, Gos. Nauch.-Tekh. Izdat., Moscow

Levin, B. J. (1964). *Distribution of Zeros of Entire Functions*, Translations of Mathematical Monographs, **5**, AMS, Rhode Island.

Makowski, I. and Neustadt, L. W. (1974). Optimal control problems with mixed control-phase variable equality and inequality constraints, *SIAM J. Control*, 12, 184–228.

Manitius, A. (1972). On the controllability conditions for systems with distributed delays in state and control, *Archiwum Automatyki i Telemechaniki*, Vol. XVII, 363–377.

Manitius, A. (1974a). *Controllability, Observability and Stabilizability of Retarded Systems*, Report H3C3J7, Université de Montreal, Centre de Recherches Mathématiques.

Manitius, A. (1974b). *On the Optimal Control of Systems with a Delay Depending on State, Control and Time*, Report CRM-449, Université de Montreal, Centre de Recherches Mathématiques, September 1974.

Manitius, A. and Olbrot, A. W. (1979). Finite spectrum assignment problem for systems with delays, *IEEE Trans. Automatic Control*, AC-24, 541–553.

Manitius, A. and Triggiani, R. (1978). Function space controllability of linear retarded systems: a derivation from abstract operator conditions, *SIAM J. Control and Optimization*, 16, 599–645.

Marshall, J. E. (1979). *Control of Time-delay Systems*, IEE, Control Engineering Series, 10, London.

Mărsik, J. (1958). Eine schnelle Berechnung des Regelungsoptimums nach Oldenbourg und Sartorius, auch für Regelkreise mit transzendenten Frequenzgang, *Regelungstechnik*, 6, 217–219.

Melvin, W. R. (1974). Stability properties of functional difference equations, *J. Math. Anal. Appl.*, 48, 749–763.

Miller, I. A. et al. (1967). A comparison of controller tuning techniques, *Control Engineering*, 14, 72–75.

Miyadera, I. (1971). Some remarks on semigroups of nonlinear operators, *Tôhoku Mathematical Journal*, 23, 245–258.

Myshkis, A. D. (1972). *Linear Differential Equations with Retarded Argument*, Nauka, Moscow (in Russian).

Nöldus, E. (1967). Analytische Untersuchung der Übertragungsverhaltens von Gleich und Gegenstromwärmeaustauschern, *Regelungstechnik*, 3, 112–117.

Oğuztöreli, M. N. (1966). *Time-Lag Control Systems*, Academic Press, New York.

Olbrot, A. W. (1972). On degeneracy and related problems for linear constant time-lag systems, *Ricerche di Automatica*, 3, 203–220.

Olbrot, A. W. (1973a). *On Controllability and Other Properties of Linear Systems with Time Lag*, Ph.D. dissertation, Institute of Automatics, Technical University of Warsaw (in Polish).

Olbrot, A. W. (1973b). Algebraic criteria of controllability to zero function for linear constant time-lag systems, *Control and Cybernetics*, 2, 59–77.

Olbrot, A. W. (1976a). Control of retarded systems with function space constraints. Part I, *Control and Cybernetics*, 5, 5–32.

Olbrot, A. W. (1976b). Control of retarded systems with function space constraints. Part II, *Technical Report*, Institute of Automatics, Technical University of Warsaw, Warsaw.

Olbrot, A. W. (1977). *Stabilizability, Detectability and Spectrum Assignment for Linear Systems with General Time Delays*, Report CRM-712, Université de Montréal, Montréal.

Pandolfi, L. (1975). On feedback stabilization of functional differential equations, *Bolletino UMI*, (4) 11, Suppl. fasc. 3, 626–635.

References

Pandolfi, L. (1976). Stabilization of neutral functional differential equations, *JOTA*, **20**, 191–204.

Parks, P. C. and Hahn, V. (1981). *Stabilitätstheorie*, Springer-Verlag, Berlin, Heidelberg, New York.

Pazy, A. (1972). On the applicability of Liapunov's theorem in Hilbert space, *SIAM J. Math. Anal.*, **3**, 291–294.

Pinney, E. (1958). *Ordinary Difference-Differential Equations*, University of California Press, Berkeley.

Plant, A. T. (1977). Stability of nonlinear FDE's using weighted norms, *Houston J. Math.*, **3**, 99–108.

Pontryagin, L. (1942). On zeros of some transcendental functions, *Izvestiya Akademii Nauk SSSR*, **6**, 115–131 (in Russian).

Popov, V. M. (1973). *Hyperstability of Control Systems*, Springer-Verlag, Berlin, Heidelberg, New York.

Pritchard, A. J. K. and Curtain, R. (1979). *Infinite Dimensional Linear Systems Theory*, Lect. Notes in Control Th., Vol. 8, Springer-Verlag, Berlin, Heidelberg, N. Y.

Prüs, J. (1984. On the spectrum of C_0-semigroups, *Trans. AMS*, **284**, 847–857.

Răsvan, V. (1983). *Absolute Stability of Automatic Control Systems with Time-Lag*, Nauka, Moscow (in Russian).

Reich, S. (1981). A nonlinear Hille–Yosida theorem in B-spaces, *J. Math. Anal. Appl.*, **84**, 1–5.

Repin, Yu. M. (1965). Quadratic Liapunov functionals for systems with delay. *Prikl. Mat. Mekh.*, **29**, 564–566.

Rolewicz, S. and Przeworska-Rolewicz, D. (1970). *Equations in Linear Spaces*, PWN, Warsaw.

Rubanik, V. P. (1969). *Oscillations of Quasilinear Systems with Retardation*, Nauka, Moscow (in Russian).

Rudin, W. (1964). *Principles of Mathematical Analysis*, 2nd ed. McGraw-Hill, New York.

Salukvadze, M. E. (1962). On the synthesis of optimal regulator in linear retarded systems, *Avtomatika i Telemekhanika*, **23**, 1595–1601 (in Russian).

Sidorov, Yu. V., Fedoryuk, M. V. and Shabunin, M. I. (1985). *Lectures on the Theory of Function of a Complex Variable*, Mir Publishers, Moscow.

Sinha, A. S. C. (1972a). Lyapunov functionals for a class of DD-systems, *Int. J. Control*, **15**, 1027–1031.

Sinha, A. S. C. (1972b). Stability of solutions of some nonautonomous delay-differential systems, *Proc. IEE*, **119**, 1375–1376.

Sinha, A. S. C. (1973). Some results on the stability of nonlinear delay-differential systems, *Int. J. Control*, **18**, 689–694.

Smith, O. J. M. (1959). A controller to overcome dead time, *ISA J.*, **6**, 28–33.

Walker, J. A. (1976). On the applications of Liapunov's direct method to linear dynamical systems, *J. Math. Anal. Appl.*, **53**, 187–220.

Walker, T. A. (1980). *Dynamical Systems and Evolution Equations. Theory and Applications*, Plenum Press, New York.

Walker, J. A. and McClamroch, N. H. (1967). Finite regions of attraction for the problem of Lur'e, *Int. J. Control*, **6**, 331–336.

Webb, G. F. (1974). Autonomous nonlinear FDE's and nonlinear semigroups, *J. Math. Anal. Appl.*, **46**, 1–12.

Webb, G. F. (1976). FDE's and nonlinear semigroups in L^p-spaces, *J. Diff. Eqs.*, **20**, 71–89.

Wexler, D. (1980). On frequency domain stability for evolution equations in Hilbert spaces via the algebraic Riccati equation, *SIAM J. Math. Anal.*, **11**, 969–983.

Willems, J. (1971). Least square stationary optimal control and the Riccati equations, *IEEE Trans. Automatic Control*, AC-16, 621–634.

Yanushevskii, R. T. (1978). *Control of Retarded Systems*, Nauka, Moscow (in Russian).

Yoshizawa, T. (1975). *Stability Theory and the Existence of Periodic Solutions and Almost Periodic Solutions*, Springer–Verlag, Berlin, Heidelberg, New York.

Yosida, K. (1980). *Functional Analysis*, 6th ed., Springer-Verlag, Berlin, Heidelberg, New York.

Zverkina, T. G. (1964). Modification of finite-difference methods for integration of ordinary differential equations with nonsmooth solutions. In: Numerical methods for solution of differential and integral equations and integral formulas, Suppl. to *Zhurnal Vychisl. Mat. i Mat. Fiz.*, 149–160 (in Russian).

Zverkina, T. G. (1965). Modified Adam's formula for integration of retarded equations, *Trudy Sem. Teor. Diff. Urav. s Otkl. Argumentom*, **3**, 221–232 (in Russian).

Index

adaptation 25
adjoint system of equations, 247, 294, 314
adjustable parameters, 201
admissible control, 27, 36, 64
approximation, 90, 91
astatic regulator, 201
asymptotic formulas for characteristic roots, 54
asymptotic stability, 101, 203
autonomous FDE, 34

basic algebraic equation, 320, 322
basis, 98
belt conveyor, 2

canonical form of linear-quadratic problem, 306
canonical system of equations, 318, 334, 335, 355, 356
Carathéodory condition, 34
causality principle, 64
characteristic equation, 54, 75
function, 179
polynomial, 75
quasipolynomial, 54, 177, 185
roots, 54, 75
closed-loop control, 30
optimal control, 305, 320, 324
commensurability, 299

control, 27, 31, 64
space, 27, 36, 37, 64
controllability, 29
, null state, 261, 264, 266
, IP-null-state, 262
, output null, 247
, output approximate, 247, 269
, R^p, 247, 251, 260
, state, 247
, spectral, 248, 283
conventional control, 200

Datko's theorem on Lyapunov equation, 126
determining equation, 259
discrete approximation of linear FDE, 80
 of linear-quadratic problem, 354
 of linear state equation, 83
 of nonlinear FDE, 85
 of optimal regulator, 356
discrete retarded equation, 73
dissipative operator, 123, 131, 133, 135, 137, 149, 171, 173
disturbances, 31, 65
domain of attraction, 101

electric transmission line, 12
equilibrium point, 25, 101

equivalent linear-quadratic problems, 306
    system, 95
estimation, 25
evolution equation, 245
evolution operators, 69
exponential asymptotic stability (EXS), 125
    stabilizability, 248, 277
extended Smith method, 238

FDE, 32, 100
FDE with distributed delays, 33
FDE with lumped delays, 33
feedback, 29
    loop, 30
    regulator, 29, 30, 31, 200
feedforward regulation, 236
filtration, 25
Fréchet derivative, 287
fundamental matrix, 262
    solution, 43, 51, 79

generalized aperiodic stability criterion, 218
global asymptotic stability, 101
Goryachenko model, 121, 175
Gram matrix, 93
Gromova–Pelevina nonlinear control system, 115

heat exchangers, 18

identification, 24, 87, 91, 92
infinitesimal generator of a linear C_0-semigroup, 123, 246
input, 63
integral error, 224
    of absolute error, 224
    of absolute error multipled by time, 225
    of squared error, 224
interpolation, 25
invariant set, 102
inverse Laplace transform, 58

Kalecki model, 6

Laplace transform, 58
La Salle–Hale invariance principle, 102
lemma of Calvarho and others, 121
    of Prüss, 126
    of Yoshizawa, 114
Lerner's diagram, 216

level stabilization, 25, 29
limit set, 102
linear C_0-semigroup, 123
    discrete equation, 74, 75
    FDE, 34, 41, 67
linear-quadratic problem of optimal control, 303
    problem with infinite time, 305, 328, 347
    problem with finite time, 304, 333
L^2-stability, 125
Lur'e control system, 177
    system of resolving equations, 108
Lyapunov functional, 102
    operator equation, 126–127
    set of equations, 145–147, 228

maintainability, 266, 267
metal rolling system, 4
Mikhaylov–Leonhard criterion for FDE, 181
Mikhaylov plot, 181
mild solution, 245, 250, 259
mismatch, 234, 238
model construction, 24
modified Adams formulas, 86
momentum method, 92
multiplicity of poles, 278

neutral FDE, 33, 35, 36, 41, 46, 134, 148, 155
non-anticipativeness, 64, 65
nonlinear semigroup, 168

observability, 30, 247, 271
observation, 25
open-loop control, 30
    optimal control, 309, 323
operator Riccati equation, 339, 340, 350
optimal absolute value criterion, 222
    control, 27, 28, 31, 285, 305
    regulator, 309, 320, 324, 336, 348
optimality conditions, 288, 289, 290, 294, 296, 297
output equation, 27, 65, 79, 247, 259, 27, 63

parametric synthesis of regulators, 215
performance criterion, 201
    functional, 28, 31
PID-regulator, 201
Plant's weighted norm, 133, 134

Pontryagin's criterion for quasipolynomials
(Theorem 6.4.7), 198
population model, 5
pole assignment, 277, 281
prediction, 25
principal term of quasipolynomial, 185, 188
principle of argument, 178
proportional action regulator, 201
  and differential action regulator, 201
  plus reset regulator, 201

reachability, 265, 266
reference input, 200
  regulation system, 233
  trajectory, 26
regression analysis, 92, 93
regulation, 30, 200
remote control, 8
resolvent, 250
retarded FDE, 33, 34, 42
Riccati set of equations, 342, 353

scalar FDE of higher order, 47
  FDE with derivatives of control, 47
semigroup of ω-quasi contractions, 169
sensitivity analysis, 242
  function, 243
series expansion of solution, 57, 59, 60
setting time, 216
Smith principle, 232
solution of FDE, 34, 36, 100–101
spectral projection, 279
space of parameters, 27
stability, 25, 101
  independently of delay, 114
  region, 202
stabilizability, 29, 306, 330, 347
state equation, 27, 65, 67, 69, 78, 79
  reconstruction, 25

trajectory, 249
  space, 27, 65, 68, 78, 100, 245
static error, 216
stationary FDE, 34
step method, 37, 49, 153, 309
strongly continuous semigroup, 246, 247, 249
strong solution of an abstract Cauchy problem, 124
sunflower equation, 104, 174, 186
system reduction, 265

tensor product of matrices, 147
terminal control, 26
theorem of Crandall, Liggett and Miyadera
  on generation of nonlinear semigroups, 169
theorem of Hale and Cruz on EXS under
  small nonlinear perturbations, 165
theorem of Hille, Phillips and Yosida, 246
'three constraints' criterion, 225
total set, 251
tracking, 26
trajectory of FDE, 102
transfer functions, 71, 80
transmittance, 71
type of a linear C_0-semigroup, 125

urban traffic, 10

variation-of-constants formula for discrete
  systems, 74, 79
  for neutral FDE, 44, 51, 160
  for retarded FDE, 43, 50
variation of performance functional, 314

Walker's version of Hille, Phillps and Yosida
  theorem, 124

Z-transform, 76
Ziegler–Nichols criterion, 225